Change is an omnipresent aspect of the environment, so that the practical problem of predicting how terrestrial ecosystems might respond in the future to large-scale human-generated changes is a major challenge for ecologists. In *Terrestrial Ecosystems in Changing Environments*, Hank Shugart describes the fundamental ecological concepts, theoretical developments and quantitative analyses involved in understanding the responses of natural systems to change.

The key ecological concepts described include the ecosystem paradigm, niche theory, vegetation–climate relationships, landscape ecology and ecological modelling. A variety of ecological models are presented, and their applications in predicting responses to change are considered. The challenge of producing ecological models capable of predicting long-term and large-area ecosystem dynamics is reviewed and several examples are provided. Finally, a review of some of the exciting new findings regarding terrestrial landscapes and their feedback with their climatic setting are discussed in the context of human land-use and global change.

D1319263

Terrestrial ecosystems in changing environments

Cambridge Studies in Ecology presents balanced, comprehensive, up-to-date, and critical reviews of selected topics within ecology, both botanical and zoological. The Series is aimed at advanced final-year undergraduates, graduate students, researchers, and university teachers, as well as ecologists in industry and government research.

It encompasses a wide range of approaches and spatial, temporal, and taxonomic scales in ecology, experimental, behavioural and evolutionary studies. The emphasis throughout is on ecology related to the real world of plants and animals in the field rather than on purely theoretical abstractions and mathematical models. Some books in the Series attempt to challenge existing ecological paradigms and present new concepts, empirical or theoretical models, and testable hypotheses. Others attempt to explore new approaches and present syntheses on topics of considerable importance ecologically which cut across the conventional but artificial boundaries within the science of ecology.

CAMBRIDGE STUDIES IN ECOLOGY

Editors

H. J. B. Birks   *Botanical Institute, University of Bergen, Norway, and Environmental Change Research Centre, University College London*

J. A. Wiens   *Department of Biology, Colorado State University, USA*

Advisory Editorial Board

P. Adam   *University of New South Wales, Australia*

R. T. Paine   *University of Washington, Seattle, USA*

R. B. Root   *Cornell University, USA*

F. I. Woodward   *University of Sheffield, Sheffield, UK*

Terrestrial ecosystems in changing environments

HERMAN H. SHUGART

Department of Environmental Sciences, The University of Virginia, USA

 CAMBRIDGE
UNIVERSITY PRESS

FRANKLIN PIERCE
COLLEGE LIBRARY
RINDGE, N.H. 03461

PUBLISHED BY THE PRESS SYNDICATE OF THE UNIVERSITY OF CAMBRIDGE
The Pitt Building, Trumpington Street, Cambridge CB2 1RP, United Kingdom

CAMBRIDGE UNIVERSITY PRESS
The Edinburgh Building, Cambridge CB2 2RU, United Kingdom
40 West 20th Street, New York, NY 10011-4211, USA
10 Stamford Road, Oakleigh, Melbourne 3166, Australia

© Cambridge University Press 1998

This book is in copyright. Subject to statutory exception
and to the provisions of relevant collective licensing agreements,
no reproduction of any part may take place without
the written permission of Cambridge University Press.

First published 1998

Printed in the United Kingdom at the University Press, Cambridge

Typeset in Bembo 11pt   [SE]

A catalogue record for this book is available from the British Library

ISBN 0 521 56342 9 hardback
ISBN 0 521 56523 5 paperback

QH
541.15
m3
.S 56
1998

Contents

Preface

For the past decade or more, I have had the opportunity to be involved with one of the most exciting research areas in ecology and the geosciences: the efforts to understand the interactions of the major Earth systems of atmosphere, oceans and the terrestrial surface. This research area is sometimes referred to as 'global change biology' or 'global change ecology'. For research scientists, interest in this area has been stirred by the actions of international scientific co-ordinating committees (notably the IPCC (the Intergovernmental Panel on Climate Change) and the IGBP (the International Geosphere–Biosphere Programme)) and by public concerns as to the eventual effects of human alterations of the composition of the atmosphere, ocean and land. My own work in this area has focused mostly on the application of ecological models in an attempt to project the possible consequences of changes in climate and other conditions on forests and other terrestrial ecosystems.

A few years ago, it occurred to me that the opportunity for today's students to gain background for studies in global change ecology was rather limited. The scientists with whom I collaborated had arrived at their interest in studying global change by different and often circuitous routes. Some were palaeontologists or palaeoecologists interested in past ecosystems. The Earth has been highly variable climatically during its recent geological history and palaeoecological research leads naturally to wanting to understand better ocean–climate–vegetation interactions. Others have been drawn into global ecology by a desire to comprehend better the patterns of plants, animals and habitats over large areas. These scientists often applied a diverse array of technologies that are unique to our times: notably, the computational power to statistically interweave larger ecological and climatological data sets, the satellite technology to allow large surveys of the Earth's surface from space, instruments to measure fluxes of water, carbon dioxide, heat, chemicals to and from ecological systems, etc.

Still others with an interest in conservation of unusual plants and animals saw the destruction and damage of essential habitat for these organisms by a wide variety of large-scale causes (climate change, human land-use, human alteration of the environment) as the motivation to understand better Earth systems function. Even with these and other diverse avenues for becoming interested in global change, the availability of a text that could provide necessary background for a student potentially interested in the area seemed to me to be lacking.

There are a large number of excellent books that are technical compendia on the subject of global change. They often feature chapters written by particular experts on topics in global change and whole Earth system dynamics. These represent a rich source of information for the advanced scholar with some background in the area, but there was still a need for a text that would provide access to the field for students with a basic background in ecology and sciences. During a sabbatical leave from the University of Virginia in the 1993–4 academic year, I had the good fortune to serve as a Visiting Scientist with Australia's Commonwealth Scientific and Industrial Research Organisation, Division of Wildlife and Ecology and, at the same time, to be a Visiting Fellow with the Research School of Biological Sciences, Environmental Biology Department, Australian National University. These two outstanding Australian institutions are among the world leaders in the ecological aspects of global change. Among old friends, I set about writing this book.

During the time in Australia and additional time working at the University of Virginia in the summer of 1994, I produced a first draft of this book. Several colleagues and students were dragooned into reading all or part of the text. I taught a course to about 50 students per year in the Department of Environmental Sciences at the University of Virginia beginning in the fall of 1994. This course ('Issues in Global Change' EVSC 493/795) was taught to third- and fourth-year undergraduate/ introductory graduate student level and has been taught every fall since. I used the draft as a photocopied text-book for this class and revised sections according to student comments. Eventually, I produced the book that follows.

This book is not solely intended to be a textbook (although it has worked as a text in my own teaching and could serve this purpose for others) and it is to some degree idiosyncratic. My own research career has moved from ornithology to plant ecology and now to global ecology. Having already mentioned that there are several rather different routes taken by research professionals to arrive at an interest in global ecology, it seems appropriate to outline the path most familiar to me – which I have done. Any lecturer using this book as a text would indubitably draw heavily from his or her own experience in developing a course syllabus. There are some larger, pedagogical points that I hoped to demonstrate in developing this text. I wanted to stress the changing, non-equilibrium nature of the biosphere. I hoped to show that global ecological studies were not a new invention, rather they derived naturally from basic ecological constructs (niche theory, plant geography, the mosaic nature of landscapes). I wanted to use ecological models as examples in a discursive manner and to discuss model formulation and model results generally. With respect to this latter point, I have included a discussion on the history and basics of ecological modelling that is strongly derived from a course taught by Professor B. C. Patten at the University of Georgia (Athens) when I was a student there. Some may find even this rather gentle taste of mathematics daunting, but modelling is an essential part of prediction of the consequences of past or future environmental changes. I also wanted to provide some of the many results that reflect the potential impact of the changes that humankind has wrought on the Earth. These are found mostly in the later chapters of the book.

Funding from grants from US Federal Agencies, particularly the National Science Foundation (DEB-90202041: *Coupling of Ecosystem Process and Vegetation Pattern Across Environmental Gradients*), NASA (NAG 5-2295: *Multidiscipline Integrative Models of Forest Ecosystem Dynamics for the Boreal Forest Biome*; NAG 5-1018: *Forest Ecosystem Dynamics*) and the Environmental Protection Agency (CR-81627-01: *Implications of Climate Change on Forests: the Development of Forest Simulation Models for Evaluating Climate Change in Global Forest Ecosystems*), supported major parts of the research work of my students and I over the years. Substantial parts of the later chapters of the book draw on this work as examples. A pair of Academic Enhancement grants from the Board of Visitors of the University of Virginia (*Global Systems Analysis Program* and *Global Environmental Change Program*) were invaluable in the development of this project and most appreciated. Travel to Australia was supported, in part, by a travel grant from the Australian CSIRO Division of Wildlife and Ecology. I have many friends and colleagues to thank for their help and patience during the development of this project. M. P. Austin, H. J. B. Birks, R. B. Carlson, F. Daria, T. E. Dennis, W. R. Emanuel, L. Gu, B. P. Hayden, B. M. McIntyre, M. W. Palace, B. R. Rizzo, G. Shao, W. L. Steffen, L. Von Schill, B. H. Walker, J. A. Wiens and F. I. Woodward read all or part of the book and provided helpful comments and encouragement. Two classes at the University of Virginia in 'Ecological Issues in Global Change' used earlier drafts of the book as a reference text. In particular and along with several others, A. W. Farmer, K. K. Caylor, W. J. Faubert, A. J. Hill, N. D. Kaufman, J. M. Owens, S. U. Seddon-Brown, R. R. Shah, P. C. Shahani, A. M. Thomson and K. E. Winterson from these classes were kind enough to provide frank advice from a student's perspective. Part of the draft text was also used in a training course on 'Modelling Land-use and Forest Dynamics' at the BIOTROP-GCTE Southeast Asian Impacts Centre, Bogor, Indonesia. Thanks to Louis Lebel, Daniel Murdiyarso, Habiba Gitay, Ian Noble and Ian Davies, who were involved with me in developing part of this and an earlier workshop. M. L. Merriam and J. R. Montambault, both undergraduate students at the University of Virginia, read an early draft for clarity. R. L. Smith, Jr drafted illustrations and Jane Ward of Cambridge University Press provided much-appreciated text editing. Finally and certainly not least, I would like to express my gratitude to my wife, Ramona Jeanne Kozel Shugart, to whom this book is dedicated.

H. H. Shugart
March 1997

Part 1
Introduction

1 · The importance of understanding ecosystem change

Semper in adsiduo motu res quæque geruntur.
(All things are caught up in ceaseless motion.)
Lucretius

We live in a technological era that is propelled by a special, and perhaps unique, form of creativity in which the imagining of difficult problems – How can we communicate over large distances? How fast can we move people or things from one place to another? Can we develop a material that transmits energy with the most efficiency? Can we eradicate small-pox? Can we map the entire genetic contents of the chromosomes of humankind? – seems to inspire technological solutions to these problems. We routinely conceive and solve problems that would have been considered in the realm of magic and witchcraft less than ten generations ago.

But for all our ability to generate, and eventually solve, the problems we present ourselves, we have stumbled in solving environmental problems. Why is this?

Modern technology excels in solving well-posed, clearly defined problems. Most people who deal with environmental problems will certify that environmental problems are rarely clearly defined or well posed. Indeed, they are often maddeningly to the contrary. In the environmental area, yesterday's logical solution can become today's problem. Unfortunately, as the human population increases, in part as a result of our ability to solve other problems such as people dying from infant mortality and curable diseases, environmental problems seem to become more difficult and larger in their spatial extent.

Since the mid-1970s, ecological problems, inherently regional and global in scale, have emerged. These involve understanding the response of ecosystems to spatially extensive environmental change. Many of these ecological problems are public concerns and, as such, have gained pen names; recent examples are 'Acid Rain' and 'The Greenhouse Effect'. Public interest in such scientific issues can be fickle. There is a news reportage of the 'Environmental Disaster of the Week' that invites the cynicism of a jaundiced readership. At the root of these and other issues is a central scientific challenge: how can we extrapolate our understanding of underlying biological, chemical and physical processes to larger land areas over long time frames? This is the kernel of the problem. If we can engage this issue, hopefully we can then pose environmental questions in a context that will allow us to apply our technological and scientific muscle towards solutions.

For example, we have a wealth of research results that have created a basic understanding of how oxidizing pollutants, such as ozone and the gaseous oxides of sulphur and nitrogen, affect the functioning of leaves. We have a degree of understanding of the analogous responses of whole plants to these pollutants. But we need to know how to extrapolate this detailed understanding to predict the potential effects (if any) of these pollutants on the species composition, productivity and functioning of the vegetation over a region. Nevertheless, it is difficult to demonstrate unequivocally that brown spots on leaves (that we can associate with air pollution) are the cause of large-scale die-backs of trees.

Similarly, we understand the responses of single plant leaves to elevated levels of CO_2 under different temperatures, light levels and humidity conditions (at least for certain species of plants). But, in a world with increases in atmospheric CO_2, how do these responses translate into alterations in forest or crop productivity? Equally difficult is the reciprocal problem, how do changes in productivity of forests (and other terrestrial systems) change the rates that atmospheric CO_2 is removed from the atmosphere? How does variation in the uptake of CO_2 by terrestrial ecosystems influence the atmospheric levels of CO_2?

Like Lucretius, quoted at the beginning of this chapter, we live in a time with an awareness of the omnipresence of changes. We are compelled by scientific interest and by the need to solve environmental problems to understand these changes better. We need to know how the Earth's systems work because we are affecting these systems through our actions and we have not yet solved the problem(s) of understanding the consequences.

The prime objective of this book is to communicate an understanding of the prediction of the 'ceaseless motion' of the Earth's vegetation: the large-scale dynamics of the terrestrial surface of the Earth in response to a changing environment. Important tools to explore such changes are computer models that can be used to predict ecosystem responses. The quantitative modelling of the dynamics of the vegetation covering the Earth's terrestrial surface under current and altered environmental conditions can be used to reconstruct the response of terrestrial ecosystems to past change and to predict responses to future changes.

One of the difficulties in treating a subject matter as broad as terrestrial ecosystems and their response to change is avoiding writing a general ecology text. There are several scientific themes that will emerge as this book develops. These are largely concerned with perspectives or points of view involved with the development of global or large-scale ecology. However, these issues are far from unique to large-scale ecology. Indeed, they are important for a wide range of scientific disciplines.

Ecologists deal with systems of great complexity and science is currently oriented towards description with an interest in detail. At times, there seems to be a fascination with the production of counter-examples to generalities. Ecological studies are often conducted at relatively small space scales. Karieva and Anderson (1988) found that 95% of the studies that they surveyed from leading ecological journals were conducted on study plots of less than 1 ha. Half of these studies used plots of 1 m^2 or smaller.

Also, ecology is strongly an observational science. Observations are the currency of the science. Ecologists traditionally argue against a particular theory (or claims for generality) by the production of counter-examples. The ecologist's seeming fascination for counter-examples is reinforced by the increased importance of the role of statistics in modern ecology and the emphasis in statistical procedures of asserting that a particular assemble of observations negates a stated hypothesis. In statistical tests, one does not 'prove' things in the sense of demonstrating them to be true. Rather, results are judged to be highly unlikely if a hypothesis called the null hypothesis, H$_0$, is true. In application, statistics is mostly about demonstrating that calculated features such as averages of samples (and other calculations) are not likely to be the same. This is not the equivalent to proving that they are different. Progress in a statistically based science proceeds by formulating appropriate hypotheses and then collecting observations to reject these hypotheses. The creative aspect of the science is to identify which hypotheses to examine.

Across ecology, an exchange of theories with examples and counter-examples drawn from different systems at inappropriate time and space scales has fuelled several acrimonious and continuing arguments. For example, the fundamental nature of succession, the importance of competition in structuring communities, or the importance of internal feedbacks in the regulation of a range of ecosystems have been debated continuously, some might say *ad nauseam*, without resolution. One factor providing fuel to these non-illuminating fires may be in the mismatch in time and space scales of the sets of observations being used to test theory.

A fundamental motif that will arise in the initial chapters of this book is that it is important to match the time and space scales of important phenomena. Time and space scales are usually expressed in terms of the ranges over time or space in which the changes in the phenomena are measurable and observable. Scale-related issues will emerge in understanding ecosystems and their spatially explicit analogue, the biogeocoenosis, and in relating environmental variables to the biology and ecology of plants and vegetation. Further, time and space scale will figure prominently in understanding the dynamics of ecosystems that occur as heterogeneous mosaics.

The importance of the appropriate blending of phenomena, time and space to gain understanding is not in the least unique to ecology. Rather, it is an integral part of most scientific investigation, theoretical, observational or experimental. For example, in particle physics, one ignores the effects of gravity because the gravitational forces are small relative to other forces; in astrophysics, gravitational forces are a central concern. Although the nature of gravity constitutes a topic of considerable interest for physicists, whether it is right or wrong to ignore gravitational forces at certain scales generally is not an issue of debate. The same type of inclusion or exclusion of important considerations in problem formulation also occurs in ecology. For example, in most ecological population models, the age-structure and sex-ratios of the populations are usually not considered explicitly. Other examples will be discussed in the development of models used to assess the consequences of large-scale environmental change.

Part 1 of the book deals with the omnipresence of change at a variety of time and space scales. This is followed by a historical review and a discussion of the ecosystem concept: a systems concept defined on problems of interest and tempered by considerations of space and time scales. One of the aims of these introductory chapters is to impress on the reader that understanding global change is an issue of both applied and basic interest. Another is to persuade the reader to adopt the use of the term ecosystem in its rich historical definition for its power in framing problems involving synthesis of information.

This is followed by Part 2, which reviews some of the basic concepts used in understanding and interpreting large-scale responses to environmental changes. This section introduces ecological modelling, a topic that will emerge in the later chapters of the book. Ecological models are a great array of physical and/or mathematical analogies to ecosystems, but generally they are manifested as computer programs thought to imitate central aspects of ecosystems. Because the time and space scales involved with some focal problems in understanding large-scale ecology are beyond the lifetimes or the logistic resources of ecologists, ecological models have developed as tools for projecting the consequences of observations or theories about how ecosystems may change over time. The degree to which these predictions can be believed actually to occur hinges to some extent on the degree to which these models can reproduce known features of ecosystems in a model-testing mode. For this reason, some effort in later chapters will be spent on issues associated with model verification and validation.

Following an introduction to ecological modelling, Part 2 then covers some of the more important ecological paradigms used in large-scale terrestrial ecology. Niche theory, the mosaic nature of landscapes and plant/environment relations at relatively large spatial scales are discussed. These are large topics. Each has been the subject of entire books and will probably continue to be so. Intentionally, the chapters in this section are relatively non-mathematical − although several of the topics (niche theory, spatial dynamics) have been advanced greatly by mathematical treatment. The emphasis is on the conceptual roots of the topics and their origins as ecological constructs, rather than on the mathematical analysis associated with these topics. The concepts in this second section are the bases for models used to simulate large-scale response of terrestrial ecosystems, which is covered in the third section of this book.

Part 3 discusses ecosystem models, the testing of these models and their implications regarding the functioning of large-scale ecosystems. The leading chapter of the section, Chapter 8, will introduce individual-based models and will emphasise the ecological features embedded in the models. Chapter 9 discusses some of the possible consequences of the dynamics of these models at the population level and for natural landscapes. Chapter 10 will introduce landscape and larger-scale models and the consequences that proceed from the application of the models to issues involving change.

Finally, in Part 4, a model-based evaluation of some of the consequences of global change will be presented. In this section, the prediction of landscape responses to changes will be developed using ecosystem models with different underlying assumptions – particularly regarding the spatial interactions and the influence of such interactions on overall system dynamics. Chapter 11 will discuss mosaic landscape models and their use in understanding change of terrestrial ecosystems. Chapter 12 will review some of the progress being made in spatially interactive landscape models, and Chapter 13 will focus on results from models that are based on the assumption of a homogeneous landscape. Dynamic changes in the structure and function of ecosystems are treated at increasingly larger spatial resolution through the text. Therefore, the final chapter (Chapter 14) will close with some of the important global issues in terrestrial system dynamics.

2 · *The omnipresence of change*

Two hundred years ago, Thomas Jefferson described the bones of an extinct ground-sloth, *Megalonyx* (Wistar 1799), that, because of its large claws, he took to be some sort of giant carnivore (*Megalonyx*, large lion). Mr Jefferson felt that the animal was still alive someplace in North America and reasoned,

> ... the bones exist: therefore the animal has existed. The movements of nature are in a never-ending cycle. The animal species which has once been put into a train of motion, is probably still moving in that train. For if one link in nature's chain be lost, another and another might be lost, till this whole system of things should evanish by piecemeal.... If this animal has once existed, it is probable that ... he still exists. (Jefferson, 1799)

Jefferson instructed Lewis and Clark, when they were sent west to explore the Louisiana Purchase, to watch for and report on large animals like the *Megalonyx*. The logic that Mr Jefferson used paralleled that of other eighteenth century intellectuals who saw extinction as an affront to the Creator. Grayson (1984), in a rich historical review of the topic of extinction, quotes Jefferson (above) and also Pope (Pope, *Essay on Man*, 1733–4),

> Vast chain of being, which from God began,
> Natures aethereal, human, angel and man,
> Beast, bird, fish, insect what no eye can see,
> No glass can reach! From infinite to thee,
> from thee to Nothing! – On superior pow'rs
> Were we to press, inferior might on ours:
> Or in the full creation leave a void,
> Where one step broken, the great scale's destroy'd:
> From Nature's chain whatever link you strike,
> Ten or ten thousandth, breaks the chain, alike.

These arguments for a constancy in nature, of an unbroken natural chain, survive in popular ideas about the 'balance of nature' and of the antiquity of certain ecological systems. However, most educated people today have little difficulty imagining that species have become extinct or that environmental character was quite different in some previous time. The ideas of an 'ice age' or a 'time when dinosaurs ruled the earth' are commonly understood phrases and, as such, are exemplars of a general appreciation of prior environmental change.

Ecologists are also aware that environments were different in the past. Given this appreciation of past differences, it is surprising that ecological theory abounds with analyses of what ecological systems should be like at equilibrium, and that ideas about ecological succession and landscape change often are posed in terms of a constant environment.* Further, the concept that a wilderness or constant primal state is 'natural' pervades many of our policies on managing parks and nature preserves.

Of course, many of these views are expediencies intended to provide answers for simplified cases and insight into more difficult cases. If one does not know what the behaviour of an ecological system is in a constant environment, then how can one hope to predict its behaviour in a changing environment? Before one can expect to appreciate the dynamic response of ecological systems, it is appropriate to know what the system is like in an unchanging state.

What is known now and is becoming clearer in its details as more information is collected is that the earth environment has been quite dynamic over almost any time scale one might consider. This chapter will provide some examples of changes at some of these time scales. Later, in Chapter 3, it will be noted that one expects changes at frequencies in time to excite responses from different ecological processes.

Long-term variations in climate

When considering the variation in climate over the period of time during which higher forms of life evolved (~800 million years ago), one can see either catastrophe or constancy. The constancy is in the sense that the Earth's temperature has remained relatively constant and in the range that supports life, despite a 25% increase in the intensity of solar radiation (Lovelock and Margulis 1974). Catastrophic events are seen in the seemingly periodic mass extinctions of forms of life. Raup and Sepkoski (1984) claim that there have been nine such extinctions with periodic

---

* Solbrig (1994) points to the 'Newtonian paradigm', that all of nature was interpretable in terms of a few laws in a 'clockwork' universe, as a contributor to the eighteenth century concept of the balance of nature. He also notes this view as an influence on most scientists until fairly recently.

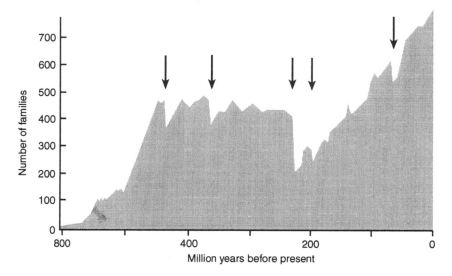

Figure 2.1. Extinction events over the past 800 million years from families of marine animals. Arrows indicate what were interpreted by Erwin *et al.* (1987) as major extinction events in the geological record. From Erwin *et al.* (1987).

components of around 33 million years and 260 million years. Figure 2.1 illustrates the number of families of marine animals living over the past 800 million years and shows several phases of extinction. The largest drop in numbers of marine animal families occurred in the Permian period (*c.* 245 million years ago) when the number of families dropped by over 50% and the number of known genera was reduced by around 80%. Such drops in the existence of higher-level taxa imply losses of species as high as 96% (Jenkins 1992). While the topic is hotly debated (Jablonski 1986), the periodicities seen in extinctions (Fig. 2.1) conform to large events on the Earth's surface (tectonic events, sea level changes, changes in the magnetic poles) and evidence for meteor impacts.

One of these extinctions of recent interest appears to have occurred at the end of the Cretaceous Period, 65 million years ago. It involved the extinction of about half of the living genera at the time (Alvarez *et al.* 1980; Ganapathy 1980). This included microscopic aquatic plants and animals of various kinds, marine and flying reptiles, and dinosaurs. However, land plants, crocodiles, snakes, mammals and many kinds of invertebrate survived (Lewin 1986). There is (and will likely continue to be) a debate as to the cause: an asteroid impact (Pollack *et al.* 1983) or volcanic eruptions have been suggested as two possible causes. Harrington (1987) observed that these cataclysmic events may obtain their periodic nature from the movements of our solar system with respect to the galactic plane of our Milky Way galaxy.

Within the million-year time frame, there have been variations in the ratios of stable isotopes of oxygen ($^{18}O/^{16}O$) taken from deep ocean sediments (Hays *et al.* 1976; Imbrie *et al.* 1984) and from glaciers in Greenland and Antarctica (Dansgaard *et al.* 1982; Saltzman 1983) that appear somewhat periodic (Fig. 2.2). The ratio of oxygen isotopes ($^{18}O/^{16}O$) varies during the formation of ice crystals. Therefore, the variation in the ratio is indicative of the amount of ice on the surface of the Earth.

The periodic changes in the ratio of $^{18}O/^{16}O$ have been related to the 105 000 year periodicity in the eccentricity of the Earth's orbit about the sun, the 41 000 year cycle in the obliquity of the orbit, and the 21 000 year cycle in the precession of the Earth's axis (Imbrie and Imbrie 1980; Berger 1984). This has been called the Milankovitch theory (proposed by Croll (1867) and derived mathematically by Milankovitch (1941)). According to this theory, the periodic cooling and warming of the Earth's climate is driven by variations in the Earth's orbit. The change in the amount of solar energy coming to the surface of the Earth as a result of these variations in the orbit is quite small. If the Earth's climate is sensitive to variations at these levels (as many Quaternary Period palaeoclimatologists believe it is), then one would expect responses in the climate from the changes in the 'greenhouse gas' concentrations in the atmosphere. These will be discussed later in this chapter.

As one considers finer and finer time slices through the recent history of the Earth's climate (obtained by interpreting different climate indicators from sea sediments, glaciers and historical observations, Fig. 2.2), the record displays variation and periodic change at almost all time scales. Along with aperiodic variation, there is also an apparently periodic variation in climatic indicators at both millennial and century time scales (Fig. 2.2c,d).

Harrington (1987) has reviewed these and other issues regarding possible mechanisms and magnitudes of climate change at a variety of time scales. Regardless of the mechanisms that cause climatic variation at different scales, there is an abundance of evidence (e.g. Fig. 2.2) for changes in the climate at multiple temporal scales. One expects terrestrial vegetation to be responsive to these changes. The relationships between climatic condition and vegetation will be discussed in Chapter 6 and the prediction of vegetation change in response to climatic and other environmental change will be the topic of several of the latter chapters.

If the climate can alter the vegetation cover and ice cover of the Earth's

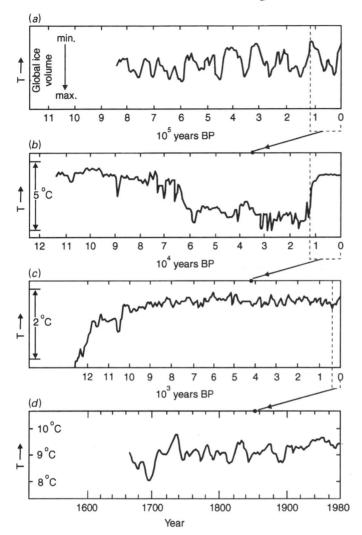

Figure 2.2. Selected climatic time series for the past 1 million years. (*a*) Global ice volume deduced from oxygen isotope variations of planktonic foraminifera in a deep sea core. (*b*) Global mean temperature variations over the past 100 000 years based on oxygen isotope variations in a long ice core from Greenland. (*c*) Global mean temperature variations over the past 10 000 years based on oxygen isotope variations in the Greenland ice core. (*d*) Thermometer readings from England for the past 300 years. From Harrington (1987); redrawn from Saltzman (1983).

surface, climate also seems to respond to changes on the Earth's surface. There is evidence for direct responses by the atmosphere to the changes on the Earth's surface. For example, the location of land masses near the poles seems to be a condition needed for continental glaciation to occur (Crowley *et al.* 1986). Volcanic eruptions that inject dust and gasses into the atmosphere can alter climate for days to years depending on the magnitude and height of the injections (Robock 1978). Variations in ice on the surface of the Earth (Fig. 2.2*a,b*) interacting with climate may introduce feedback cycles with a range of periodic components.

Therefore, it seems possible that there may be significant feedbacks between the terrestrial surface and the atmosphere. It is possible that a change in climate might produce changes in the Earth's vegetation and these surface changes might then produce climatic alterations etc. The nature and strength of such feedbacks is a topic of current interest and will be discussed in Chapter 14. One of the current scientific challenges is to understand if the changes that we are currently making to the land cover of the terrestrial surface are of a significant magnitude to alter the surface interaction with climate, and thus to cause a global change in the environment.

Changes in the Quaternary Period

The present geological period, the Quaternary Period, is divided into two epochs, the Recent or Holocene Epoch (which goes back from the present about 10 000 years) and the Pleistocene Epoch (from 10 000 years ago to ~2 million years ago). In the Pleistocene Epoch, there were periodic formations of continental-scale glaciers, giving it the popular name the Ice Age.

The Quaternary Period has been a particularly active period in the sense of climate. It has shaped the current distributions and patterns of plants and animals on the terrestrial surface to a great degree.

A variety of clever analyses of a diverse array of data has been developed to obtain an understanding of ecosystem change in the Quaternary Period, particularly the events of the past 100 000 years. This time frame is particularly relevant to our understanding of present-day distribution of the biota and the composition of modern terrestrial communities (Delcourt and Delcourt 1991). The Quaternary Period also provides several important lessons for our understanding of modern ecological systems. Examples of these will be discussed in the sections that follow and include: the formation of ecological systems not currently extant, a documentation of the degree to which some modern ecosystems appear to be relatively recent constructions, the extinction of a number of large mammals over continents with associated landscape effects, and the modification of the terrestrial surface by prehistoric and historic human societies.

Novel ecosystems and novel patterns in the Quaternary Period

There have been several documentations of the tendency for the individual species of dominant plants on terrestrial landscapes to change their ranges independently over time. In England during the periods when glaciers had advanced, there were no trees because of the severe climatic conditions and the presence of an ice cap (West 1977). During the interglacials (when the glaciers were in retreat), tree species migrated from locations south of the Pyrenees and the Alps (Huntley and Birks 1983) to the British Isles. West (1977) documented the changes in major tree taxa during four such interglacials (Fig. 2.3).

There are several differences in the patterns of vegetation change for each of these interglacials. Species arrive at different times in the sequence. Some species are absent in certain interglacials. The order in which different species become predominant varies among interglacials (Fig. 2.3). There is a wide range of possible causes of these differences (many of which do not exclude others), including differences in the seasonal contrast of temperatures and precipitation altering competitive abilities of species, chance events involving seed dispersal, timing of the sea level rise cutting off the British Isles from the rest of Europe, and others (West 1961, 1977; Davis 1976; Wright 1977; Watts 1988; Birks 1989; see Delcourt and Delcourt 1991).

Davis (1976, 1981a,b, 1983, 1986) used dated sequences of fossil pollen from lake sediments to develop maps of change in the ranges of eastern North American tree taxa. These maps clearly document an independence in the pattern of dynamic change of the ranges of major tree species as the species migrated across eastern North America with the retreat of the continental glaciers. Comparable maps have been developed for Europe (Huntley and Birks 1983; Huntley 1988). Webb (1988) compiled similar maps illustrating the systematic changes in major plant taxa at 2000 year intervals for eastern North America.

Figure 2.4 is extracted from Webb's (1988) maps and shows the changes in the major species that make up the boreal forest for the past 18 000 years. Considering today's boreal forest both in Eurasia and North

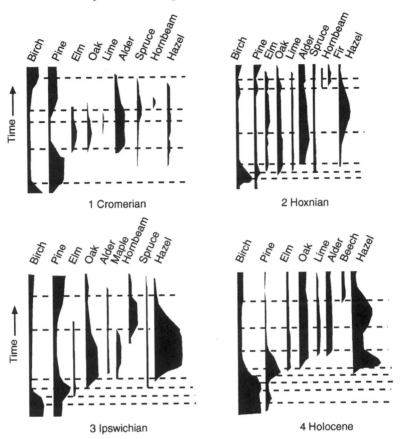

Figure 2.3. Pollen diagrams representing the general sequence of changes in forest composition in eastern England during four interglacial cycles from oldest (1) to the youngest (4). The second youngest, the Ipswichian, is thought to have reached a temperature maximum about 120 000 years ago. The width of the lines corresponds to the relative abundance of pollen grains in sediments deposited during the interglacial cycles (West 1977). From Wright (1977).

America, one is struck by the great similarity in the structure and landscape pattern of the systems (Shugart *et al.* 1992a). Birches (*Betula* spp.) and aspens (*Populus* spp.) are found in successional stands following fires; mature stands of spruce and fir (*Picea* and *Abies* spp.) form dark conifer stands, and sandy or drier locations support pines (*Pinus* spp.). There are exceptions to the comparative similarity in the global boreal forest (the Siberian larch, *Larix sibirica,* forest of Siberia being one obvious example), but the larger patterns on the landscape are quite similar. A Russian ecologist can inspect a Canadian landscape and correctly read with confi-

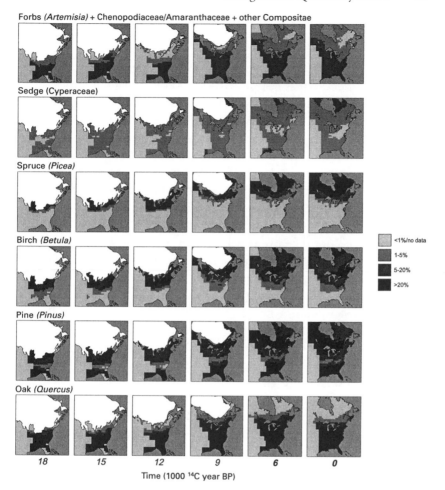

Figure 2.4. Changes in the major tree taxa that make up the boreal forest of North America mapped at 2000-year intervals for eastern North America for the past 18 000 years. The white areas indicate ice, which recedes through the sequence with glacial retreat. Shading indicates the apparent presence of each taxa through the time sequence. Species that are now mixed to produce the landscape and successional pattern of the modern boreal forests vary in distribution, relative abundance and degree of range overlap through the time sequence. From Webb (1988).

dence the history of wildfires across that landscape for the past several hundred years.

On inspection, the global boreal forest appears to offer a unity of pattern – an ecological system adapted to fire and cold climate. Webb's palaeoecological reconstruction of the changes in the components of the forest provides a different view. The relative abundances of the species that make up the forest have changed greatly over the past 18 000 years. Indeed, many of the combinations that co-occur and behave so predictably in their successional responses today were virtually disjunct in the past. *Picea* was in one location, *Pinus* in another. Alders (*Alnus*) seemed to show a greater presence 8000 years ago. Birch (*Betula*) was in much lower abundance in the past. What one finds in these reconstructions are unique mixtures of trees and other plants relative to today. Analogous cases can be demonstrated in tropical (Haffer 1987; Whitmore and Prance 1987; van der Hammen 1988), temperate (Delcourt and Delcourt 1991) or arctic (Lamb and Edwards 1988) present-day climates.

At a higher level of detail, one can also consider the changes in ecological systems at a locality using palaeoreconstruction. For example, the material in the middens (or nests) collected by a variety of small mammals in arid environments (Betancourt *et al.* 1990) can be used to reconstruct local habitat patterns. Packrats (*Neotoma* spp.) build nests from material that they collect within a range of about 30 m from middens. Packrat middens can be dated and since they provide pieces of plants one can identify the species that were near the midden at a given time. Cole (1982, 1985) examined the changes in the distributions of woody plants at different elevations in the Grand Canyon of Arizona (Fig. 2.5). He was able to challenge the idea that the plant communities in the western USA moved up and down the elevational gradient as units. What he found, when he examined the contents of packrat middens of different ages and from different locations, were differences in patterns of zonations in response to environmental change over the past 24 000 years: communities in existance today that were not in evidence in the past and vice versa. The vegetation zones were not constant in time. Elements of the vegetation of each mountain zone had a tendency to change with a degree of independence from other elements.

The extinction of a megafauna

A striking change in terrestrial landscapes in the past 20 000 years has been the extinction of a large number of the large animals in Eurasia, the Americas, Australia and elsewhere (Martin and Wright 1967; Martin and Klein 1984). The causes of these extinctions are a subject of discussion (see Martin and Klein (1984) for a collection of different views and three different summaries of the available data). Climate change, actions of pre-

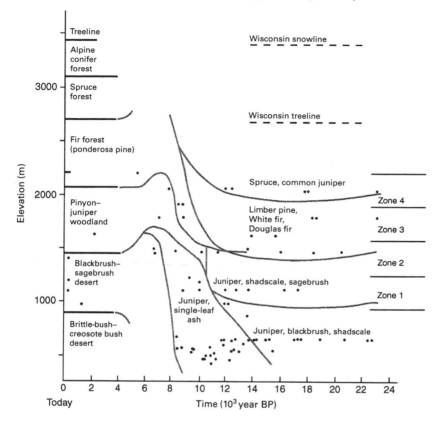

Figure 2.5. Changes in northern Arizonan vegetation zones during the past 24 000 years based on plant macrofossil data from radiocarbon-dated packrat middens for the Grand Canyon. Note that the zones do not shift symmetrically, that zones present today were absent in times in the past, and vice versa. From Cole (1985).

historic people and coincidence of the evolutionary turnovers of species are all candidate causes (Martin and Klein 1984; Owen-Smith 1988), but these could work in concert with one another.

Regardless of the cause of these extinctions, accounts of these species (Nilsson 1983; Anderson 1984) stir the imagination. Large carnivores – huge bears, sabre-tooth tigers, cave lions, dire wolves, dogs of various sorts – were distributed through Europe and North America. A diverse array of herbivores – long-necked camels, giant beavers the size of modern bears, horned giraffes, giant armoured armadillos, huge ground sloths the size of small elephants – grazed vegetation in North America and Europe that today support nothing resembling such creatures. A zoo full of the species of large mammals from the Pleistocene Epoch would need to be over twice the size of today's zoos. Since about 10 000 years ago, almost 40 genera of mammals, most of which are large mammals, have become extinct (Webb 1984).

For example, today there are two elephants (subfamily Elephantinae), the African elephant (*Loxodonta africana*) in Africa and the Asiatic elephant (*Elephas maximus*) in Asia. In the late Pleistocene Epoch, there were these two species in Africa and Asia, respectively, but the African elephant occurred in Europe along with another elephant, the woolly mammoth (*Mammuthus primigenius*). North America had three elephants, the woolly mammoth, the imperial mammoth (*M. imperator*) and the Columbian mammoth (*M. columbi*). Asia also had the three elephants including the woolly mammoth and the modern elephant (Maglio 1973; Agenbroad 1984). Along with these elephants, there was also a diverse array of related genera from other subfamilies of elephant-like creatures, notably the mastodon, *Mammut* (Fig. 2.6).

We know from direct observations in Africa and Asia that the effects of elephant herds and other large herbivores (Owen-Smith 1988) on the vegetation structure and composition can be highly significant. Elephants in Africa have been recorded as a major factor in altering the dominant tree species in the vegetation. This can be seen in elevated levels of tree mortality and damage in areas of relatively high elephant abundance and in conversion of grassland to closed-canopy woodland in areas from which elephants are excluded.

It is likely that the effect of at least some of these creatures on the vegetation in the past was equally substantial. Guthrie (1984) suggests that, in Alaska, the Pleistocene landscape was structured by large herbivores that had a generalist diet and inhabited a range of habitats. Further, the local diversity of the mammal grazers was higher than in the Holocene Epoch and there were more large herbivores that were habitat and dietary specialists. Guthrie proposed major changes in plant community organisation, habitat diversity, landscape pattern and trophic levels associated with the end of the Pleistocene Epoch.

Changes in the Holocene Epoch

It is apparent from the earlier discussions in this chapter concerning the Quaternary Period in general that there have been significant variations in climatic conditions and in the vegetation in the past 10 000 years (i.e. the Holocene Epoch). There were apparent variations in the climate (Fig.

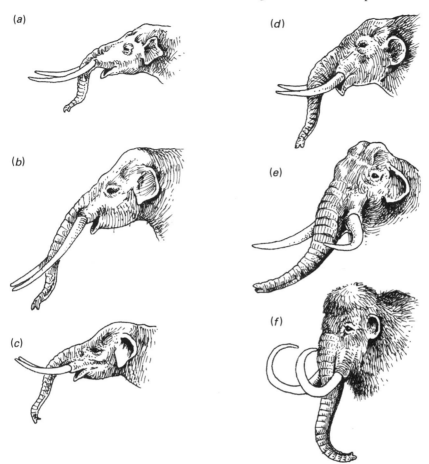

Figure 2.6. Extinct proboscideans from the late Pleistocene Period. (*a*) *Cuvieronius*; (*b*) *Stegomastodon*; (*c*) *Haplomastodon*; (*d*) *Mammut*; (*e*) *Mammuthus meridionalis*; (*f*) *Mammuthus primigenius*. The last records for the *Cuvieronius, Stegomastodon* and *Haplomastodon* spp. are from South America between 11,000 and 13,500 BP. *Mammut* (the mastodon) was last recorded in sites from North America with dates between 9000 and 12,000 BP. Similar dates (*c.* 11,000 BP) hold for most recent records of mammoths (*Mammuthus* spp.) in Eurasia and North America. (Dates from Anderson (1984). Illustrations redrawn from Anderson (1984).)

2.2). Important species of plants have migrated over continental ranges at different rates and with differing patterns of movement (Fig. 2.4). Terrestrial ecosystems have changed in their character with some ecosystems disappearing and others forming locally (Fig. 2.5) and over large areas. The rate of change in vegetation in a given area may be relatively

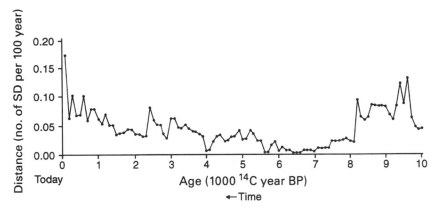

Figure 2.7. Vegetation change in central Minnesota (Billy's Lake) through the past 10 000 years BP based on pollen changes in 100 year intervals. The distance on the *y*-axis increases as the rate of change in the vegetation. From Jacobson and Grimm (1986).

slight for long periods and then increase during periods of relative instability (Fig. 2.7).

Fire and insect outbreaks

There have also been changes in the patterns of major external factors that shape terrestrial landscapes. Over the past 10 000 years in the Atherton Tableland in northern Queensland (Fig. 2.8), one sees a transition from sclerophyll forest (dominated by species of *Eucalyptus*) to the rain forest species that are found there today. This started in about 6800 BP (before present) and was associated with a reduction in the accumulation of charcoal (indicating a reduction in the number of wildfires, which is associated with an increase in rainfall). Similar variations in the distribution of rain forest, savanna and dryer vegetation types have been recorded elsewhere in the tropics (e.g. Bartlett and Barghoorn 1973; van der Hammen 1974, 1985).

There are records of reductions in the abundance of major species that appear to be caused by the outbreak of diseases at continental scales. One example is the abrupt demise of eastern hemlock (*Tsuga canadensis*) occurring about 5000 years ago (Fig. 2.9). This decline has been attributed to the infestation of the insect pest the hemlock looper (*Lamdina fiscellaria*), spreading throughout the range of a tree species largely lacking a resistance to the insect (Davis 1981a). The insect may have spread from the west where it is resident on western hemlock (*Tsuga heterophylla*) and

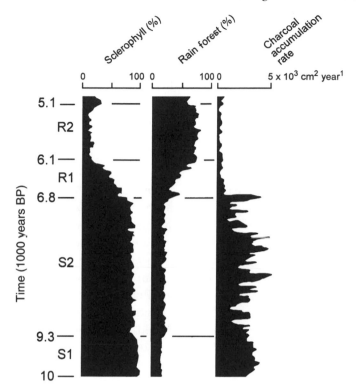

Figure 2.8. Changes in the relative frequencies of pollen of plant species characteristic of sclerophyll woodland dominated by *Eucalyptus* and rain forest during the early to middle Holocene Epoch at Lake Barrine on the Atherton Tableland in northern Queensland, Australia. The decrease in charcoal around 6800 years ago implies a decreased fire frequency. From Walker and Chen (1987).

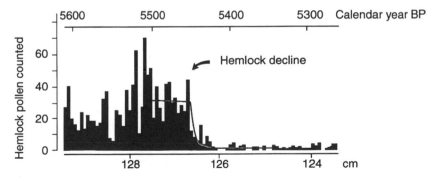

Figure 2.9. Number of pollen grains of eastern hemlock (*Tsuga canadensis*) per unit sediment from Pout Pond, New Hampshire, USA. Note the severe regional decline in eastern hemlock in the mid-Holocene Epoch. Hemlock stages a slow recovery over the next 1000 or so years. From Allison *et al.* (1986).

erupted in the eastern hemlock forest where natural controls (disease or predators) were lacking (Davis 1981b).

One particularly significant source of change in the Holocene Epoch has been the increasing level of human alteration of landscapes and eventually the global environmental condition. It has been postulated that the pronounced extinction of Pleistocene large animals (Martin and Klein 1984) was the result of the increased hunting pressure from prehistoric peoples. This would extend the measurable global effects of human alterations of the environment far back into prehistory. Similarly, human alterations of fire frequencies by either accidental or intentional means have also been implicated as instances where prehistoric people altered the character of relatively large areas of land (Singh *et al.* 1981; Nicholson 1981). Barker (1985) sees a considerable prehistoric alteration of vegetation in Europe in Neolithic times, between 6000 and 4500 BP, presumably associated with increased human densities and developing agricultural technologies. Delcourt (1980) postulated that the change from deciduous forests to coniferous forest in the mid- to late-Holocene Epoch was the result of Archaic Indians setting fires to drive game and to encourage the quality of herbivore browse. Similar practices are speculated to have occurred in Australia as far back in time as 40 000 BP (Nicholson 1981).

Change in atmospheric CO_2 levels in historical time

A striking change associated with modern human society has been the alteration in the level of atmospheric CO_2 associated with the increased burning of fossil fuels (coal, petroleum, natural gas) since the industrial revolution (Baes *et al.* 1977). In 1957, Keeling (e.g. Keeling 1983) began measuring the concentration of CO_2 in the atmosphere at Mauna Loa, Hawaii. The initial observations indicated an annual fluctuation in the amount of CO_2. Over time, these observations also confirmed that the amount of CO_2 in the atmosphere was increasing exponentially and was associated with the levels of consumption of fossil fuels (Baes *et al.* 1977). Figure 2.10 shows the pattern of variation in CO_2 measured at four different stations (Barrow, Alaska; American Samoa; Mauna Loa, Hawaii; South Pole) from 1973 to 1983 (Harris and Bodhaine 1983). While there is a considerable variation in the degree of annual oscillation of these data (this appears to be the result of photosynthesis and respiration of the terrestrial surface; see Chapter 13), there is a clear tendency for the measured levels of CO_2 to increase regularly throughout this record in all cases.

Because CO_2 is an essential component of plant photosynthesis, systematic, multiple-year change in CO_2 levels immediately suggests that

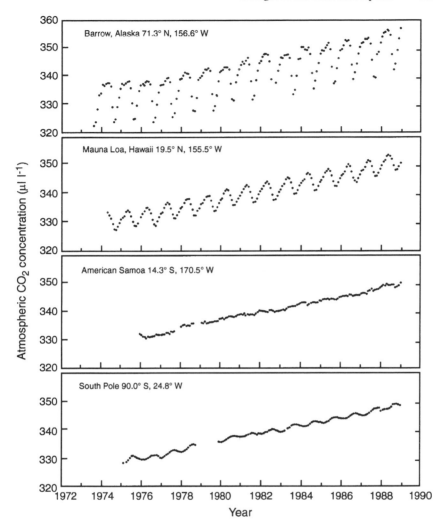

Figure 2.10. Selected monthly mean CO_2 concentrations from continuous measurements at National Oceanic and Atmospheric Administration/Geophysical Monitoring for Climate Change (NOAA/GMCC) stations at four locations (Barrow, Alaska; American Samoa; Mauna Loa, Hawaii; South Pole) from 1973 to 1983. From Harris and Bodhaine (1983).

these changes may in some way be altering the way plants function. If the functioning of plants is changed in some way, what other changes might ensue if the CO_2 in the atmosphere continues to increase?

There is evidence that the number of stomata per unit area on plant leaves has varied inversely in relation to the change in CO_2 concentration since the industrial revolution (Woodward 1987a,b; see Fig. 2.11). Stable-

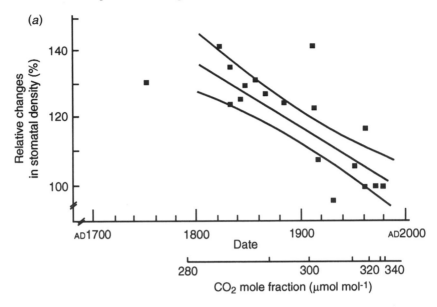

Figure 2.11. (*a*) Historical relative values of abaxial (lower surface) stomatal densities of herbarium stored leaves of *Acer pseudoplatanus, Carpinus betulus, Fagus sylvatica, Populus nigra, Quercus petraea, Q. robur, Rhamnus catharticus* and *Tilia cordata* from samples from the Botany Department Herbarium, University of Cambridge. Only leaves on reproductive shoots from herbarium specimens (five leaves of each species sampled from different dates back to 1750 AD) were sampled. All specimens were from collections made in the Midlands of England. Absolute stomatal densities varied by a factor of about two across all species. The graph shows the percentage stomatal density relative to recent collections (made between 1970 to 1981). The lower line shows reconstructed atmospheric CO_2 mole fraction (from Friedli *et al.* (1986), based on CO_2 concentration in air trapped in bubbles in an Antarctic glacier), scaled temporally to match collection dates of herbarium specimens. The linear regression line (shown with 95% confidence limits) shows a 40% reduction in the ratio of stomatal densities over the past 200 years ($r = -0.828$). (*b*) Experimental results from changes in the CO_2 mole fraction for individuals of *Acer* in *A. pseudoplatanus* (●), *Quercus robur* (◆), *Rhamnus catharticus* (▲), and *Rumex crispus* (■) grown in the laboratory under different concentrations of CO_2. Data from herbarium material (○) for these species are plotted according to the glacial CO_2 mole fraction as reconstructed in (*a*). The decrease in stomatal density seen in the herbarium specimens appears to be reproducible in the laboratory by manipulating CO_2 concentrations. Further, the magnitude of the laboratory responses is at what appears to be appropriate levels. Figures from Woodward (1987a).

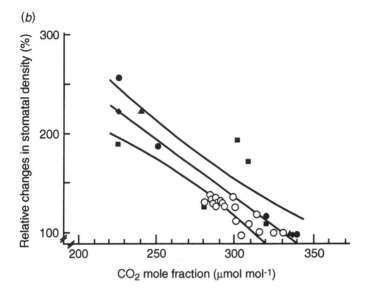

(b)

isotope analysis of carbon isotopes and calculation of the ratio of $^{13}C/^{12}C$ as an index of the leaf water-use efficiency (the ratio of carbon fixed by photosynthesis to water used for transpiration) for the same herbarium leaves (Woodward 1993) implies that the water-use efficiency has increased since the industrial revolution. This may be the result of an increase in photosynthesis and a decrease in stomatal transpiration, which is predicted by the reduction in stomatal density since the industrial revolution, and the associated increase in atmospheric CO_2.

The potential for anthropogenic climate change

An important consequence of an increased ambient level of CO_2 in the atmosphere is related directly to the role of this relatively rare atmospheric constituent in the heat balance of the Earth: CO_2 is a greenhouse gas. The Earth radiates shortwave radiation from the sun as longer wavelength radiation. If the Earth had no atmosphere, the surface would attain a temperature at which the heat radiated to space as longwave radiation would balance that received from the sun. This would occur when the Earth's surface has a temperature of 256 K. The average temperature of the Earth's surface (288 K) is 32 K warmer because of the effect of the atmosphere (Sarmiento and Bender 1994).

The CO_2 in the atmosphere allows light to enter and becomes heated by exiting longwave radiation (Fig. 2.12). Clouds absorb longwave radiation

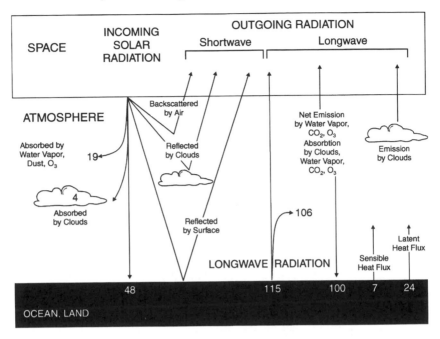

Figure 2.12. Schematic diagram of the global average components of the Earth's energy balance. From MacCracken and Luther (1985).

(a greenhouse function) but also reflect incoming radiation back to space (a cooling function). Other atmospheric components are also active in the Earth's energy balance, notably CH_4, O_3, NO_x, chlorofluorocarbon compounds and water vapour, which act as greenhouse gasses. Some of these other atmospheric components are increasing at rates that exceed that of CO_2. An estimate of the possible effects of these atmospheric components based on projected increases from 1980 to 2030 are shown in Fig. 2.13 (Bolin *et al.* 1986). These computations are based on relatively simple models that do not attempt to take into account the interactive nature and the feedback effects among the components of the atmosphere (Dickinson 1986).

These feedback effects are incorporated into massive computer models (General Circulation Models or GCMs) in an attempt to understand the global climate system and to assess the effects of changes in the atmosphere such as those that might result from the observed change in atmospheric CO_2 (Dickinson 1986). The resultant climate models are complex. At present, even the largest and fastest computers are unable to solve the equations at a fine spatial level (solutions are typically for an Earth divided

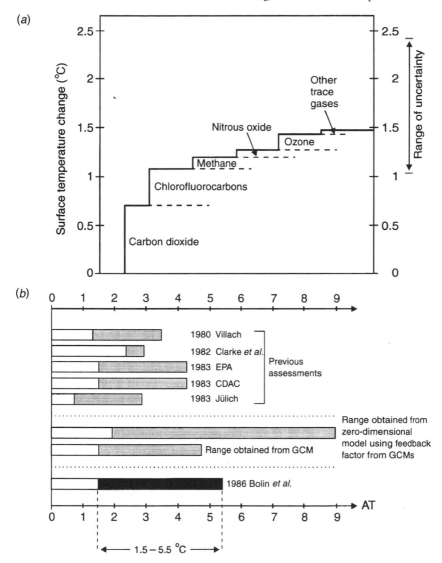

Figure 2.13. (*a*) Cumulative surface temperature warming caused by an increase in CO_2 and other trace gasses from 1980 to 2030 as computed by a one-dimensional model. In these calculations the atmospheric concentration of CO_2 is expected to double in this time period. From Ramanathan *et al.* (1985).
(*b*) Estimates of GCM models that include dynamic feedbacks for the global warming expected from a doubling of CO_2. AT, globally averaged surface temperature. For details on Villach (1980), see World Climate Programme (1981); EPA 1983 is reported in Seidel and Keyes (1983); Jülich (1983) is reported in Volz (1983). Figure is from Bolin *et al.* (1986).

into large blocks *c.* 2° latitude by 2° longitude) or to represent ocean currents that change dynamically as the climate changes (see Chapter 14).

The GCMs vary with respect to the way that processes important to the Earth's climate are depicted in a given formulation. They also are formulated with different assumptions about how best to approximate conditions that cannot be simulated in the models directly (e.g. the formation of clouds, the energy exchange between the atmosphere and the ocean, or the formation of ice at sea). Despite these model differences, the models converge in that they do predict an increase in the Earth's temperature as a consequence of increased atmospheric CO_2. The models differ considerably in the details of their predictions even at the coarse scales at which they have been developed. In some cases, they produce substantial differences in the patterns of temperature change in space. The models also differ significantly from one another with respect to patterns of precipitation (Dickinson 1986).

There is a considerable level of uncertainty in the predictions of the rates of increase of atmospheric CO_2 over the coming years, in the reliability of the GCMs, in the effects of the behaviour of the oceans under a changed climate and so on. Nevertheless, there is a need for a clearer understanding of the response of vegetation to climate change. This is true both with respect to the effect of climate on the vegetation and with respect to the role of the terrestrial biota in the carbon cycle of the Earth.

Anthropogenic changes on natural landscapes

The concern over possible climate change should not overshadow other major changes being wrought by humans on the Earth's surface. Conversion of land from native vegetation to agricultural and urban uses has greatly altered the landscapes of Europe and parts of North America, Australia and Asia. Land conversion is continuing to alter the landscapes of developing nations in the tropics and subtropics of Asia, Africa and Central and South America. There is also a considerable degree of forest clearing in the Russian and Canadian boreal forests. As the patterns of natural landscapes are changed, habitats for plants and animals are altered and certain species are eliminated while others are favoured. There is a great deal of proper concern as to what these changes mean for the biotic diversity of our planet. In addition, land conversion releases CO_2 into the atmosphere and is, therefore, a source of an important greenhouse gas. Further, land cover changes alter the surface properties of the earth surface in ways that can also alter local and global climates.

In the chapters that follow, ecological models as tools for evaluating and predicting the consequences of environmental changes will be introduced and discussed. Many of these large-scale models have been applied to assess initially the effects of climate change on ecological systems. In part, this is because climate change has been a topical issue while many of these models were being developed. Also, it is often more straightforward to apply models to evaluate the effects of an external stress (such as climate change) rather than to study the more complex problem of evaluating the consequences of internal changes (such as the extinction of species). The models and concepts that have been developed to deal with evaluating environmental change can provide useful insights into the consequences of changes in plant and animal habitats resulting from human land-use changes.

Concluding comments

There has been a rich and complex pattern of changes in important environmental conditions that affect the vegetation of the terrestrial surface of the Earth. These changes have occurred over almost any time scale that one might select. Terrestrial vegetation should be in a state of dynamic change in response to significant environmental changes over multiple time and space scales. Unfortunately, there are almost no direct observations of landscapes over sufficiently long periods to indicate the longer-term changes that may result from what we feel are likely to be important longer-term environmental changes.

One course of action to predict how terrestrial vegetation might change in response to environmental change is to understand better the basic processes and interactions controlling dynamic ecosystem function under a range of time and spatial dimensions. A. S. Watt (whose work will be discussed in more detail in later chapters) termed this to understand 'the plant community as a working mechanism' (Watt 1947). The magnitudes and scales of the changes in important environmental variables are sufficient to imply a dynamic change in the terrestrial ecosystems that cover the Earth at multiple scales of observation.

Cowles (1899), one of the first American ecologists to study ecological succession, characterised succession as 'a variable approaching a variable'. This perception that landscape dynamics are those of a system perturbed away from – but moving towards – an equilibrium that is itself changing seems the appropriate paradigm for terrestrial ecosystems varying under conditions of omnipresent change.

Appreciating the degree to which the environment has changed, from the geological past through the historical past and into the present, highlights the inappropriateness of the notion of constancy suggested by Jefferson and Pope and discussed in the introduction to this chapter. Of course, this idea of constancy extends well beyond Mr Jefferson and Mr Pope: it is a theme echoed by a long list of ecologists and conservationists in catch phrases such as the balance of nature, the wilderness concept or the virgin forest and prairie. We manage parks, nature reserves and conservation preserves with these concepts as a basis or as a philosophical underpinning for our actions.

One can imagine the difficulty of a series of Pleistocene park superintendents trying to preserve the natural character of their park as the continental glaciers retreat, populations of cold-adapted species become locally extinct and exotic new species invade from the south. Of course, we may have, or be having, the identical types of problem ourselves. Conservation of species and the preservation of nature are desirable, even though the validity of the scientific basis of some of the concepts behind these goals has paled since the eighteenth century. As knowledge about ecological systems and their lack of constancy increases, it is important to understand better the dynamics of natural systems. We need to understand both the nature of change and the responses of natural systems to change, as the press of human activities alters the environment on local, regional and even global scales.

The following chapter deals with one element of understanding the response to change, the ecosystem concept. The ecosystem will be presented as arising from historical roots as a flexible concept intended to focus on dynamic change in ecological systems. In this view, a wilderness is a sort of ecosystem (perhaps one that has been exposed to only a moderate level of external change in the environment or one that covers a large area), but a managed pine plantation is also an ecosystem. The challenge is to understand how such systems change dynamically and how we can learn from each to help in predicting the response of either.

3 · *Temporal scale, spatial scale and the ecosystem*

Ecology is in a transition from being a descriptive science to becoming a predictive science that can answer questions involving dynamics and change of natural systems. The ecosystem concept was an initial, important step in this transition. The implications of the word ecosystem and its continuing development as a scientific concept are important beyond an interest in simply defining the term. The word, its genesis and its implications reflect several basic issues in ecology, as well as other sciences.

The ecosystem concept involves understanding the appropriate levels of abstraction with which the processes and parts of natural systems can be treated to predict better their dynamics. This is central to understanding the responses of the Earth's surface to environmental changes, particularly novel, large-scale changes, and, therefore, to the theme of this book. How should one abstract the complex and highly interactive webs of physical, chemical and biological processes and entities in appropriate ways to solve particular problems? Abstraction is an essential aspect of science, and the most appropriate way to abstract ecological systems is very much a topic still under development.

Roots of the ecosystem concept

Ecosystem is a sufficiently common term in scientific writings as well as in everyday use that many find it surprising that it has a recent origin. In what has become a classic paper, A. G. Tansley used and simultaneously defined the term ecosystem in 1935. This neologism was in a section of a paper treating the use and misuse of terms in community ecology. He defined an appropriately abstract term to describe interactive ecological systems and intended to define the central concepts involved in scientific abstraction in ecology:

. . . the more fundamental conception is, as it seems to me, the whole *system* (in the sense of physics), including not only the organism-complex, but also the whole complex of physical factors forming what we call the environment of the biome – the habitat factors in the widest sense. Though the organisms may claim our primary interest, when we are trying to think fundamentally we cannot separate them from their special environment, with which they form one physical system.

It is these systems so formed which, from the point of view of the ecologist, are the basic units of nature on the face of the earth. Our natural human prejudices force us to consider the organisms (in the sense of the biologist) as the most important parts of these systems, but certainly the inorganic 'factors' are also parts – there could be no systems without them, and there is constant interchange of the most various kinds within each system, not only between the organisms but between the organic and the inorganic. These *ecosystems*, as we may call them, are of the most various kinds and sizes. They form one category of the multitudinous physical systems from the universe as a whole down to the atom.

The ecosystem concept had its roots in theoretical concepts regarding the organisation and dynamics of natural systems. F. E. Clements (1916) and, particularly, John Phillips (1934, 1935a,b) were ascribing to ecological communities the attributes of a super-organism: a highly organised and co-evolved assemblage of plants and animals interacting in a dynamic system analogous to the manner that cells embryologically interact to produce an organism. The underlying conviction was that evolution and internal interactions would produce homogeneous regional 'climax' vegetations or communities of uniform species composition (Rich 1988). It was Tansley's negative view of this interpretation of the community as a super-organism that inspired his development of the ecosystem concept. Tansley's original development of the word ecosystem was as an alternative to the word community. After Tansley's introduction of ecosystem the term community still continued to be used in ecology. This usage continues today. In present usage, a community is an assemblage of interacting plant and/or animal populations with some degree of interdependencies. (The degree of interdependency and/or co-evolution in the structure of communities will be discussed in Chapter 5 in the context of the ecological niches of species.)

Important modifications of the use of ecosystem as a concept developed as ecology developed after the mid-1930s. One of the most important of these is associated with the work of Lindeman (1942), who provided one of the first examples of an ecosystem (a lake) and placed an emphasis on understanding the material flows into and out of components of the system.

The trophic–dynamic concept

According to McIntosh (1985), a now famous paper, *The Trophic–Dynamic Aspect of Ecology*, published posthumously by Raymond Lindeman in 1942, represents the birth of ecosystem ecology. In this paper, Lindeman gave a ringing endorsement to Tansley's ecosystem concept with: 'The *ecosystem* may be formally defined as the system composed of the physical–chemical–biological processes active within a space time unit of any magnitude, i.e., the biotic community *plus* its abiotic environment. The concept of the ecosystem is believed to be of fundamental importance in interpreting the data of dynamic ecology.' The Lindeman (1942) publication also gave a tangible demonstration to Tansley's idea that the ecosystem had a '...constant interchange of the most various kinds within each system, not only between the organisms but between the organic and the inorganic.' Lindeman asserted that a lake could best be considered as an ecological unit in its own right. A lake was an entity beyond the plant aggregations (phytoplankton, marginal emergent plants on the lake margin, etc.) that might be considered by a botanist, or the animal communities that might be studied by a zoologist.

Lindeman synthesised data on the biomass of different feeding or trophic levels from three lakes in Minnesota and characterised a lake in terms of the transfer of energy from one part of the system to another (Fig. 3.1). According to Cook (1977), a reviewer of this work from a historical perspective, Lindeman's principal contributions to ecology were:

1. To stress the major role of trophic function and to emphasise the quantification of trophic functions to determine succession
2. To establish the validity of a theoretical orientation in ecology
3. To identify a fundamental dynamic process (energy flow) with which the seasonal trophic relations of animals could be integrated into the long-term process of community changes.

Elements of Lindeman's work had been developed and discussed earlier (McIntosh 1985). The importance of energetics (Lotka 1925), the progressive removal of energy through the steps of a food chain on the basis of thermodynamics (Semper 1881; Petersen 1918), the lake as an interactive system (Forbes 1897; Thienemann 1918; Allee 1934) were all developed prior to Lindeman. The uniqueness of the Lindeman paper (1942) was a product of the creativity and theoretical richness of his work and of the timing of the publication.

Even so, Lindeman's contribution was not extensively influential until

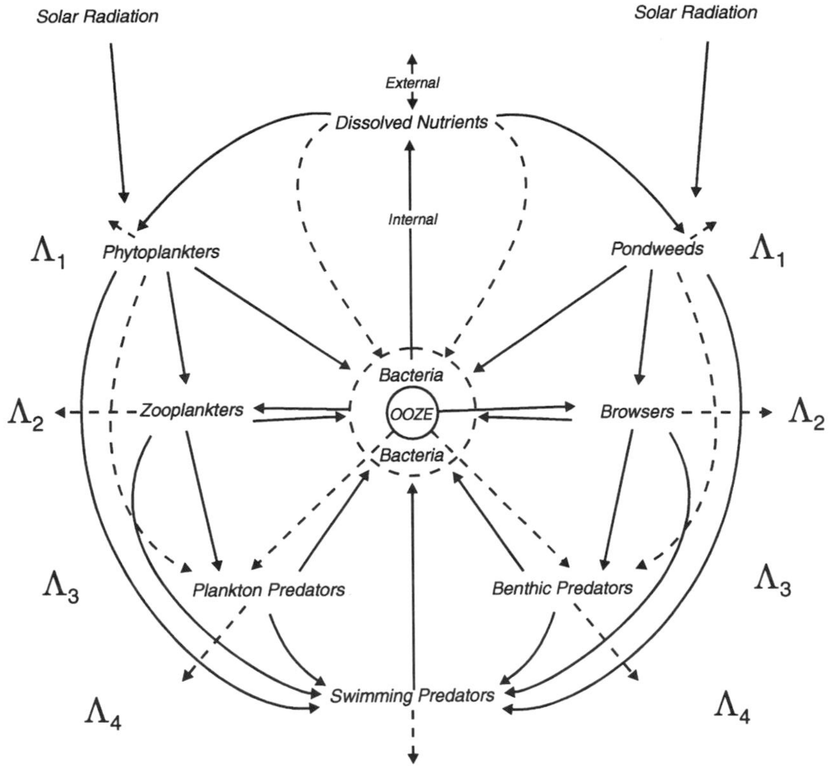

Figure 3.1. Generalised lake food-cycle relationships in a lake (from Lindeman 1942). Transfers of energy are indicated by arrows, and the Greek lambdas indicate trophic levels (producers, consumers, etc.).

the 1950s (Wiegert 1988). At that time, E. P. Odum (1953) produced an ecology textbook organised around the concepts of ecosystems and their structure and function. By the 1960s, there was an explosive development in ecosystem ecology (McIntosh 1985). Because of the central importance of understanding the trophic-dynamics of ecosystems (ecoenergetics), Odum in 1968 could state, 'Ecoenergetics is the core of ecosystem analysis'.

Rich (1988) maintains that there have been two explanations of the root causes of the structure of exchanges of energy and matter that Lindeman (1942) described as trophic-dynamics. These views are focused on the issue of whether or not it is the properties of individuals or the attributes of interactive ecosystems that are at the root of determining cause of observed trophic structure.

The first of these views is based on the theory of evolution. In this view, evolution is driven by natural selection on heritable variation in the too numerous offspring of individuals. These offspring are produced because excess fertility is an integral property of organisms. This intrinsic tendency for over-production of offspring is sometimes called the Malthusian axiom. Thus, predation (or decomposition), the next step in the trophic pyramid, can occur because there is an excess of prey produced in natural systems (Rich 1988).

The second explanation for the cause of trophic structure originates in the holistic philosophy of Herbert Spencer, a contemporary of Darwin, who, like Darwin, was interested in evolutionary deism (Rich 1988). Spencer was a remarkable scientific innovator and is considered by many to be a founder of sociology and a co-founder (with Freud) of psychology. A Spencerian explanation of the root of trophic structure would be that evolution develops organised configurations with increased internal organisation and mutual interdependencies. In this view, prey are seen to have a surfeit of offspring because they have evolved in a system in which these additional offspring will be eaten by the predators of the system.★

The importance of an evolutionary view of population dynamics versus a holistic view of system dynamics (or patterns) arises repeatedly in ecology. Are regularities in successional change the products of individuals found in a particular location or are they an ecosystem feature arising inevitably from the dynamics of ecologically organised complex systems? Are the patterns of species abundance the consequences of internally organised communities or are the observed patterns of rareness and commonness simply a product of chance? Can we predict change at the ecosystem level without resolving these seemingly fundamental issues?

Food chains, food webs and element cycles

When the numbers of animals in a community are grouped according to size, by the trophic level at which the various species feed, or according to the amount of energy transferred by feeding relations, a pyramid structure often results. In a collection of animals from a fixed area, there are generally fewer large animals than small (for all sizes). There is a greater biomass of

★ It is significant in understanding the emphases in ecosystem studies to note that A. G. Tansley was a co-author of the 1899 edition of Spencer's *Principles of Biology*, the leading English biology text of the late nineteenth century. Tansley's definition of the ecosystem and his lack of acceptance of Clements' and Phillips' holistic views that the ecological community was a coherent, organism-like entity was not a rejection of a holistic view of ecology, it was a debate among holistic ecologists (Rich 1988).

Table 3.1. *Area, net primary productivity[a], average biomass and percentage of primary production consumed by herbivores for terrestrial ecosystems*

Ecosystem type	Area (10^6 km^2)	Mean net primary productivity (g m^{-2} year^{-1})	Mean biomass (kg m^{-2})	Primary production consumed by herbivores (%)
Tropical rain forest	17.0	2200	45	7
Tropical seasonal forest	7.5	1600	35	6
Temperate evergreen forest	5.0	1300	35	4
Temperate deciduous forest	7.0	1200	30	5
Boreal forest	12.0	800	20	4
Woodland and shrubland	8.5	700	6	5
Savanna	15.0	900	4	15
Temperate grassland	9.0	600	1.6	10
Tundra and alpine meadow	8.0	140	0.6	3
Desert scrub	18.0	90	0.7	3
Extreme desert, rock, ice and snow	24.0	3	0.02	1
Cultivated land	14.0	650	1	1

Note:
[a] Information on net primary productivity from this table will be discussed later in Chapter 13 in terms of the problem of estimating the productivity of the Earth using satellite-based observations.
Source: From Whittaker and Likens (1975).

plants than herbivores and of herbivores than carnivores in most terrestrial ecosystems. Further, the transfer of chemical energy from the feeding on plants by herbivores is greater than the energy transferred from herbivores to carnivores. There are exceptions to these general patterns, but trophic pyramids can be explained in terms of the Second Law of Thermodynamics (DeAngelis 1992). Across a broad range of terrestrial ecosystems, transfer of material from plants to animals seldom exceeds 10% over large areas or over several years (Table 3.1). Feeding relations among plants and animals are often depicted as food chains (in the least complex cases) and food webs (in cases in which there are multiple pathways through the system).

The importance of the circulation of energy and matter in ecological systems remains a dominant theme in ecosystem ecology. The processing of material and forms of energy through terrestrial systems and the residence time of materials in different ecosystems links terrestrial ecosystems directly to other global systems (atmosphere and oceans).

Part of the initial interest in ecosystem energetics and one of the problems that was focal in Lindeman's initial formulation of the trophic–dynamic concept stemmed from the observation that there are seemingly fundamental patterns in the plants and animals constituting ecological communities. Both Shelford (1913) and Elton (1927) were instrumental in emphasising the importance of these patterns although the basic ideas were known earlier (see McIntosh 1985).

Ecological succession

The fundamental concept in succession theory is that, following an event that changes the system, there is a regular and predictable replacement of one sort of community with another over time until a final community develops which perpetuates itself and is called the climax community (Clements 1916, 1928). As a theory for how ecological systems change over time, succession theory continues to be a hotly debated topic in the ecological sciences (Glenn–Lewin *et al.* 1992). Among the best examples of such succession in the USA are observations on the patterns of vegetation on abandoned fields that over time become forests and the pattern of vegetation change on new substrates (sand bars, receding glaciers, new lava flows).

Debates in succession theory included arguments as to whether or not the communities that occur in a sequence do so because the earlier communities are necessary to make it possible for the next community to grow at the site (Clements 1936; Egler 1954). There are other debates as to the climax ecosystem that might occur in a given locale (Drury and Nesbit 1973). In a broader context, there is a debate as to whether the climax system even exists or if natural systems always change over time (Connell and Slatyer 1977).

There is a division of opinion between scientists who view succession as an ecosystem-level property (to one degree or another) and those who view the process of succession as arising from the attributes and interactions of individual organisms (McIntosh 1981). (I noted earlier (Shugart 1984) that both sides are partially correct and that the arguments result in part from the differences in point of view, particularly with regard to time and spatial scales of interest.) The two points of view can be illustrated by examples.

1. *Whole system theories.*   Figure 3.2 was developed by Van Cleve and Vierick (1981) to illustrate what they felt are the important features in understanding the dynamics of the forests in the boreal forests in the

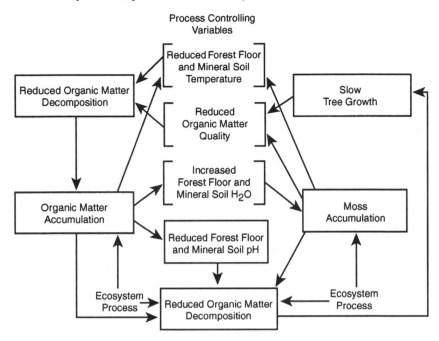

Figure 3.2. The factors that control the dynamics of forests in the northern part of North America according to Van Cleve and Vierick (1981). Note that the species of plants do not enter into this diagram of important processes controlling boreal forest dynamics.

vicinity of Fairbanks, Alaska, USA. The area lies near the Arctic Circle in one of the remaining wildernesses on the North American continent. The diagram does not mention plant species at all in its explanation of the dynamics of the forest. Instead it identifies processes, tree growth and organic matter break-down and physical conditions such as soil temperatures. The successional dynamics of the ecosystem are conceptualised as having their origins in a complex web of interactions that involves physical and chemical processes. Because these processes are strongly related to the soil temperature, one finds very different patterns of ecological succession following a fire on south-facing slopes (that have a greater exposure to the sun and are warmer, Fig. 3.3) than one finds on a north-facing slope (Fig. 3.4). This same importance of physical factors has also been identified in the boreal forests of what was the USSR (Sukachev, 1968a).

2. *Individual organism-oriented theories.* The explanation of the pattern of successional dynamics shown in Figs. 3.2 to 3.4 (and presumed to be

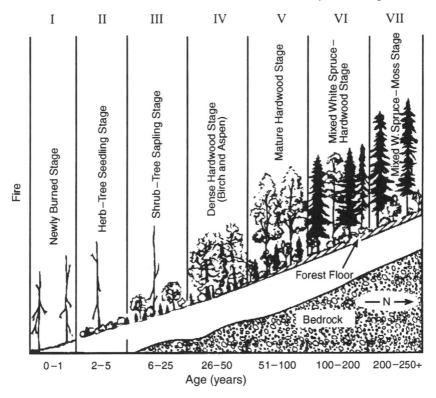

Figure 3.3. Ecological succession on south-facing slopes in the vicinity of Fairbanks, Alaska. From Van Cleve and Vierick (1981).

correct for the boreal forests of the northern hemisphere) takes no explicit account of the biology of the species. However, one can also think of other systems in which the plant biology at the individual plant level and the interactions among the species can have great importance in the understanding of the ecosystem dynamics (Connell and Slatyer 1977; Noble and Slatyer 1978, 1980; Grime 1979a,b). Pickett *et al.* (1987) reviewed the causes and processes in vegetation dynamics and identified a multi-levelled set of interacting processes and factors that are strongly related to ecological differences among species (Table 3.2). For example, Fig. 3.5 shows a strangler fig, a frequently encountered type of tropical tree. A strangler fig goes through a life cycle starting as an epiphyte in the rain forest canopy. It then drops roots to the ground and encases the host tree in which it is found. Stranglers can eventually become one of the dominant canopy species in the forest. The successional dynamics of a forest that has strangler figs (*Ficus* spp.) as an important component are

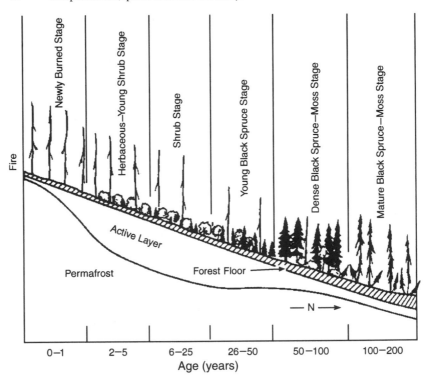

Figure 3.4. Ecological succession on north-facing slopes in the vicinity of Fairbanks, Alaska. Note the great differences compared with Fig. 3.3 all resulting from changes in ecological processes related to the ecosystem interactions shown in Fig. 3.2. From Van Cleve and Vierick (1981).

strongly influenced by the interesting biology of the species. Figs are often very large both in height and crown spread. They frequently are a major source of food for birds and mammals as well as other less conspicuous animals. The unusual habits of these figs allow them to compete success-fully with other, non-strangling types of tree. Strangler figs can join multi-ple host trees together and can be quite a hindrance to foresters attempting to harvest trees in the rain forest, particularly in Malaysia, Indonesia and the Solomon Islands (Whitmore 1975). An ecologist ori-ented toward an individualistic view of ecosystems dynamics might claim, with justification, that the unique features of this plant life form make it necessary to include their presence, absence and abundance in an under-standing of how tropical forests function.

The understanding of the factors that control succession and of the level at which ecological succession operates is limited by the nature of

Table 3.2. *A summary of the causes, processes and factors of vegetation dynamics*

General cause of succession	Contributing processes or conditions	Modifying factors
Site availability	Coarse-scale disturbance	Size, severity, time, dispersion
Differential species availability	Dispersal	Landscape configuration, dispersal agents
	Propagule pool	Time since last disturbance, land-use treatment
	Resource availability	Soil condition, topography, microclimate, site history
Differential species performance	Ecophysiology	Germination requirements, assimilation rates, growth rates, population differentiation
	Life history	Allocation pattern, reproductive timing, reproductive mode
	Environmental stress	Climate cycles, site history, prior occupants
	Competition	Hierarchy, presence of competitors, identity of competitors, within community disturbance, predators, herbivores, resource base
	Allelopathy	Soil chemistry, soil structure, microbes, neighbouring species
	Herbivory, predation and disease	Climate cycles, predator cycles, plant vigour, plant defences, community composition, patchiness

Source: From Picket *et al.* (1987).

the data concerning the long-term dynamics of ecosystems. Most of the information on ecosystem successional dynamics is obtained indirectly (Whittaker 1975). Ecological data on succession are obtained by collecting information regarding the pattern of vegetation at a certain age following a disturbance and then looking for general patterns when several

Figure 3.5. A stage in the life cycle of a strangler fig (*Ficus cunninghamii*) from the tropical rain forests of Queensland, Australia. The roots of the fig, which has started as an epiphyte in the top of the canopy, have wrapped around the host tree. Eventually, the host tree will be killed and replaced by the strangler. Photograph by H. H. Shugart.

such cases are sequenced according to their age. In some cases, these sequences are actually created by unusual physical processes. For example, in some areas glaciers have been receding at a steady rate over time. In these areas, the vegetation becomes progressively older as one walks away from the glacier and the sequence of plant communities physically reflects the expected successional pattern.

Most studies of ecological succession are not in such idealised conditions as the vegetation sequence at the foot of a receding glacier. The sequencing of the plant communities is more arbitrary and the patterns in the data are often identified to a degree according to the theories of succession that are prevalent at the time. The arrangement of the data is subject to the point of view of the scientist who is organising the data (and is not an ideal way to test theories).

The biogeocoenosis and the ecosystem

Odum (1971) points out that one of the roots of the ecosystem concept, the fundamental idea of a unity of organisms and their environments, is a sufficiently venerable concept to appear as far back in history as one may care to look. Woodward (1987b) traces the relationship to the peak of Ionian philosophy (third to fifth centuries BC) and the writings of Menestor in the fifth century BC (Morton 1981). The work of Theophrastus (c. 370 to 285 BC), sometimes referred to as the 'father of ecology' and a pupil of both Aristotle and Plato, identifies 'sympathetic relationships' between plants and their environments. Theophrastus was an ingenious observer of relationships among plants and their environment and even outlined transplantation experiments to investigate these relationships that presage experiments of today (Woodward (1987b), drawing on Morton (1981) and on translations of Hort (1916)).

More modern influences in the formulation of the ecosystem concept were rooted in the late nineteenth century with the works of Möbius (1877), describing an oyster reef as a 'biocoenosis', and of an American, Forbes (1897), describing a lake as a 'microcosm'. In Russia, Dokuchaev (1889) expanded Möbius' ideas to define a 'biogeocoenosis', a joining of biotic and abiotic elements.

The biogeocoenosis is a concept with many similarities to the ecosystem. Even today there is a tendency for English speakers to use the term ecosystem and for Germanic or Slavic language speakers to use biogeocoenosis (or geobiocoenosis) in ways that are quite similar. The terms are similar, in some senses, confusingly so. The concepts differ significantly with regard to whether or not the area of the system considered is fixed (the biogeocoenosis) or of any size (the ecosystem).

A biogeocoenosis can be defined (Sukachev 1968a) as:

> ...a combination on a specific area of the earth's surface of homogeneous natural phenomena (atmosphere, mineral strata, vegetable, animal and microbic life, soil, and water conditions), possessing its own specific type of interaction of these components and a definite type of interchange of their matter and energy among themselves and with other natural phenomena ...

The principal difference between the biogeocoenosis and the ecosystem is in the emphasis on a fixed or specific area of the former and on arbitrary dimensions of the latter. This is not a trivial distinction. Many supporters of the biogeocoenosis concept interpret the ecosystem as a nebulous approximation of reality (e.g., Sukachev 1968a,b; Gilmanov 1992).

Along with generating a relatively elaborate terminology (Fig. 3.6), the biogeocoenosis concept has embedded in it some relatively strong assumptions about the intrinsic structure of ecological systems. Important among these assumptions is that there is a degree of homogeneity in the interactions among the biota, geology and atmospheric conditions that can be expected at a certain size dimension. This is the size dimension of the biogeocoenosis. It is this definable size that makes the biogeocoenosis a natural unit in nature, one that can be identified independently of the views of the observer. In this sense, the biogeocoenosis begins to resemble the super-organism concept associated with Clements (1916), Phillips (1934, 1935a,b) and others (see McIntosh (1985) for a review). To clarify the emphasis in ecosystem versus biogeocoenotic views, it is helpful to review the ecosystem definition (as originally proposed) with more detail with respect to spatial considerations.

Recent interpretations of the ecosystem

Tansley's original definition of the ecosystem was an abstract concept free of intrinsic considerations of space or time boundaries. The current uses of the word ecosystem have evolved from Tansley's fundamental ('... the basic units of nature on the face of the earth'), general ('... the whole system ...') and dimension-free concept ('... of the most various kinds and sizes.'). Some of the evolution in the use of the term reflects adoption of older holistic ideas about interactions among organisms and their environments. These include the ideas that species were supposed to be interconnected as an 'unbroken natural chain' (see Chapter 2 and pp.

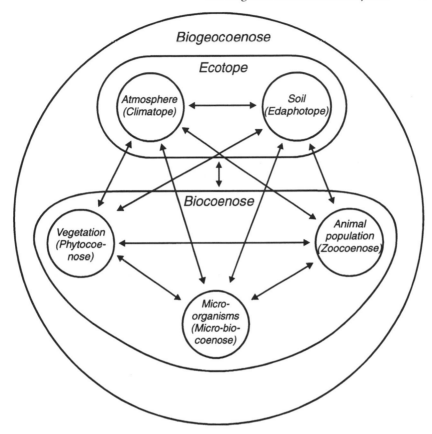

Figure 3.6. Interactions of the components of a biogeocoenosis. From Sukachev (1968a).

457–464 in Boorstin, 1983), or that communities of great antiquity were thought to be the natural cover of the landscape (the 'wilderness concept' or the 'forest primeval').

Recent textbooks and introductions to the ecosystem concept have had a tendency to make more specific Tansley's intentionally abstract definition. Recent re-definitions of the ecosystem have moved from Tansley's original definition and toward the concepts historically associated with the super-organism version of the community concept. One sees this in such statements as,

> Likewise, community in the ecological sense (sometimes designated as 'biotic community') includes all of the populations occupying a given area. The community and the nonliving environment function as an ecological system or ecosystem. (Odum 1971)

> We have, however, to fix some limits to the spatial size of the ecosystem. . . . First, the ecosystem must be large enough for all of the functional processes to be represented. . . . Secondly, the boundaries should be so sited that the inputs and outputs across the boundaries are relatively easily measured. (Usher 1973)

> An ecosystem is basically an energy processing system whose components have evolved together over a long period of time. (Smith 1980)

> Ecosystems are communities of plants, animals, fungi and monera, with different degrees and kinds of interdependence among the component species. (Burrows 1990)

These (and other) views of the modern concept of the ecosystem do not adhere strictly to the generality in Tansley's original definition.* Instead, they (and others) tend to elaborate the importance of having a size dimension based on some criteria, such as representing all of the internal flows of nutrients or energy in the system, or having sustainable populations of all the major species. These are points of view better associated with the biogeocoenosis concept (with its emphasis on identifiable boundaries), rather that the ecosystem concept (at least as it was originally defined). Indeed, as was mentioned above, the lack of an essential size of an ecosystem in Tansley's original definition is a standard criticism of the ecosystem concept by those preferring the biogeocoenosis concept (e.g., Sukachev 1968a; Gilmanov 1992). The usual expression of the modern ecosystem concept has tended, to degrees, to presuppose that a reference size is associated with any particular ecosystem. From this view, the standard use of ecosystem is converging on the standard use of biogeocoenosis. It is not clear that this convergence is desirable. A biogeocoenosis is a special case of the ecosystem. Namely, it is an ecosystem of a particular size.

The ecosystem as a system of definition

The ascribing of size to an ecosystem probably should not proceed in an ad hoc fashion, especially when issues involving environmental change and the alteration of ecosystems are involved. In practice among ecologists, the term ecosystem has come to have two somewhat different uses: one with relatively explicit assumptions, the other with an implicit assumptive structure.

The first use of ecosystem is as a system of definition: an arbitrary

---

* These comments are not intended to be critical of these authors, who have taken on the difficult task of writing textbooks that communicate ecological concepts to students at an introductory level. The point here is that these textbook definitions reflect the current use of the ecosystem concept – which has drifted from the original definition of Tansley (1935).

system defined by the specific considerations for a particular application. Therefore, if one were interested in the ecology of packs of wolves in the taiga of Siberia, size of the ecosystem might correspond to the several thousands of square kilometres over which a wolf pack hunts. One with an interest in the movement of a radioisotope from a forest landscape to a small stream might assign the stream's watershed of a few hundred hectares as the ecosystem size. The ecosystem is defined relative to the objectives of a given study. An ecosystem could be defined to be the interactions of the microbiota in a drop of water hanging from a leaf or it could be defined to be the entire biosphere of the Earth. This purposefully abstract use of ecosystem is probably closer to what is implied in Tansley's original definition.

The value of an abstract ecosystem concept is that it restricts the definition to the important processes of the particular case of interest. The definition of which aspects of a system are important (or unimportant) to a particular problem is stated explicitly. The ecosystem, as an abstract term, is closer to systems concepts in other sciences. Among other things, this makes the application of analytical procedures from other branches of science more straightforward.

The second modern use of the term ecosystem is a traditional understanding among workers in a particular branch of ecology as to what they mean when they speak of the ecosystems for which they have a particular interest. For example, limnologists studying small lakes might see the lake margin as a natural boundary of their ecosystems. Ecologists interested in element cycling might say ecosystem but actually focus on the watershed of a small stream. Forest ecologists would take a forest stand (perhaps 10 to 100 ha) as the ecosystem. In this usage, ecosystem is an idiom for a commonly held view of what should be considered the object of study. This latter use of ecosystem is also as a system of definition, but the details of the definition are largely implied in terms of a given context or shared research experience.

It is clear that in the history of the ecosystem concept and at present ecologists have tended to operate in a spectrum of opinion regarding the 'wholeness' of assemblages of plants and animals interacting with their environment. At one extreme is to consider such assemblages to be systems of definition in a most abstract sense – 'call them ecosystems, if you like'. At the other extreme is to regard some of these assemblages to be natural units with emergent properties. An example of this latter view would be to assume that the movement of energy through an ecosystem is analogous to the metabolism of an organism, or to liken ecological succession to the development of an embryo. More subtle is the tendency found in ecology and biology textbooks to follow and often espouse a natural, organisational hierarchy starting from the molecule, then to the cell, to the tissue, to the organism, to the population, to the community, to the ecosystem, and on to higher levels (O'Neill *et al.* 1986).

Superficially, some ecosystems appear to represent a greater degree of abstraction than others. A lake appears to have a clearly definable margin and, therefore, may seem a natural unit. A watershed, the land area that provides the water that flows past a given point in a stream, is a slightly more abstract unit. A forest stand is an area of more-or-less homogeneous vegetation dominated by trees of similar species and sizes and is an even more abstract system. However, as has been discussed in Chapter 2, ecologists are realising that the systems with which they deal have been subjected to a varied history of change at all time scales. The constancy of what, in the present time sense, appear to be natural landscape units tends to be diminished when one considers longer time scales.

The current use of the ecosystem concept, its historical origins and the development of related concepts all illustrate a tendency for ecologists interested in dynamic changes among plants, animals and the environment either to think of the ecosystem as an entirely abstract construct (in one extreme) or to think of it as a natural ecological unit (in the other extreme). These same considerations are also embedded in discussions around the issue of whether it is (or is not) appropriate to ascribe an area of particular dimensions to an ecosystem.

Temporal scale, spatial scale and the ecosystem

One reason that there is concern about the definition of spatial scale in the ecosystem theory is that particular phenomena appear to be more or less important at different scales in time and space. The issue of understanding space and time scale in ecological systems has been identified as a necessary preamble to understanding how ecosystems will respond to large-scale environmental change (O'Neill 1988). Further, the experience in building interdisciplinary research teams indicates that an attention to space and time scales may not guarantee success; but to ignore such factors seems to enhance the likelihood of failure (Shugart and Urban 1988). This attention to scale has recently been highlighted in the development of the hierarchy theory in ecology (Allen and Starr 1982; O'Neill *et al.* 1986) and may be its most important contribution.

While the idea of phenomenological scale may seem unfamiliar, we are all familiar with phenomena that have a particular range in time or space at which they are operative (or observable) as part of everyday experience. One can dip one's hand into hot water or touch a hot iron for a fraction of a second without damage, but a longer exposure would be painful. The rapid motion of an aeroplane propeller renders it invisible to the eye. The movement of the second hand of a watch is observable, while the slower movement of the hour hand is not. A dog will answer a whistle that is so shrill that the human ear cannot hear it. These examples are part of an overall feature of dynamic systems in general. Phenomena have time (and space) scales over which they are observably dynamic and others in which they are not.

System response in the frequency domain

The response of dynamic systems to inputs of various frequencies can be used to exemplify scale of response. The above example of dogs' ears hearing sounds at higher frequencies than humans' implies that relatively similar systems (ears) can differ significantly with respect to frequency of input (sound). In any linear dynamic system given an input to the system in the form of a sine wave, one will observe three broad types of response (Fig. 3.7). (Formal definition of what is meant by linearity will be found

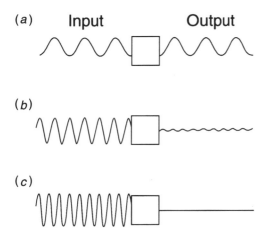

Figure 3.7. Response of a linear dynamic system to an input sine wave. (*a*) Low-frequency response. (*b*) Intermediate-frequency response. (*c*) High-frequency response. A linear system will attenuate high-frequency input signals, will modify intermediate input signals and will follow low-frequency input signals. The difference among the relative terms (high, intermediate and low) is dependent on the rates of dynamic responses in the linear system.

in the latter sections of Chapter 4.) At sufficiently high frequencies of input (Fig. 3.7c), the response of the system is to attenuate the input signal. At sufficiently low frequencies of input (Fig. 3.7a), the system response will follow the variation in the input signal. There is an intermediate frequency of input at which the response of the system is to delay the signal (the peak in the input signal differs in timing from the peak in the output signal) and to alter the amplitude of the input signal. The difference among the relative terms of high, intermediate, or low frequencies in this example depends on the rates of dynamic responses in the linear system. What is a high-frequency input to one system may be an intermediate-frequency input to another.

Woodward (1987b) illustrated this concept in considering the responses of different plant processes to various components of the characteristic cycles of variation in climate (Fig. 3.8). Woodward did not consider the impact of rapidly varying extremes in developing this spectrum of responses because these might exceed biological thresholds and thus could exert large effects. The frequency and the amount on variance in the climate signal associated with that frequency (Fig. 3.8b) were aligned with the plant processes that responded to those frequencies (Fig. 3.8a).

For example, expansion and contraction of the ranges of different types of plant (late successional trees, herbaceous perennials, annuals) responded to annual or longer frequencies of climatic variation (Fig. 3.8a). Woodward (1987b) felt that late successional trees would require climatic cycles of multiple centuries to induce a contraction in their distribution. The responses of other types of plant were more rapid. One might expect range expansion of a population to be driven by frequencies of variation that were at slightly higher frequencies than range contractions. However, responses in species ranges were not felt to be influenced by the monthly to daily variations in the climate – even though this scale of variation accounts for a large proportion of the variance in climate. The daily to monthly periodic variations, which seem to be a relatively less important consideration in understanding range changes in species, excite a different aspect of the plant response (flowering, germination). Further, the significant variations in the climate signal at minute to second periods are most strongly involved with still another aspect of the plant response: tissue-level responses and plant physiology (stomatal opening, leaf gas exchange).

Time scale, space scale and the ecosystem

Time and space scale are of significant importance in understanding system dynamics. Several authors have attempted to categorise phenomena

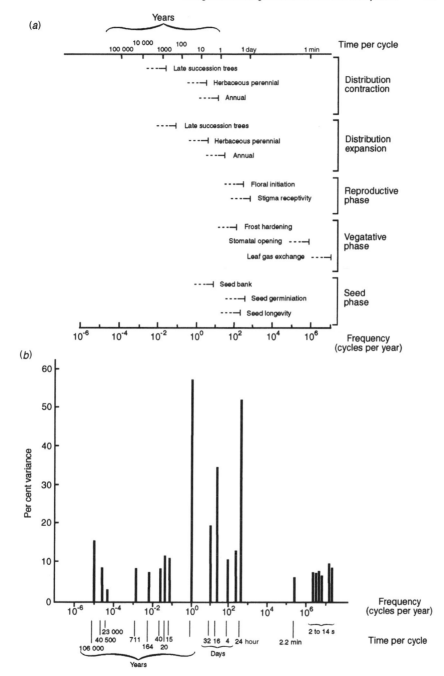

Figure 3.8. Concurrence of response times of (*a*) plant processes and
(*b*) characteristic cycles of climate (Woodward 1987b).

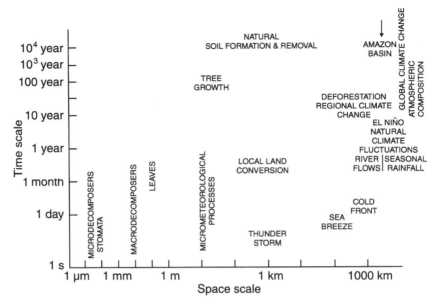

Figure 3.9. Scales of the atmospheric processes linked to surface processes. From Dickinson (1988).

of interest in the time and space domain. In attempting to understand how the terrestrial surface might respond to large-scale changes in the environment, ecologists are being required to consider new environmental conditions operating on biological processes at unusual time and space scales with respect to the typical study in ecology (Karieva and Anderson 1988). Indeed, the potential scales of consideration are quite broad. Dickinson (1988) illustrates scales of surface processes that interact with atmospheric processes across a space scale of sub-millimetre to thousands of kilometres, and for times from as short as a few seconds to millennia (Fig. 3.9).

Delcourt *et al.* (1983) considered the time and space scale of different disturbance factors, the ecological mechanisms that are excited by these phenomena and the patterns produced by the interactions between the disturbances and the ecological mechanisms. They illustrated this in a three-part diagram (Fig. 3.10) in which the disturbances, biotic responses and resultant patterns are indicated at the space and time intervals over which they were typically measured. In the sense of the earlier discussions, disturbances can be thought of as inputs to the systems.

Different disturbances operate on the vegetation systems at different frequencies. Just as in the Woodward (1987b) example above, inputs of

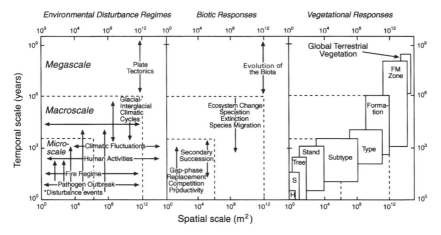

Figure 3.10. Environmental disturbance regimes, biotic responses and
vegetational patterns viewed in the context of space–time domains in which the
scale for each process or pattern reflects the sampling intervals required to observe
it. The time scale for the vegetational patterns is the time interval required to
record their dynamics. The vegetational units are mapped as a nested series of
vegetational patterns. ★Examples of disturbance events: wildfire, wind damage, clear
cut, flood and earthquake. H, herb; S, shrub. From Delcourt *et al.* (1983).

different frequencies in time or space induce responses from different
vegetation processes and can be thought of as producing different patterns. The time and space scales of plate tectonics in separating, moving
and melding the continents over millions of years are an active disturbance on the evolution of the biota and on the diversity of the vegetation formations for continental-scale zones. Plate tectonics is sufficiently
slow to be a non-consideration in understanding the response of ecological succession to decade-to-century variations in the incidence of wildfires, producing differences in local vegetation composition (types and
subtypes) (Fig. 3.10).

One useful feature of the Delcourt *et al.* (1983) diagram is that it illustrates some of the factors involved in considering the space scale in the
earlier comparison between the ecosystem and the biogeocoenosis. If
one stacks the three panels (environmental disturbance regimes, biotic
responses, vegetational patterns) from Fig. 3.10 and sticks a pin in the
stack at the bottom-left corner, then the resultant conjunction of frequent disturbances (fire outbreak etc.) interacting with the biotic processes of competition and productivity would be the important processes
to consider in understanding the patterns of a forest stand ecosystem. A
different insertion of the pin at the top-right of the stack would generate

a different set of important disturbances working on a different array of processes to be considered for a different kind of ecosystem. To a degree, the specification of time and space domains isolates some of the choices in the way one formulates the system of definition. What is important to consider is a scale-dependent feature of ecosystems.

Concluding comments

In this chapter, it has been emphasised that the ecosystem concept has been an abstract, whole-system-oriented concept from its inception. The ecosystem is an intellectual construct. Ideally, ecosystems are defined in ways to make clear the solution to a particular problem or to improve one's ability to predict changes in natural systems. Some conceptualisations of ecosystems are more successful than others in producing understanding of the functioning of natural systems. The typical problems central to defining an ecosystem stem from determining how particular processes interact in nature to produce observable ecological patterns over time or space.

Success in formulating ecosystems can be attributed to wise selection of processes and responses for the time and space scales of interest. Investigation of certain sets of interactions appears to be more productive than investigations of others, especially when used as the basis of defining ecosystems and understanding the behaviour of dynamic systems. Different processes are operative at certain scales in time and space and are less obvious at others. Historically, there has been a struggle between the abstract ecosystem concept and particular definitions of ecosystems that, at least initially, seem so efficacious that they become the ecosystem in use among a segment of the ecological scientists. It is always useful to remember that while a watershed can be defined as an ecosystem, not all ecosystems are watersheds.

The dynamic nature of the factors that influence terrestrial ecosystems operates at several different time scales (Chapter 2). It is appropriate to think of terrestrial landscapes as being excited by a range of factors operating simultaneously over different extents in time and space. While the complexity of hierarchical systems driven simultaneously by changes in different time and space domains is daunting, the situation is not impossible.

As will be seen in later chapters, ecologists have had considerable success in predicting the responses of ecosystems at a range of scales. The class of systems to which terrestrial landscapes belong has an extremely wide range of potential behaviours; terrestrial systems appear to be surprisingly predictable in some cases. The challenge is to determine the degree to which this predictability can be generalised or applied to novel conditions.

Part 2
Basic concepts

4 · *Ecological modelling*

A diverse array of physical objects and abstract constructs are referred to as models (Table 4.1). In the usual usage, however, ecological models are mathematical expressions developed to be analogous, in some sense, with an ecosystem of interest. The models of principal interest here are those that are used to integrate information and to produce predictions of responses of ecosystems to change. This chapter provides a basic background on the mathematical formulations, concepts and terms commonly used to model the dynamics of ecological systems. It is intended to be a preamble to the chapters that follow (which are oriented toward the large-scale predictions and ecological consequences of models at several temporal and spatial scales). The focus is also on models that are explicitly dynamic. For readers interested in an expanded but accessible treatment of these topics, Beltrami (1987) reviews the mathematics of dynamic models used in a variety of sciences including ecology.

With an abruptly increased availability of computer power (in terms of speed, cost and magnitude of computation) since the 1970s, there has been an explosive development of the application of computer models in ecology as well as in other sciences. New computer software allows scientists to explore complex dynamic equations using small (but powerful) computers in the same manner that an earlier generation of ecologists used paper as 'scratch pads' to sketch data patterns and relations.

Models are playing an ever-increasing role in the development of ecological theory at several scales, from understanding the mechanisms of carbon fixation (Farquhar and Sharkey 1982; Farquhar and von Caemmer 1982) and plant water balance (Cowan 1982, 1986) through scaling of physiological processes to whole plant function (Reynolds *et al.* 1986) to exploring how ecosystem processes of carbon and nitrogen cycling operate at continental to global scales (Emanuel *et al.* 1984, 1985a,b). An important role of modelling is in exploring phenomena that occur at

Table 4.1. *Examples of different types of model commonly used in ecosystem analysis*

Type of model	Description
Conceptual models	Diagrams or descriptions of the important connections among the components of an ecosystem
Microcosms	Small (usually small enough to fit on a laboratory bench) physical and biological analogues to a larger ecosystem of interest
Population models	Usually systems of differential or difference equations that compute the change in the numbers of individuals in a population
Community models	Often of similar structure to population models but including terms that involve interactions with other populations
Compartment models	Usually systems of differential or difference equations that follow the transfer of elements, energy or other material through an ecosystem
Multiple commodity models	Compartment models that treat the interactive transfers of several different materials through an ecosystem
Individual-based models	Models in which the dynamic changes in the individuals in an ecosystem are used as a basis to understand larger system dynamics

spatial and temporal scales at which extensive direct observation and experimentation are prohibitive, if not impossible. Examples include investigating spatial and temporal variation in competition on ecosystem functioning (Sharpe *et al.* 1985, 1986; Wu *et al.* 1985; Walker *et al.* 1989), extrapolating the processes of carbon fixation and water balance to the landscape scale to link ecosystem models with remotely sensed data (Running and Coughlan 1988; Running *et al.* 1989), and exploring the implications of the evolution of plant adaptations to varying environmental conditions on current patterns of ecosystem structure across environmental gradients (Tilman 1988).

The mathematical techniques used in developing and analysing ecological models have been treated in several books (e.g. Caswell *et al.* 1972; Smith 1974; Odum 1983; Jørgensen 1986; Beltrami 1987; Yodzis 1989) and are the focus of several ecological journals. The sections that follow are intended to provide the reader with an initial exposure to some of the concepts and methods used in systems ecology and to serve as an introductory background for those interested in approaching the modelling literature directly. Some of the material used in this chapter is derived from an earlier *Primer for Systems Ecology* by Patten (1971).

Finite-state automata

A common procedure used in studies of ecological succession is to develop what is called a 'space-for-time substitution' to piece together an expectation of the long-term dynamics. For example, if one had identified a number of vegetation types called A, B, C, D, E, F, G, H and I and found that over an interval of time, the successional changes were:

A B C D E F G H I
⇓ ⇓ ⇓ ⇓ ⇓ ⇓ ⇓ ⇓ ⇓
B C E G I D I F I

One might assume that the successional sequence was:

$$H \Rightarrow F \Rightarrow D \Rightarrow G$$
$$\Downarrow$$
$$A \Rightarrow B \Rightarrow C \Rightarrow E \Rightarrow I$$

This procedure has often been used to piece together the stages of a successional sequence from a set of observations collected over a relatively short time interval. There are two critical assumptions in using this approach to obtain a dynamic representation of vegetation change. The first assumption is that the vegetation can be categorised into a finite number of states, usually by defining a small number of vegetation types. The second is that the processes that change the vegetation types from one to another at different times or different locations are the same. It is this assumption that allows the pattern found in space (at different locations) to be interpreted as a temporal replacement sequence. Generally, systems conforming to this last condition are said to be ergodic. An approach to modelling systems that have a finite number of states and are ergodic is to use what is called a finite-state representation of the systems.

For example, consider a simple system, the light switch. One can think of a light switch as having two states, 'on' and 'off', which can be denoted by s_{on} and s_{off}, respectively. The dynamics or changes in the state of a light switch amount to changes in the system from s_{on} to s_{off} and vice versa. These changes occur when an agent of change, say 'your hand', operates on the light switch to change its state. The 'your–hand' operator could be denoted τ_{hand}. The rules for how this simple dynamic system works can be captured by a simple diagram:

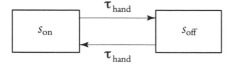

or as a matrix of the rules of change (a **state transition matrix**):

State of the system	New state when operated on by τ_{hand}
S_{on}	S_{off}
S_{off}	S_{on}

This extremely simple example can be used to introduce some definitions. The system as described is a finite-state system (or finite-state automata) in that it has a finite number (two) of possible system states. There is one operator that changes the state of the system. The system is deterministic, in that if one knows the state of the system at any instant one can predict the current state of the system uniquely by knowing the history of application of the operator. If this were not the case, for example if the response of the system in state s_i in response to an operator was to go to state s_k with a probability of 0.8 and to go to state s_j with a probability of 0.2, the system would be stochastic.

The operator on a dynamic system is usually time. However, it is possible to have any number of operators, and to have any number of states in a particular system. The light-switch example is intentionally simple, but it is important to realise that simple systems can be elaborated to systems of great complexity. The digital computer, for example, is a finite-state machine whose state is one of the combinations of the millions of on-or-off 'switches' that make up the memory of the machine. A digital computer has a large number of states, but this number is finite.

Consider a somewhat more complex system with a few more system states (Fig. 4.1). In this case, some of the states are 'transient states' in that the system will change to another state over the application of the operator (s_1, s_2, s_3 and s_5 in Fig. 4.1a; r_1, r_3 and r_4 in Fig. 4.1b). Other states are 'steady states' in that once the system arrives in the state (s_4 in Fig. 4.1a) it with remain in that state. States can also be configured as cycles (r_2, r_5 and r_6 in Fig. 4.1b).

Finite-state automata have applications in ecological modelling that are

(a)

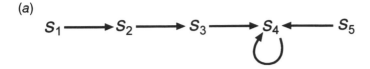

(b)

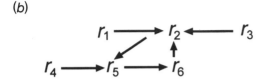

Figure 4.1. Example dynamics of finite-state systems. (*a*) System *S* with states s_i, $i = 1$ to 5. (*b*) System *R* with states r_j, $j = 1$ to 6.

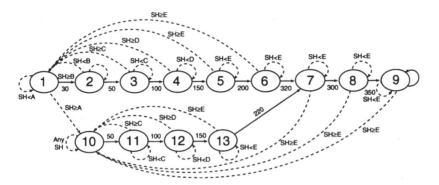

Figure 4.2. Successional pattern for 13 Montana habitat types. Transitions in the absence of disturbances are determined by stand ages (in years). Transitions in response to fire are a function of the fire scorch height (SH) (Van Wagner 1973). Forest types are shown in Table 4.2. From Kessell and Potter (1980).

important beyond their use here as examples. Figure 4.2 illustrates landscape dynamics for habitat types in Montana under succession and fire disturbance (Kessell 1976, 1979a,b; Cattelino *et al.* 1979; Potter *et al.* 1979; Kessell and Potter 1980). The state-transition matrix for this finitestate system is shown in Table 4.2. Also, the species composition and relative abundance for each system state (s_1 through s_{13}) are shown in Table 4.3. The operators in this system are the passage of time and the occurrence of wildfires of different scorch heights ranging from relatively mild fires of scorch heights less than height A, to devastating fires with scorch heights greater than height E. Overall, there are seven operators in this system. Notationally the passage of time without a wildfire is τ_{time}, and

Table 4.2. *A state-transition matrix for successional changes in 13 Montana habitat types. The vegetation composition for the 13 types is presented in Table 4.3. The system has seven operators. One operator is the passage of time without a wildfire (τ_{time}). The six others are based on the intensity of fires, $\tau_{<A}$ for fires with a scorch height less than A, τ_{AB} for fires with scorch heights between A and B, etc.*

System states	Operators						
	τ_{time}	$\tau_{<A}$	τ_{AB}	τ_{BC}	τ_{CD}	τ_{DE}	$\tau_{>E}$
S_1	S_2	S_{10}	S_1	S_1	S_1	S_1	S_1
S_2	S_3	S_3	S_2	S_1	S_1	S_1	S_1
S_3	S_4	S_4	S_3	S_3	S_1	S_1	S_1
S_4	S_5	S_5	S_4	S_4	S_4	S_1	S_1
S_5	S_6	S_6	S_5	S_5	S_5	S_5	S_1
S_6	S_7	S_7	S_6	S_6	S_6	S_6	S_1
S_7	S_8	S_{10}	S_{10}	S_{10}	S_{10}	S_{10}	S_7
S_8	S_9	S_{10}	S_{10}	S_{10}	S_{10}	S_{10}	S_8
S_9	S_9	S_{10}	S_{10}	S_{10}	S_{10}	S_{10}	S_9
S_{10}	S_{11}	S_{10}	S_{10}	S_{10}	S_{10}	S_{10}	S_{10}
S_{11}	S_{12}	S_{11}	S_{11}	S_{11}	S_{10}	S_{10}	S_{10}
S_{12}	S_{13}	S_{12}	S_{12}	S_{12}	S_{12}	S_{10}	S_{10}
S_{13}	S_7	S_{13}	S_{13}	S_{13}	S_{13}	S_{13}	S_{10}

the six other operators based on the intensity of fires are: $\tau_{<A}$ for fires with a scorch height less than A, τ_{AB} for fires with scorch heights between A and B, etc.

The finite-state system shown in Fig. 4.2 and Table 4.2 can be used to predict the state of a small plot of land over successional time by applying a history of different operators producing a sequence of dynamic changes. The finite-state systems such as the example (Fig. 4.2) have been implemented as digital computer programs. These programs produce dynamic maps for use by park managers in national parks in the USA and in Australia to forecast changes in the park landscapes following wildfires Kessell 1976, 1979a,b; Cattelino *et al.* 1979; Potter *et al.* 1979; Kessell and Potter 1980). In these applications, a finite-state model was applied to every element of a digitised map of the particular park. One of the advantages of finite-state automata for such applications is that they can be solved rapidly, even using small computers. Development of these and

Table 4.3. *Relative abundances of species in successional states used to develop the finite-state automata for Montana habitat types. The successional state numbers correspond to those shown in Fig. 4.2*

	Importance values[a] for system states												
	s_1	s_2	s_3	s_4	s_5	s_6	s_7	s_8	s_9	s_{10}	s_{11}	s_{12}	s_{13}
Trees >1.4 m in height													
Pinus contorta			3	2	1								
Pinus ponderosa			2	2	2	2	1	1			3	3	2
Pseudotsuga menziesii			4	4	5	5	5	6	6		4	4	5
Trees <1.4 m in height													
Pinus contorta	1	1											
Pinus ponderosa	1	1								1			
Pseudotsuga menziesii	1	1	1	1						1	1	1	
Total tree cover			4	4	4	4	4	4	4		4	4	4
Shrubs and subshrubs													
Juniperus communis			2	1							2	1	
Potentilla fruticosa				1	1	1	1	1	1		1	1	1
Ribes spp.	2	2											
Rosa spp.			1	2	2	2	2	2	2		1	2	2
Sheperdia canadensis			2	1	1	1	1	1	1		2	1	1
Spirea betulifolia			2	2	2	2	2	2	2		2	2	2
Symphoricarpos albus	1	1	2	2	2	2	2	2	2	1	2	2	2
Berberis repens			1	1	1	1	1	1	1		1	1	1
Total shrub cover	2	2	4	4	4	4	4	4	4	2	4	4	4
Forbs and grasses													
Calamagrostis rubescens	1	1	3	1	1	1	1	1	1	1	3	1	1
Carex geyeri	2	2	3							2	3		
Festuca scabrella	1	1	1		1					1	1		1
Gramineae[b]	1	1		1	1					1		1	1
Arica spp.			2		1	1	1	1	1		2		1
Astragalus spp.			1	1	1	1	1	1	1		1	1	1
Fragaria spp.	2	2	1		1	1	1	1	1	2	1		1
Total forb and grass cover	5	5	3	2	1	1	1	1	1	5	3	2	1

Notes:
[a] Importance values are expressed on a seven point scale where: blank = <1%; 1 = 1–5%; 2 = 6–25%; 3 = 26–50%; 4 = 51–75%; 5 = 76–95%; 6 = 96–100%.
Importance values for trees greater than 1.4 m in height are based on their relative densities. All other importance values are based on absolute cover.
[b] All grasses not listed by species.

other landscape-level simulators will be discussed in more detail in Chapter 10.

State variable representations of dynamic systems

As one can infer from their name, finite-state automata have finite numbers of states. In the examples that have been provided so far, the states have been abstract. In the last example, the states were forest community types of a complex, multiple-pathway successional system. Change in a dynamic system over time can also be expressed as changes in variables (called state variables) whose values represent the state of the system. Dynamic change in the system is manifested as changes in the system's state variables.

Two related mathematical formulations are often used to predict the change in the state variables of a system. In the first, the difference equation, the change in the state variable at a given time t (denoted Δx_t) over a fixed interval of time (Δt) is used to compute the new value of the state variable at the next time $t + \Delta t$ (denoted $x_{t+\Delta t}$). The resultant equations are of the form:

$$x_{t+\Delta t} = x_t + \Delta x_t \qquad (4.1)$$

or

$$\Delta x_t = x_{t+\Delta t} - x_t$$

Each of the state variables in a system could be simulated by such an equation.

In a difference equation, the incremental change of the state variable can be measured over the interval Δt as:

$$\frac{\Delta x}{\Delta t} = \frac{x_{t+\Delta t} - x_t}{\Delta t} \qquad (4.2)$$

If the interval of time (Δt) is made smaller until it approaches zero (denoted $t \rightarrow 0$), then the change in x over the time interval is the slope of x versus time at time t. This is denoted dx/dt and is the differential equation for the instantaneous change of x with time. Differential equations are the second mathematical formulation used to simulate changes in the state variables of a dynamic system.

As was the case for the finite-state automata, both difference and differential equations are deterministic. If one knows the values of the state variables of the system (and any external factors from outside the system that operate to change these state variables), then one can predict the future state of the system. Difference and differential equations were initially developed to understand the dynamics of the systems of interest in classical physics. It is the deterministic features of such equations (and the belief that the universe behaved according to such equations) that led Laplace to posit,

> Given for one instant an intelligence which could comprehend all the forces by which nature is animated and the respective positions of the beings which compose it, if moreover this intelligence were vast enough to submit these data to analysis, it would embrace in the same formula both the movements of the largest bodies in the universe and those of the slightest atom; to it nothing would be uncertain, and the future as the past would be present to its eyes. (Pierre Simon de Laplace in *Oeuvres*, vol. VII, *Théorie Analytique des Probabilités* 1812–20)

While the universe may not be as deterministic as Laplace thought it was, certainly the application of difference and differential equations to the understanding of physical and other systems has a rich history. There is a legion of procedures, analytical techniques and solutions for certain formulations. In the sections that follow, a geometrical representation of system dynamics will be used to illustrate some of this heritage.

Change in state space: a geometrical representation of system dynamics

In a state-space representation of system change, the dynamics of the system are represented as a movement of a point in an abstract space whose axes are a system's state variables. The value of the state variables that represent the system state are the co-ordinates at a given time. In Fig. 4.3, the dynamics of a simple system with two state variables are shown in terms of changes in the state variables over time (Fig. 4.3*a*) and in state space (Fig. 4.3*b*). State-space representations of systems are particularly useful in illustrating systems concepts because they are geometric inter-pretations of systems concepts that otherwise require a considerable mathematical background. To illustrate state-variable approaches to dynamic systems, it is necessary to have a system of equations that can be used for examples. Probably the best known ecological equations that can be used for such purposes are for the dynamics of idealised populations.

Population dynamics models

Familiar state-space interpretations of system dynamics are the graphical analyses of the Lotka–Volterra equations for competition. The Lotka–

(a)

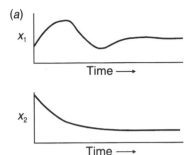

(b)

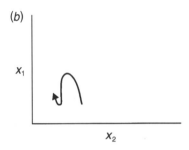

Figure 4.3. State-space representation of the dynamics of a system. In this case the system has two state variables (x_1 and x_2). (*a*) Change in x_1 and x_2 over time. (*b*) Change in the systems of interest in state space.

Volterra equations were part of the mathematical interpretation of the outcomes of the classic competition experiments of Gause (Chapter 7) and have been a mainstay in population ecology since the mid-1930s. The equations are derived from simple models of population dynamics and for this reason are a useful introduction to population dynamics models.

Initially, one can consider a simple population in a relatively unchanging environment. The state variable of interest is the population density of the population, N. In this simple case, one could predict the changes in the population by keeping track of the births and deaths of the individuals that make up the population over some time interval. One can represent the number of individuals in the population by a simple difference equation,

$$N_{t+\Delta t} = N_t + B_t - D_t \tag{4.3}$$

where $N_{t+\Delta t}$ is the population at time $t+\Delta t$; B_t is the number of individuals born between time t and time $t+\Delta t$; and D_t is the number of individuals dying between time t and time $t+\Delta t$.

If instead of the change of the population over an interval Δt, we consider the rate at which the population is changing (the slope of the curve when one plots population change against time), the differential equation most frequently used to represent population change is:

$$\frac{dN}{dt} = (b-d)N \text{ or } \frac{dN}{dt} = rN \qquad (4.4)$$

where b is the proportion of the population giving birth; d is the proportion of the population dying; and r is $(b-d)$ and is called the intrinsic rate of population increase.

The rate at which the number of individuals are born into a population is a function of the size of the population (larger populations tend to have more births than small populations etc.) and the simplest relationship that expresses this is to assume that the number of individuals born is a constant proportion of the population. A similar situation holds for the death of individuals over a time interval, with a simple assumption being that the number dying is a constant proportion of the population.

This representation (Eq. 4.4) of change in populations has been known for quite some time. The increase in a population implied by Eq. 4.4 over time can be solved analytically to obtain:

$$N_t = N_0 e^{rt} \qquad (4.5)$$

where N_0 is the size of the population at time 0 and N_t is the size of the population at time t. It is the exponential increase of populations that this formulation yields that was noted by Malthus (1798, *An Essay on the Principle of Population*) and that prompted him to note, 'Population, when unchecked, increases in a geometric ratio. Subsistence increases only in an arithmetic ratio. A slight acquaintance with numbers will show that immensity of the first power in comparison of the second.'

That populations tend to increase as an exponential function (what Malthus called geometric increase) was central to Darwin's thoughts when he formulated his theories on the evolution of species. It is the potential for population over-production that provides the surplus individuals to be eliminated (or not) by natural selection.

Recall that in Eq. 4.4, the intrinsic rate of increase of the population r was defined as the difference in the birth rate constant and the death rate constant ($r=b-d$). This implies that the rate of population birth and death is a fixed proportion of the number of individuals in the population. As populations get large with respect to space or resources available to them, one would expect the proportion of individuals dying to increase owing to such effects as stress, shortages of food or space, etc. One might also expect a depression in the proportion of females in the population giving birth to diminish for much the same reasons. These factors are termed density-dependent factors in that they are conditions that alter the rate of population growth as a function of the population density.

A simple case of how the births and deaths in a population might be affected by density of the population would be to assume that the decrease in birth rate is a linear function of population density. A similar assumption could be made for the population death rate. The population growth as a function of time under these assumptions would be:

$$\frac{dN}{dt} = [(b - b_{dd}N) - (d - d_{dd}N)]N \tag{4.6}$$

where b is the proportion of the population giving birth; b_{dd} is the density-related change in the proportion of the population giving birth; d is the proportion of the population dying; and d_{dd} is the density-related change in the proportion of the population dying.

One can rearrange the terms of this equation to obtain:

$$\frac{dN}{dt} = [(b - d) - (b_{dd} + d_{dd})N]N \tag{4.7}$$

If the expression $b_{dd} + d_{dd}$ is replaced by a new expression so that:

$$b_{dd} + d_{dd} = \frac{r}{K}$$

then the differential equation describing the change of the population becomes:

$$\frac{dN}{dt} = \left(r - \frac{r}{K}N\right)N \tag{4.8}$$

$$\frac{dN}{dt} = rN\left(1 - \frac{1}{K}N\right) \tag{4.9}$$

Population ecologists refer to this equation as the logistic equation or the Verhulst–Pearl equation. The dynamics of the equation are considerably different from the equation for the exponential increase of a population (Eqs. 4.4 and 4.5). A population behaving as described by the logistic equation will rise or fall to K over time (Fig. 4.4). K is called the carrying capacity and is associated with concepts that are well ingrained in the vocabulary of population biology and wildlife management – so much so

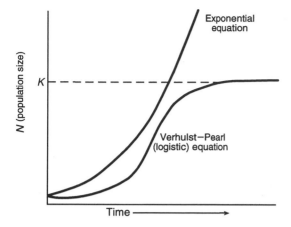

Figure 4.4. Dynamics of a population under equations for exponential growth and the Verhulst–Pearl logistic equation.

that it is taken to be a tangible property of a given habitat or location. For example, wildlife managers are involved with assessing the carrying capacity of a given area with regard to a deer herd or other managed animal population. The carrying capacity concept applied in wildlife biology is derived directly from the logistic equation (Eq. 4.9). There are important limitations in the behaviour of populations implied by the logistic equation. Important among these is that the expected state of the population over time is an unchanging population at the carrying capacity.

For the present purpose, a valuable feature of Eq. 4.9 is that the effect of density on the population is equal to $(1/K)N$, which expresses the negative effect on the birth rate or the increase in the death rate (or both) as a consequence of increased density. Following the same logic, if one had two populations N_1 and N_2, the effects of one on the other could be expected to be a similar formulation $(\alpha/K)N_2$. In this formulation, α is the effect of density of a competing population relative to the effect of the population's density. The resultant equations for two competing populations are:

$$\frac{dN_1}{dt} = r_1 N_1 \left(1 - \frac{1}{K_1}N_1 - \frac{\alpha}{K_1}N_2\right) \tag{4.10}$$

$$\frac{dN_2}{dt} = r_2 N_2 \left(1 - \frac{1}{K_2}N_2 - \frac{\beta}{K_2}N_1\right) \tag{4.11}$$

where N_1 and N_2 are the densities of populations 1 and 2, respectively; α and β are constants that represent the negative effect of each population on the other; and r_1, r_2, K_1 and K_2 are the intrinsic rates of increase and the carrying capacities of each population.

According to Eq. 4.11, the value of α is indicative of the relative negative effect on population N_1 of adding an individual of population N_2. The equivalent relative negative effect of adding an individual of the same (N_1) population is 1. In many populations, the negative effect of the addition of a further individual of either the same or the competing species is strongly a function of the size of the particular individual rather than just the tally of numbers. This is particularly the case in sessile organisms, such as plants. Alternative models of population dynamics (individual-based models) that consider the size of each individual in competing populations are discussed in Chapter 8. However, Eqs. 4.10 and 4.11 are the classic equations for the competition between two populations. They were initially developed by Lotka (1925) and Volterra (1926). Gause (1932, 1934, 1935) demonstrated experimentally that laboratory populations appeared to behave as the equations predicted (see Fig. 5.2, p. 111).

A state-space analysis of the dynamics of the Lotka–Volterra competition equations

The Lotka–Volterra equations are particularly useful because they can be solved graphically to determine the expected dynamics of a system of competing populations. The analysis is based on solving for the conditions under which each of the populations can be expected to not change. Mathematically, this amounts to determining the set of conditions where dN_1/dt or dN_2/dt is equal to zero. When the differential equation describing the dynamics of a population is zero, the population is constant. To solve for these conditions for the N_1 population, one initially sets the differential equation to zero. Setting $dN_1/dt = 0$ one finds that:

$$r_1 N_1 \left(1 - \frac{1}{K_1} N_1 - \frac{\alpha}{K_1} N_2 \right) = 0 \qquad (4.12)$$

This equation can be true if $r_1 N_1 = 0$, but this implies that the population is extinct (because $N_1 = 0$) or that it has no reproductive potential (because $r_1 = 0$). The other condition under which Eq. 4.12 can be true is when:

$$\left(1 - \frac{1}{K_1} N_1 - \frac{\alpha}{K_1} N_2 \right) = 0 \qquad (4.13)$$

This latter case is the case of interest. One can rearrange the terms to find:

$$K_1 - N_1 - \alpha N_2 = 0 \tag{4.14}$$

$$N_1 + \alpha N_2 = K_1 \tag{4.15}$$

$$N_1 = K_1 - \alpha N_2 \tag{4.16}$$

Equation 4.16 is the equation for a line on a plane with a co-ordinate system with axes of N_1 and N_2. The two state variables in the population system that is being considered are N_1 and N_2. Thus, the N_1 and N_2 plane is the two-dimensional state space for the population system (see Fig. 4.5 for example dynamics of the Lotka–Volterra equations in time and in this two-dimensional state space). The condition under which population N_1 shows zero change ($dN_1/dt = 0$) in this state space is a straight line (according to Eq. 4.16) and is called the N_1-isocline. The intercepts of this line are:

1. Intercept when $N_1 = 0$ is $N_2 = K_1/\alpha$
2. Intercept when $N_2 = 0$ is $N_1 = K_1$.

One can repeat this procedure to determine the conditions under which N_2 shows no change ($dN_2/dt = 0$) and obtain a second line in the state space for this condition ($N_2 = K_2 - \beta N_1$) with intercepts:

1. Intercept when $N_1 = 0$ is $N_2 = K_2$
2. Intercept when $N_2 = 0$ is $N_1 = K_2/\beta$.

The N_1- and N_2-isoclines can be drawn in the system state space to determine the pattern of dynamics and the behaviour of a two–population competition system over time but with different starting points (see Fig. 4.5).

The Lotka–Volterra equations for competition have four cases of dynamic behaviour depending on the relative values of the intercepts of the N_1- and N_2-isoclines. Recall that in the initial derivation of the Lotka–Volterra population equations (Eqs. 4.10 and 4.11), the α parameter scaled the relative effect of adding a member to the competing N_2 population compared with the negative effect of adding another member to the N_1 population (and vice versa for the β parameter).

The four cases of competitive outcomes for these equations and the dynamic responses that are associated with them are shown in Fig. 4.6. There are two cases of competitive exclusion, in which one species eliminates the other no matter what the initial values of the two populations (Fig. 4.6a,b). In each of these two cases there is an asymmetry in the

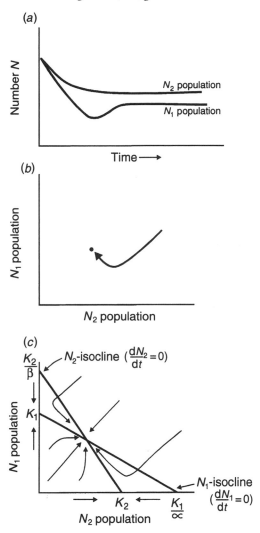

Figure 4.5. Dynamics of the Lotka–Volterra competition equations. (*a*) Dynamic response of the two populations with time. (*b*) Response of the populations plotted in state space. Change in the populations over time is indicated by change in the locus of the N_1 and N_2 co-ordinates in the state space. (*c*) Response of populations with several initial starting conditions and in relation to the N_1- and N_2-isoclines.

competition between the species, with the negative effect of one species on the other being large but the reciprocal negative effect being small. In a third case, one species or the other survives depending on the starting point of the initial populations. In this case (Fig. 4.6*c*), each population has

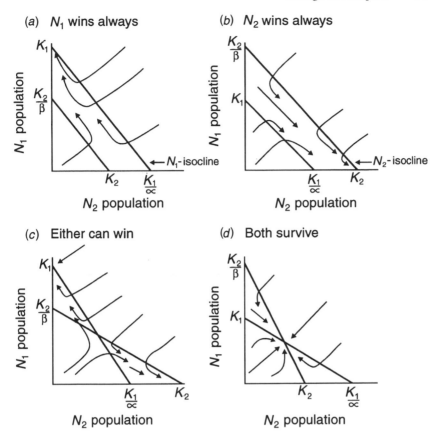

Figure 4.6. Outcomes of competitive interactions predicted by the Lotka–Volterra equations for four different cases of isocline intercepts. The relative magnitudes of the intercepts are controlled by the value of K and competition terms (α and β) in the coupled equations.

a strong negative effect on the other. In the fourth case, each species has a relatively mild negative effect on the other and both species survive regardless of the initial starting point (Fig. 4.6d). This last case is also the one illustrated in Fig. 4.5.

The point at which the two isoclines cross is an equilibrium. In the case shown in Figure 4.6c, it is an unstable equilibrium point. Any small perturbation in the populations will perturb them away from the point and they will move away to some other value. The equilibrium point can be likened to a ball perched at the top of a hill that slopes away on all sides. Any small change in the ball's position causes it to roll down the hill away from the top. The case shown in Fig. 4.6*d* illustrates a stable equilibrium point towards which the internal dynamics will move the population numbers. This equilibrium point can be likened to a ball resting in the lowest point of a depression. When the ball is displaced, it will roll back to its original position.

One of the most appealing features of the Lotka–Volterra equations for competition is that they can be solved easily in state space (Figs. 4.5 and 4.6). The positions of the isoclines in the N_1–N_2 state space determine the behaviour of the system and the outcome of competition between populations. Experimental demonstrations of such population dynamics in laboratory populations were developed by Gause (1932, 1934, 1935) and tied by him to Elton's (1927) concept of the ecological niche. Competitive interactions appear to be a regular and repeatable feature in laboratory populations (Arthur 1987). Sometimes, in long-term laboratory experiments, the losing population does not become extinct but remains at a low level for a long period of time (and in some cases 'regroups' to win the competition). Curved isoclines have been observed in some laboratory populations (Ayala *et al.* 1973), implying that the dynamic equations representing the competitive outcomes should be more complex than those used in the Lotka–Volterra equations. That the isoclines in some observed laboratory experiments (Ayala *et al.* 1973) appear to be curved rather than straight (as implied by a direct analysis of the Lotka–Volterra equations provided in the section above) may imply difficulties in the equations representing even relatively simple laboratory populations. However, any model is likely to have simplifications that reduce its applicability in some cases or at certain levels of detail. The identification of the possibility of curved isoclines in population state space for competing populations leads to the possibility of using theoretical arguments about the position and shape of the isoclines as a basis for understanding system dynamics (as discussed in the next section).

Developing theoretical derivations of system dynamics in state space

DeAngelis *et al.* (1986) developed a state space analysis of mutualism based on arguments about the shapes and positions of the population isoclines. Mutualism is one of six interactions to be found among pairs of interacting populations (Table 4.4) and is the condition in which each population has a beneficial effect on the other. One can derive an impression of the shape of the isoclines for one of the two populations in a mutualistic interaction by initially considering the isocline for one of the

Table 4.4. *Interactions among pairs of populations*

Type of interaction	Population		Description of the interaction
	A	B	
Competition	−	−	Each population inhibits the other
Predation, parasitism, Batesian mimicry	+	−	Population A (the predator, parasite or mimic) kills or exploits population B (the prey, the host or the model)
Neutralism	0	0	Neither population affects the other
Obligatory or facultative mutualism, Müllerian mimicry	+	+	Interaction is favourable to both populations
Commensalism	+	0	Population A is benefited; population B (the host) is not affected
Amensalism	−	0	Population A is negatively affected by interactions with population B; B is not affected by A

Source: From Pianka (1994).

populations (say the N_1 population) in state space under the condition that there is no interaction whatsoever with the other species (N_2). If the population behaves in a manner approximately like that described by the logistic equation, one would expect the N_1-isocline to be a line parallel to the N_2 axis with an intercept at K_1 (Fig. 4.7a). With no interaction with the other species, the N_1 population should approach its carrying capacity (K_1) regardless of the population density of the N_2 population.

If the interaction of N_1 with the N_2 population was as an obligatory mutualism, then one would expect the N_1 population to be unable to increase unless the N_2 population exceeded some critical value. Thus the N_1-isocline would cross the N_2 axis at the level of N_2 needed to allow survival (Fig. 4.7b). One might expect the isocline to increase towards what would a resource-base-determined carrying capacity (K_1) as the N_2 population is increased. This would cause the N_1-isocline to increase along the N_2 axis in the state space (Fig. 4.7b). Reciprocally, one could expect the isocline for the other population (the N_2-isocline) to have an analogous shape if it was also an obligatory mutualist with respect to N_1.

(a)   Logistic growth

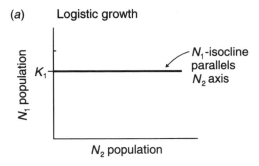

(b)   Logistic growth

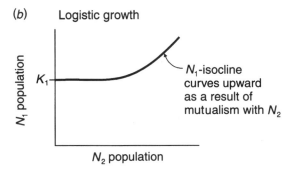

Figure 4.7. Isoclines for the N_1 population in a system in which the N_1 and N_2 populations are mutualists. (*a*) Shape of N_1-isocline with no interaction with the N_2 species. (*b*) Shape of the N_1-isocline with a mutualistic interaction with the N_2 population.

For two mutualist populations with curved isoclines, as in Figure 4.7*b*, there are two cases of population outcomes from the interaction. These cases are dependent on the positions of the isoclines – specifically on whether or not the isoclines cross one another (Fig. 4.8). When the two isoclines do not cross, regardless of the initial starting points, the numbers in the two populations take on infinitely large values as the two populations mutually elevate one another's carrying capacities (Fig. 4.8*a*). Presumably, in the real world, such a positive feedback interaction would eventually be stopped by some resource limitation.

When the two isoclines do cross (Fig. 4.8*b*), the populations have a rather complex interaction that features a stable equilibrium at one of the crossing points. The line called the separatrix (shown by a dotted line in Fig. 4.8*b*) separates the areas of the state space where starting points go to

(a)

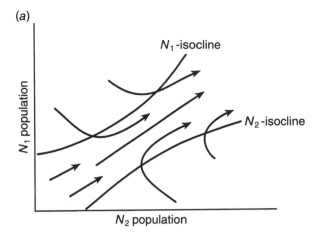

(b)

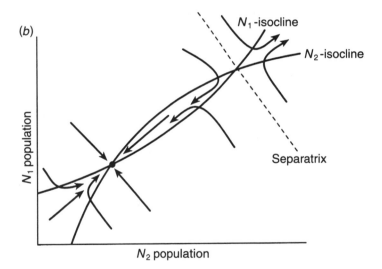

Figure 4.8. State-space dynamics for two cases of mutualistic interactions with isoclines as exemplified in Fig. 4.7. (*a*) Case 1 in which the populations have no equilibrium (the isoclines do not cross) and the populations take on infinitely large values regardless of the starting point. (*b*) Case 2 in which the populations either seek an equilibrium or take on infinitely large values depending on the starting points. The dotted line indicates the separatrix between areas with starting points that correspond to the two eventual outcomes.

the stable equilibrium and those that take on infinite values. Whether the species survive or increase owing to mutual reinforcement depends on the initial starting point. DeAngelis *et al.* (1986) present an analogous case for a pair of obligatory mutualists. In this case, the survival of the two species required a large initial innoculum of at least one of the mutualists as a starting condition.

Observational data displayed in state space to test theories of system dynamics

A third way one can use the pattern of system change in state space is to test theories about the pattern of system behaviour. Such a procedure was mentioned earlier for the experiments of Ayala *et al.* (1973) in which competitive interactions in laboratory fruit fly populations were graphed to inspect the isocline shape. Rosenzweig and MacArthur (1963) developed a theoretical argument on the shape of the prey isocline in predator–prey interactions. The essential aspects of this argument were:
1. The isocline should run through zero when there are no prey
2. In the vicinity of the origin of the state space (where there are low densities of prey and predators), the prey should increase
3. The isocline should also cross zero at the high prey densities (near the carrying capacity).

The conclusion they developed was elaborated as a graphical theory of predation by Rosenzweig (1969). In this paper, Rosenzweig also pointed out that the logistic population growth equation (Eq. 4.9) implied that the prey isocline should have a maximum between zero and the carrying capacity. This is because the rate of population growth is the greatest at an intermediate density.

To demonstrate the theory that Rosenzweig and MacArthur (1963) had derived, Rosenzweig (1969) plotted the results of a classic laboratory study of predator–prey interactions by Huffaker (1958) in predator–prey state space (Fig. 4.9). In Huffaker's study, populations of two species of mites, one of which was a predator on the other, were studied in experimental systems composed of oranges (the food of the prey mites). Huffaker designed a series of microcosms composed of arrangements of oranges of varying complexity of connections and physical structure. He found that the two populations both survived as the physical structure of the experimental system became more complex. Figure 4.9 (from Rosenzweig (1969) and based on Huffaker's most complex experimental microcosm) shows a pattern of increase and decrease in pairs of preda-

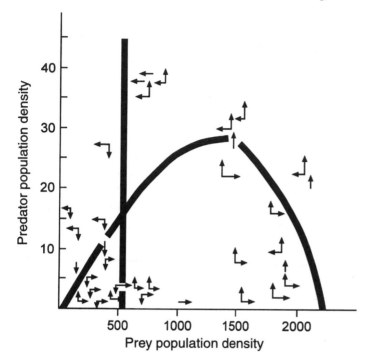

Figure 4.9. Dynamics of a predator–prey system maintained under laboratory conditions by Huffaker (1958) displayed in a state. The pairs of arrows indicate the directions of change for both the predator and prey for observed pairs of population densities observed in the population dynamics. The pattern of these changes in the populations allowed Rosenzweig (1969) to infer the location of the predator and prey populations' isoclines. From Rosenzweig (1969).

tor–prey observations that agrees with the responses expected from the isoclines drawn in the state space.

The display of data in state space provides insight into the system dynamics and can be used to test theories about the shapes of the isoclines in the state space. In actual practice, these sorts of model test are more likely to be used in a qualitative manner. In such cases, one might assemble predictions from a model that under a given set of conditions one state variable should increase while another increased, under another state of conditions both state variables should increase, etc. Such predictions could be used to collect a 'fingerprint' of observations that would support a model as an explanation of the dynamics of a system. Such qualitative tests can be developed by inspecting the state-space dynamics of a system and predicting patterns in observations.

Compartment models and material flow

Food chains and food webs initially emphasised the energy-related aspects of feeding relations in ecological systems. The understanding of transfers of energy in and out of plants and animals was (and is) an important topic for physiologists and physiological ecologists. There was an important cross-seeding of techniques developed in the 1930s by physiological researchers. These mathematical techniques were intended to quantify the dynamics of various substances moving through the vertebrate body (Shugart and O'Neill 1979). These systems-oriented methods evolved from laboratory experiments involving following the movements of tracers. Initially, these tracers were dyes. Radioisotope tracers were increasingly used following the 'atomic-age' availability of isotopes for experimental purposes in the late 1940s. A typical tracer experiment might involve injecting a dye into an artery and noting the time required for the dye to reappear in an adjacent vein. The reason for such an experiment might be to estimate the rate of blood circulation. As the dye dilutes and is mixed through the entire body, one might also use the degree of dilution to estimate the total volume of blood.

A mathematical theory was developed for the tracer experiments (Teorell 1937; Zilversmit *et al.* 1943) that eventually became known as tracer kinetics or compartment modelling (Solomon 1949; Sheppard and Householder 1951; Hearon 1953). Mathematical aspects of this topic will be discussed in this chapter in the context of ecological modelling (also see Shugart and O'Neill (1979) for a compilation of classic examples). In tracer kinetics, the body is divided into a mutually exclusive set of compartments (e.g., blood, bone, liver, etc.). The change in the material in each component is related to changes in the other components as flows of material from one compartment to another.

The analogy between this approach and conceptualisations of energy or nutrients flowing through the components of an ecosystem is the basis for a large body of ecological modelling.

At present, an area of great importance in ecosystem studies is determining the magnitudes and the controlling factors for ecosystem element cycles (or food webs). In such studies, the important parts of the ecosystem are represented as compartments. The changes in compartments are the result of the fluxes of material out (to other compartments or out of the system) or in (from other compartments or from outside the system). In the sense of the previous chapter, the compartments are the state variables of the ecosystem. The changes in these state variables are each represented by one of the equations in an interacting set of differential equations.

The fluxes of material from one compartment to another can, in some cases, be extremely difficult to measure. Often the fluxes are estimated using procedures that amount to assuming that the material going in and out of a compartment is equal. This is true when the amount of material in the compartment is constant. In other cases, fluxes are measured by assuming that rates of transfer noted in laboratory experiments or in small field experiments can be applied to larger systems.

The formulation of element cycling compartment diagrams (or models) involves subtle, underlying assumptions about the constancy of ecosystems and the extrapolation of observations to larger spatial scales. For these reasons, the scientific importance of particular ecosystems is great for ecosystems, such as watersheds, in which some of the fluxes can be measured directly (Bormann and Likens 1979a).

Formulation of compartment models for ecosystem studies

The usual application of compartment models in ecology is in representing the internal and external transfers of nutrients and energy in ecosystems. This section will provide background on mathematical formulation of compartment models and the definition of terms used in ecosystem studies that arise from compartment models.

A straightforward introduction to these models is through the terms and procedures associated with estimating the parameters of the model. The fundamental unit of a compartment model is the single compartment and the associated flows of material or energy in and out of the compartment:

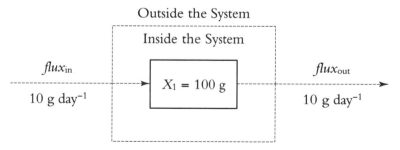

This representation is typical of that used in compartment modelling in general. In the case shown above, the compartment is in a state of equilibrium: the amount of material going in and out is the same. Therefore, the amount of material in the compartment is not changing. The lack of change in the state of the system under these conditions is the origin of the term steady state. Also, the balance of inflows and outflows inspires the term equilibrium. In the usual case, the material flowing through the system is conserved (neither created nor destroyed in the process), and the two terms (equilibrium, steady state) describe an equivalent condition.

The compartment (X_1) contains 100 g of the material of interest. The flux into the example compartment from outside the system is $10 \, g \, day^{-1}$ and the flux from the compartment to outside the system is also $10 \, g \, day^{-1}$. Note that the dotted line defines what is considered to be inside and outside the system. This defines the system of interest and amounts to an ecosystem definition. A differential equation representing this compartment is:

$$\frac{dX_1}{dt} = flux_{in} - flux_{out} \tag{4.17}$$

In simple compartment models, the loss from a compartment is taken to be a constant proportion of the amount of the material in the compartment and this proportion is called the transfer coefficient or the rate constant. It is important to point out the difference between the rate of change and the rate constant. The rate of change has the dimensions of units of material per unit time (in the example: $g \, day^{-1}$); the rate constant has dimensions of reciprocal units of time (in the example, day^{-1}). In compartment model notation, the form of the differential equation used to model material flows is:

$$\frac{dX_1}{dt} = f_1 - \lambda_{11} X_1 \tag{4.18}$$

where: f_1 is the input to compartment X_1 $(g \, day^{-1})$ from outside the system and λ_{11} is the rate constant for compartment X_1 losses (day^{-1}). Note that the input from outside the system is a rate of input dimensioned in $g \, day^{-1}$ and the loss from compartment X_1 is the product of the contents of the compartment (g) and a rate constant (units of day^{-1}) and, therefore, as a product is also dimensioned in $g \, day^{-1}$. The value of the rate constant in this equation can be calculated from the flux out of the system $(10 \, g \, day^{-1})$ and the equilibrium value of the compartment (100 g). Because $\lambda_{11} X_1 = 10 \, g \, day^{-1}$ and $X_1 = 100 \, g$, then $\lambda_{11} = 10 \, g \, day^{-1}/X_1 = 10 \, g \, day^{-1}/100 \, g = 0.1 \, day^{-1}$.

Even a simple system (such as the one described in Eq. 4.18) provides some useful concepts for understanding the flow of materials through a system. One important measure is the time constant (or turnover time): the time needed for the amount of material flowing through the compartment at equilibrium to equal the amount of material in the compartment. The time constant is often denoted by T (for an entire system) or T_i (for the ith compartment in a system).

Complexity in a compartment model arises in the linking together of compartments representing parts of the system. A simple case, the linking of two compartments, might be represented as:

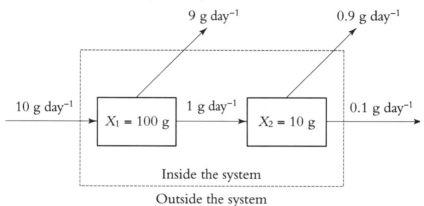

$$9 \text{ g day}^{-1} \qquad 0.9 \text{ g day}^{-1}$$

$$10 \text{ g day}^{-1} \qquad X_1 = 100 \text{ g} \qquad 1 \text{ g day}^{-1} \qquad X_2 = 10 \text{ g} \qquad 0.1 \text{ g day}^{-1}$$

Inside the system

Outside the system

with the corresponding pair of differential equations being:

$$\frac{dX_1}{dt} = f_1 - \lambda_{11} X_1 \tag{4.19}$$

$$\frac{dX_2}{dt} = \lambda_{12} X_1 - \lambda_{22} X_2 \tag{4.20}$$

where f_1 is the input to compartment X_1 (g day^{-1}) from outside the system or the forcing to compartment X_1; λ_{11} is the rate constant for compartment X_1 losses (day^{-1}); λ_{12} is the rate constant for the transfer of material from X_1 to X_2 (day^{-1}); and λ_{22} is the rate constant for compartment X_2 losses (day^{-1}).

As an example of the dynamics associated with compartment model representation of ecosystems, consider the classic study of the energy flow in the Silver Springs ecosystem developed by Odum (1955). In this study, the transfers of energy in a spring in Florida were quantified with respect to the amounts of energy in different parts of the system (compartments) and the flows of energy into, within and out of the system. The pattern of energy flow is shown as a compartment model in Fig. 4.10. (Note that in

Fig. 4.10 and the text that follows, kilocalories are used to conform to the original paper. 1 kilocalorie=4184 joules.) The system of differential equations representing the fluxes of energy flow is:

$$\frac{dX_1}{dt} = f_1 - \lambda_{11}X_1 \tag{4.21}$$

$$\frac{dX_2}{dt} = f_2 + \lambda_{12}X_1 - \lambda_{22}X_2 \tag{4.22}$$

$$\frac{dX_3}{dt} = \lambda_{23}X_2 - \lambda_{33}X_3 \tag{4.23}$$

$$\frac{dX_4}{dt} = \lambda_{34}X_3 - \lambda_{44}X_4 \tag{4.24}$$

$$\frac{dX_5}{dt} = \lambda_{15}X_1 + \lambda_{25}X_2 + \lambda_{35}X_3 + \lambda_{45}X_4 + \lambda_{55}X_5 \tag{4.25}$$

where: f_i is the input to compartment X_i (kcal m^{-2} yr^{-1}) from outside the system; λ_{ii} is the rate constant for compartment X_i losses (yr^{-1}). When $i \neq j$, λ_{ij} is the rate constant for the transfer of energy from X_i to X_j (yr^{-1}).

The notation used allows the rate constants to be expressed as a matrix. Using the notation $dx_i/dt = \dot{X}_i$, the system of equations can be expressed in matrix form as:

$$\begin{bmatrix} \dot{X}_1 \\ \dot{X}_2 \\ \dot{X}_3 \\ \dot{X}_4 \\ \dot{X}_5 \end{bmatrix} = \begin{bmatrix} \lambda_{11} & 0 & 0 & 0 & 0 \\ \lambda_{12} & \lambda_{22} & 0 & 0 & 0 \\ 0 & \lambda_{23} & \lambda_{33} & 0 & 0 \\ 0 & 0 & \lambda_{34} & \lambda_{44} & 0 \\ \lambda_{15} & \lambda_{25} & \lambda_{35} & \lambda_{45} & \lambda_{55} \end{bmatrix} \begin{bmatrix} X_1 \\ X_2 \\ X_3 \\ X_4 \\ X_5 \end{bmatrix} + \begin{bmatrix} f_1 \\ f_2 \\ 0 \\ 0 \\ 0 \end{bmatrix} \tag{4.26}$$

or using the values for rate constants calculated from the information in Fig. 4.10:

$$\begin{bmatrix} \dot{X}_1 \\ \dot{X}_2 \\ \dot{X}_3 \\ \dot{X}_4 \\ \dot{X}_5 \end{bmatrix} = \begin{bmatrix} -6.08 & 0 & 0 & 0 & 0 \\ 1.01 & -15.81 & 0 & 0 & 0 \\ 0 & 1.79 & -6.19 & 0 & 0 \\ 0 & 0 & 0.33 & -2.11 & 0 \\ 0.84 & 5.14 & 0.74 & 0.66 & -191.58 \end{bmatrix} \begin{bmatrix} X_1 \\ X_2 \\ X_3 \\ X_4 \\ X_5 \end{bmatrix} + \begin{bmatrix} 20810 \\ 486 \\ 0 \\ 0 \\ 0 \end{bmatrix} \tag{4.27}$$

In the matrix representation of the model, the empty or zero entries in the rate constant matrix make up what is referred to as the structure of the model.

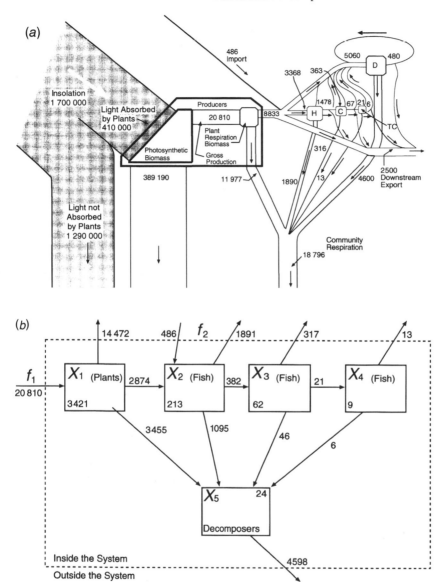

Figure 4.10. Energy flow in a clear spring in Florida as represented by Odum (1955). (*a*) Flows of energy in the spring as represented from Odum (1955). (*b*) Compartment model of energy flow in the system. Values in the compartments represent the steady state or equilibrium values of energy in each compartment determined by Odum in units of kcal m^{-2}; values beside each arrow represent the fluxes of energy in the systems (kcal m^{-2} yr^{-1}). Numerical values follow Patten's (1971) computation of energy fluxes based on the original Odum (1955) paper.

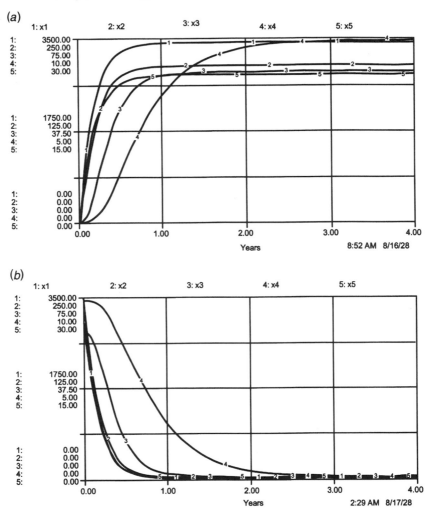

Figure 4.11. Dynamic responses of the Silver Springs compartment model. (*a*) Forced response. (*b*) Free response. (*c*) Impulse response.

Commonly, compartment models are solved by using numerical techniques on a digital computer. Numerical methods for solving the differential equations representing the material or energy flows in a compartment model involve determining the values of the derivative at some starting point. These derivatives are then applied to determine an approximate value for the state variables a short interval later, recalculating the derivatives at this new point in time and approximating the state variables over the next short interval. There are several methods for cor-

(c)

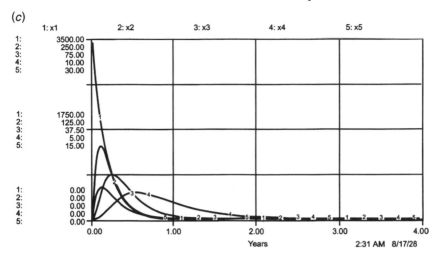

recting for inaccuracies in this approximation and to vary the calculation interval to optimise computational speed to solve with the
appropriate and desired accuracy of approximation (see Patten 1971;
Beltrami 1987).

Figure 4.11 illustrates the dynamic response of the Silver Springs
model under several diagnostic conditions and based on the parameters
from Eq. 4.27. Figure 4.11a shows what is termed the forced response
of the system. In this case, the dynamics of the system are determined
for a case in which the initial values or initial conditions of the
compartments are all set to zero. The inputs of the system or forcings
are set to their normal values. The dynamics that are shown are analogous to a case in which the material (or energy) going into a system is
labelled in some way from a particular time onwards and the dynamics
of this new, labelled material as it replaces the old unlabelled material is
followed.

Figure 4.11b illustrates the free response of the Silver Springs model. In
this case, the compartment state variables begin the simulation with the
values that one would expect them to have at equilibrium, and the input
terms in the equations are set to zero. The dynamics are analogous to
those expected if all material in the system were labelled and the rate at
which this labelled material was flushed out of the system by the input of
new, unlabelled material is followed. Figure 4.11c is the impulse response
of the system. This response has its analogue in a rapid injection of
labelled material into an otherwise, unlabelled system.

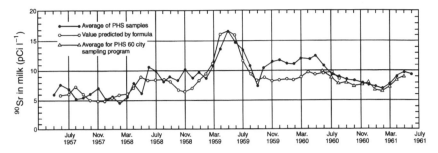

Figure 4.12. Comparison of predicted and observed concentrations of ^{90}Sr in milk from 1957 until 1961. (From Eisenbud (1973), original data from Knapp (1961)). PHS, Public Health Service; 1 Ci = 3.7 × 10^{10} Bq.

Applications of compartment models

Compartment models are appealing for ecological studies for several reasons. First, they are mathematically tractable. The procedures for estimating the parameters for such models are well developed in several fields (applied mathematics, physics, engineering). Historically, these procedures were transferred to ecosystem studies. Compartment models also appeared to predict certain aspects of ecological systems, particularly the movement of radioactive materials through a variety of aquatic and terrestrial ecosystems. Following the development of the atomic bomb and the nuclear industry, concern about radioactive pollution inspired considerable interest in the movement of radioisotopes through ecosystems. The experimental (and in some cases, accidental) application of isotopes to laboratory and field systems ('radioecology') was an important aspect of ecology in the 1950s and 1960s.

As an example of both the problem and the application of these models, Fig. 4.12 shows the levels of ^{90}Sr in milk for a number of cities across the USA predicted by a simple compartment model. With respect to its hazard to health, ^{90}Sr was a major isotope released into the atmosphere during the era of testing of nuclear weapons until the practice of open-air testing was banned by international treaty in 1963. Potentially, it is a problem isotope in health hazards from nuclear reactors. Strontium tends to move in the environment and in an organism's metabolism in ways that are analogous to calcium. It is an alkaline earth element that tends to concentrate in the bone. Its half-life (the time needed for half the material in a compartment to be lost) in bone is relatively long, ~1800 days. A major environmental pathway of ^{90}Sr to humans is through the atmosphere→soil→cows→milk→food product transfer pathway. The model of movement of ^{90}Sr into milk supplies (Fig. 4.12) duplicates the dynamic response of the milk in the different cities and, as an example of model performance, is indicative of the reasons for interest in this class of model in ecosystem studies. For materials such as radioisotopes, and trace substances of various kinds, compartment models have proved to have reasonable predictive ability (given an appropriate parametrisation).

Common concepts used in ecological modelling

Ecological modelling is a 'concept-rich' representation of the dynamics and interactions among system components. In recent years, systems and mathematical ecologists have appropriated analysis procedures and concepts from other fields (physics, mathematics and engineering) and applied them to a range of ecological problems. Indeed, the variety of such techniques used in ecological contexts prohibits treating all of these topics in an introductory chapter such as this one. Since the concepts of sensitivity analysis, non-linear systems dynamics and stability concepts will reoccur in the latter sections of this book, these topics will be briefly discussed below.

Linear systems

One very important feature of a system is whether or not it is linear. For ecologists, system linearity is a topic that has inspired a considerable level of debate involving ecosystem linearity. However, system linearity sometimes is not well understood by ecologists. There is a tendency to associate non-linearity with complexity. In some discussions, it is assumed that linear systems are related with linear regression in statistics and that linear systems have dynamics that are represented by straight lines. To avoid some of this confusion, it is appropriate to define what is meant by a linear system. It is useful to develop some notation to accompany this definition.

If $z_{(t)}$ is defined as an input to a system M and the response from system M to this input is $y_{(t)}$, then denote the relationship as $M[(z_{(t)})] \rightarrow y_{(t)}$. The system M is a mathematical function that relates an input to an output. Note that $z_{(t)}$ and $y_{(t)}$ may be single values or they can be a vector of values (such as the history of inputs to the system or the history of responses). A system is linear if for $z_{1(t)}$ and its response $y_{1(t)}$, for $z_{2(t)}$ and its response $y_{2(t)}$, and for b_1 and b_2, which are scalars (numbers), then:

$$M[(b_1(z_{1(t)})) + (b_2(z_{2(t)}))] = b_1 M[(z_{1(t)})] + b_2 M[(z_{2(t)})] \qquad (4.28)$$

If the system does not always conform to this condition, it is non-linear.

For example, a system whose responses are the integral of its inputs is linear. The demonstration of the linear nature of the integrator system provides a useful example. Consider a system M in which:

$$M[(z_{(t)})] = \int z_{(t)} dt = y_{(t)} \qquad (4.29)$$

For the system to be linear, it must be superposable (Eq. 4.28) or the equality

$$\int (b_1 z_{1(t)} + b_2 z_{2(t)}) dt = b_1 \int z_{1(t)} dt + b_2 \int z_{2(t)} dt \qquad (4.30)$$

must be true. Solving the left-hand side of the equation,

$$\int (b_1 z_{1(t)} + b_2 z_{2(t)}) dt = \int b_1 z_{1(t)} dt + \int b_2 z_{2(t)} dt = b_1 \int z_{1(t)} dt + b_2 \int z_{2(t)} dt \quad (4.31)$$

one can demonstrate that linking the output of one linear system so that it is the input to a second linear system in series produces a linear system. Linking species in parallel, or as cycles, also produces aggregated systems that are linear. Therefore, compartment models composed of linear compartments linked together are linear. Since one can develop linear compartment models of great elaboration in terms of model structure, there is not a necessary connection between linearity and system complexity.

Linear systems have input/output relations that follow the principle of superposition (DiStefano et al. 1967). Using the notation used in the earlier definition of linearity (Eq. 4.28), superposition is the condition that the response of a system to several inputs ($z_{1(t)}$, $z_{2(t)}$, etc.) is equal to the sum of the responses of each input acting alone, or:

$$M\left[\sum_{i=1}^{n} z_{i(t)}\right] = \sum_{i=1}^{n} M[z_{i(t)}] \qquad (4.32)$$

This definition follows directly from the earlier definition of linearity (Eq. 4.28).

An important implication of linearity and superposition is that if one knows the response of a system to a conveniently shaped input function, then one can perform various scaling and adding manipulations to produce a variety of differently shaped input functions. Using the principle of superposition, the responses to these new input functions are created by performing the equivalent manipulations on the response to the original function. One set of 'conveniently shaped' input functions for such

Table 4.5. *Examples of singularity functions. These functions are related to one another by numerical integration or differentiation. For linear systems, the response to singularity functions can be used to reconstruct the expected responses to a variety of inputs*

Name of function	Graph of function	Description of function
Unit ramp function		The unit ramp function has a value of zero when t is $<t_0$. At t_0, it increases constantly with a slope of 1 from 0. It is denoted as the integral of the unit step function (below) as: $$\int_{-\infty}^{t} u(\tau-t_0)\mathrm{d}\tau$$
Unit step function		The unit step function has a value of 0 when t is $<t_0$. At t_0 and thereafter, the function is equal to 1. The unit step function is denoted: $$u(t-t_0)$$
Unit impulse function		The impulse function $\delta(t)$ is obtained by: $$\delta(t)=\lim_{\Delta t \to 0, \Delta t>0}\left[\frac{u(t)-u(t-\Delta t)}{\Delta t}\right]$$

procedures are called singularity functions (Table 4.5). Singularity functions are related to one another by numerical integration or differentiation.

There are experimental procedures that can be performed on different systems that approximate the response to a singularity function. For example, the injection of a dye or an isotopic tracer into an ecosystem resembles the impulse singularity function, or an increase in the supply of nutrients to an ecosystem is like a step singularity function. In many fields, determining the response of a system to a particular singularity function is part of the normal protocol for analysing data about system response. In hydrology, for example, the response of a stream to a step increase in precipitation input (the 'unit hydrograph') is used to characterise and compare stream dynamics.

In some cases, pollutant releases, accidental spills and other events represent unplanned experiments that can be capitalised upon to understand better ecosystem dynamics. For example, the rapid increase in ^{14}C in the atmosphere resulting from open-air testing of fusion weapons in the early 1960s and the abrupt drop in the input of ^{14}C following treaties in 1964 represent an atmospheric input resembling an impulse and is one test of global models of carbon cycling.

Sensitivity analysis

The basic aim of sensitivity analysis is to document the pattern of variation in the state variables or system responses to small changes in the model parameters. The usual objective of sensitivity analysis is to determine which parameters in a model are 'important' with respect to developing better measurements. Sensitivity analysis also provides insights into theories on system structure and control of system behaviour. The basic concept of determining how a system changes in response to changes in the parameters is intuitively appealing. It mimics the sort of experimental manipulations that one would like to be able to perform to understand how a system works.

For a model parameter, p, one can imagine observing the change in a system state variable, X, after a small change (Δp). The change with respect to an incremental change in p can be denoted $\Delta X/\Delta p$. The size of this change increases as Δp becomes large. There is some value in inspecting the sensitivity of a system component as Δp becomes arbitrarily small ($\Delta p \rightarrow 0$). The resultant partial differential equation representation of sensitivity $\delta X/\delta p$ can be calculated directly from the model equations in some cases – in particular the formulations used for compartment models.

The change in the system with a change in a parameter does not have to be a state variable but can also be other indices of system response. In general, parameters that are large tend to have relatively larger sensitivities, and state variables that are numerically large also tend to have large sensitivities. For this reason, the 'absolute' sensitivities ($\delta X/\delta p$) are sometimes scaled by the size of the state variable (($\delta X/\delta p)(1/X)$, the normalised sensitivity), or by the parameter and the state variable (($\delta X/\delta p)(p/X)$, the relative sensitivity). Another complication is that in dynamic systems, the sensitivities can vary in size and in relationship to one another at different stages of the system's transient response (Brylinsky 1972). A state variable can be insensitive to changes in a particular parameter at one stage in a transient response and quite sensitive to the parameter in another stage of the response. This inspires a notation expressing the time when a particular sensitivity applies to the system response (e.g. $(\delta X/\delta p)(t)$ for the absolute sensitivity at a particular time, or $(\delta X/\delta p)(t \rightarrow \infty)$ for absolute sensitivity as time becomes large). These notational conventions reveal one important feature of sensitivity analysis: with the potential to compute the sensitivity of each state variable to each parameter both at equilibrium and at any time, sensitivity analysis can produce a copious quantity of numbers to be evaluated and can leave the analyst awash in a sea of sensitivity indices.

There are several related concepts to sensitivity analysis. Important among these is what has been called 'error analysis'. An error analysis approach to evaluating a model might be to determine (or theorise) the amount and pattern of the variability in a model parameter and then propagate this variability through the model. For example, if one determined that the value used for the first forcing (f_1) in the energy model for Silver Springs (Eqs. 4.21–4.25) varied as a normal distribution with a standard deviation of 300 (the value of the forcing) when it was sampled, then one could draw several thousand parameters from a distribution with a mean of 14 472 and a standard deviation of 300, put each of these parameters in the model and solve the model. Finally, the amount of variation that one might expect in a state variable originating with variation in the f_1 forcing could be computed to determine the error that one might expect from the model owing to sample variation.

The procedure just described is called a Monte Carlo procedure and is a frequent approach to developing error analyses. Error analysis procedures are similar to sensitivity analyses in that one can compute the variation in response at any time, for all combinations of parameters versus state variables. Several parameters can also be allowed to vary. In this case, the Monte Carlo procedure would be to draw sets of parameters from some statistical representation of the variation and co-variation among all the parameters. As is the case with sensitivity analyses, the volume of information arising from such procedures is considerable. Error analysis probably finds its best applications on relatively practical problems using well-established models.

Stability analysis

One case in which it is important whether a system is linear or non-linear involves determining the system stability. The fundamental concern in analysing the stability of a system is the long-term response of the system to an external change or a perturbation. Clearly, this is a topic of interest in considering the response of ecosystems to changes in the environment. Recall that in the earlier discussion of the Lotka–Volterra competition equations, cases of stable and unstable equilibria were identified in terms of whether or not the system would return to a location in state space after a small perturbation.

Historically, the problem of system stability concerned understanding the stability of the solar system (Timothy and Bona 1968). The question involved understanding whether or not the interactions of the planets in their orbits around the sun would destabilise one another. Could the planets be expected to remain in orbit forever or might their interactions cause one or another planet to lose its orbital position and either fall into the sun or hurl off into space?

There are two rather different mathematical traditions and associated methods for determining stability of dynamic systems. The first is appropriate only for linear systems and originates from some of the great thinkers of physics and mathematics (Newton, Euler, Laplace, Lagrange, Maxwell). The second applies to linear and non-linear systems and was the doctoral dissertation of a Russian mathematician, A. M. Lyapunov (Timothy and Bona 1968).

Linear stability analyses are involved with the response of the system to perturbations or changes in inputs. In linear systems, a system is stable if its response to an impulse function approaches zero as time approaches infinity (DiSephano et al. 1967). There are a number of methods that can be used to determine the stability of linear systems. Linear compartment models in which the material being transferred from one compartment to another is conserved (e.g. Eq. 4.27) are stable – because of the structure of the matrix of parameters for the models (the negative diagonal terms are all greater than or equal to the sum of the off-diagonal terms).

One method that has been applied in several ecosystems to determine stability of systems is the procedure known as 'loop analysis' (Levins 1974, 1975; Yodzis 1989). Loop analysis is derived from criteria for determining the stability of linear systems, called the Routh–Hurwitz criteria (Timothy and Bona 1968). The appealing aspect of loop analysis is that it allows one to determine from the structure of interactions whether a linear system will be unstable. This finding is independent of the particular parameter values (other than their signs) that are involved. One cannot use the analysis to prove that a system is stable (which can depend on the magnitudes as well as the signs of the parameters). Applications of loop analysis to non-linear systems are usually couched on the assumption that the system behaves in a linear manner, at least under the conditions being considered.

The second approach to the problem of systems stability was that of A. M. Lyapunov. In this case, stability concepts could be applied to non-linear (as well as linear) systems. Lyapunov recognised two sorts of stability. The first sort of stability in a Lyapunov sense concerned whether or not the system would return to the vicinity of its former state following a small perturbation (Fig. 4.13). In a system that is stable, any (small) perturbation on a system within ϵ of the state of the system at some time (t_0) produces a response from the system $(x_{r(t)})$ that is bounded within a distance δ of the unperturbed response $(x_{(t)})$ of the system (Fig. 4.13a). In a system that is asymptotically stable, the difference between the perturbed system dynamic and the unperturbed system dynamic goes to zero after sufficiently long time (Fig. 4.13b).

Concluding comments

This chapter provides a basic introduction to some of the considerations in developing ecological models. The initial concepts introduced involve state variables and the concept of system state and examples are given of the broad classes of model. The treatment of many important topics in the area of ecological modelling has been intentionally brief. Understanding the dynamics of terrestrial ecosystems subjected to altered environmental conditions is a model-rich topic, but this book is not intended as a treatise on ecological modelling. This chapter does not represent an exhaustive treatment of ecological modelling, but is in the spirit of a primer: a brief introduction of concepts and terms. There are several texts on the topics of systems and mathematical approaches in ecology, and several of these are noted as citations in the text. The chapters that follow provide several examples of different sorts of model applied to problems involving dynamic ecosystem response to environmental changes.

The principal examples used in this presentation have been differential equations used to model relatively simple interaction systems such as simple vegetation change, the interactions between two populations, or the movement of materials through ecosystems. These examples were chosen both to demonstrate the dynamics of these interactions and to illustrate development or analysis of a dynamic model. For population interaction models, a combination of types of population interaction (competition, predation, etc.) were paired with different example models of analysis. The combinations were:

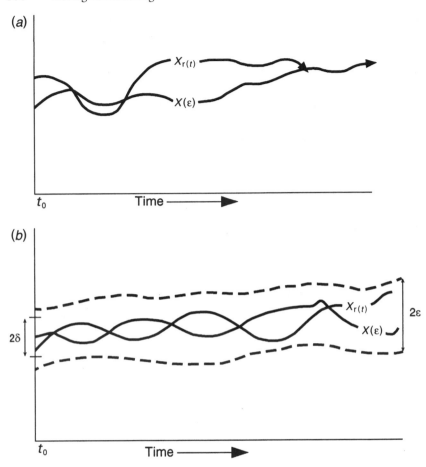

Figure 4.13. Lyapunov stability concepts. (*a*) In a system that is stable, any (small) perturbation on a system within ϵ of the state of the system at some time (t_0) produces a response from the system ($x_{r(t)}$) that is bounded within a distance δ of the unperturbed response ($x_{(t)}$) of the system. (*b*) In a system that is asymptotically stable, the distance between the perturbed system dynamic and the unperturbed system dynamic goes to zero over sufficiently long time.

1. Competition and graphic analysis of system dynamics
2. Mutualism and general theory on dynamics in state space
3. Prey–predator interactions and comparisons with experimental data.

The examples certainly do not exhaust the topic of ways to analyse or use dynamic models for simple ecological systems but are intended as a sampler. Population interactions of the type developed in these examples will be treated again in the next chapter.

Additional concepts dealing with system linearity are developed using the compartment model as a familiar ecological example. Compartment models were not initially developed by ecologists, but they are used to a great degree by ecologists – particularly those concerned with the topics of energy flow in ecological systems, nutrient cycling and chemical dynamics of larger systems such as watersheds or lakes. Indeed, compartment models are so associated with 'ecosystem ecologists' that they are sometimes taken as 'ecosystems models' that are in some sense contrasted with the models used in population studies. This is a false dichotomy in that all of these are ecosystem models. Their formulation involves defining an ecological system of interest and then proceeding to consider the dynamics of the resultant ecosystems.

Compartment models have been used in studies involved in energy transfers through ecosystems and in studies of element cycling through ecosystems to a sufficient extent that the model structures and the differential equations that they imply are a standard feature in ecosystems analysis. The use of compartment models in other fields (engineering, pharmacology, physiology, hydrology, etc.) was strongly enhanced by the predictive abilities of this relatively simple class of model. Compartment models have not been used as extensively in ecosystem studies in a predictive fashion as in other fields. This may be because many of these ecosystem studies are often at the 'biogeocoenosis' level; they are focused (typically) on understanding the responses of homogeneous units of 10 to 100 ha in size with respect to the dynamics of element cycles that require tens to thousands of years to turnover. These longer time scales of interest and consequently relative short observational time-series of data tend to proscribe purely predictive applications of linear compartment models.

The best applications intended for prediction and based on compartment models are probably in the studies of radioisotopes in the environment. In these cases, following trace amounts of isotopes through systems is aided by the ability to measure changes in these isotopes. Much of the evaluation of the transport of radioactive isotopes from various sources, through the environment to humans is made using linear compartment models as bases.

In most ecosystem studies, compartment models have taken on more of a synthesis role: the codification of observations and changes in systems into compartment models force a systems' level view of what is (or is not) known about a particular ecosystem. Formulation of data into compartment models interweaves diverse sorts of observation and provides ways to determine research priorities when the representation of a system in a complete sense is the objective of the researchers.

Even though this introduction to modelling and mathematical application in dynamic systems has been intentionally brief, several of the approaches introduced here will appear in later chapters as models to be used to assess the response of ecosystems to environmental change. Finite-state automata are used as models of landscape change in Chapter 10, as are Markov models. The FATE model (Moore and Noble 1990) is essentially a procedure for developing finite-state automata by grouping species into functional groups according to their life history, response to the environment and response to disturbances. A wide variety of Markov models of vegetation dynamics are also introduced in Chapter 10 (Horn 1975a,b, 1976; Noble and Slatyer 1978, 1980; Moore and Noble 1990, 1993). The population dynamics of the Lotka–Volterra equations have not been used to any great degree in assessing the responses of communities to change, but theoretical derivations from these equations and their 'descendants' (improved community models) have had a profound influence on the evaluation of community organisation and dynamics. Newer modelling approaches ('individual based models', Huston *et al.,* 1988; DeAngelis and Gross 1992) to be introduced in Chapter 8 are oriented towards merging population modelling, as introduced in this chapter, with an understanding of the response of individual organisms to their environment.

This chapter has presented some of the more important dynamic system concepts used in ecology. The text errs in the direction of conceptual (as opposed to formal) treatment of the models used to understand dynamic changes in the terrestrial surface. In this spirit, the three chapters that follow in this section will treat some of the ecological concepts that are often imbedded in models of terrestrial change (niche theory, vegetation pattern in response to environmental variables, landscape mosaic dynamics). Modelling will be revisited as a topic in Chapter 8, which treats a modelling approach that is still developing in ecological studies: individual-based models of relatively complex and large systems. The application of models in understanding ecosystem change at large spatial scales will then be considered in the last part of the book, following Chapter 9.

5 · Niche theory

People familiar with the ecological requirements of a given animal (or plant) frequently have an uncanny ability to locate individuals even when the species is rare, the individuals are few and the area to be searched is large. This ability, often a product of years of experience, is a point of pride among ecologists, wildlife scientists and other field researchers. It is also the mark of a good hunter, guide or angler. Indeed, knowledge of the needs of animals (or plants) and the association of these needs with the distribution of what the landscape provides is as old as innumerable, pre-historic discussions around campfires. The aspects of the biology of a species necessary for predicting the distribution and occurrence of individuals constitute the understanding of the niches and habitats of species.

Species niches also are important in the understanding of how species respond to a mosaic, heterogeneous landscape. In later chapters, niche concepts will be used to develop individual–organism-based computer models intended to predict the dynamic changes in natural ecosystems (Chapters 8 and 9). The understanding of the application of our ecological knowledge of a species to predicting the response of the species to change (its survival, change in geographical range, etc.) is essential in studies of the consequences of large-scale environmental change on biotic diversity.

The niche concept, involving the understanding of how plants and animals are distributed both locally and geographically, was initially formalised in the first third of the twentieth century. This formalisation is far from complete and, perhaps for this reason, the terminology is confusing. Nevertheless, the ideas and progress in predicting where different species are found and the challenging questions that are associated with predicting the structure of ecological communities give importance to the concepts associated with the niche and habitat of species. This chapter treats two traditional views of the species niche: the one involving the ways species interact with one another to produce patterns in ecosystem structure, and the other concerning the ways a species attributes control its distribution.

Grinnell (1904, 1917a,b) developed the niche concept with the intention of explaining how the attributes of individuals determined the manner in which they would fit into a range of environmental conditions (Table 5.1). Such relations can be aggregated to explain the potential distributions of a species in time and space. The Grinnellian niche was applied to the problem of understanding the distribution of single species. It subsequently developed largely as an individual-organism-based concept.

At approximately the same time and yet different from Grinnell's niche concept, the Eltonian niche (Elton 1927) was developed as a community-centric concept. A niche to Elton was what a species did in an ecological community, its function, how it fitted with other species. If the aim of the Grinnellian niche is to understand where individuals constituting a species population might be found, the aim of the Eltonian niche is to understand the way populations come together to form communities and, thus, to understand the seeming regularities in the patterns of sizes, of numbers and of biotic diversity in a community. Giller (1984) notes four central issues in community ecology that derive from this latter point of view:

1. How do species fit together to form a community?
2. What determines the number of species making up different communities?
3. How might the interactions among species populations set an upper limit on the number of species in a community?
4. What are the implications of differences in relative abundances of species in a community?

Eltonian niche theory (and its intellectual descendants) focuses on the theoretical basis for understanding these and related questions. Usually, the community concept typically used in Eltonian niche theory is relaxed from the 'community as a super-organism' view (Chapter 3) that motivated Tansley (1935) to define the ecosystem as an alternative. A community, in this context, is a local assemblage of plants and animals. Nevertheless, some of the issues and explanations in Eltonian niche theory involve evolutionary mechanisms and, therefore, invoke a more co-ordinated and co-evolved notion of the ecological community (Table 5.1).

Niche in its use as an ecological term has to do with the 'fit' of biotic entities to their environment and with one another. But at what level of organisation is this fitting most appropriately applied? This remains an issue of debate. In the following sections, the implications of the niche for species distribution (the Grinnellian niche) and for community interactions (the Eltonian niche) are discussed.

The Grinnellian niche

In 1917, Grinnell considered the factors that influenced the distribution of the California thrasher (*Toxostoma redivivum*), a bird found in the chaparral vegetation in California (Grinnell 1917a). He used the term 'niche' to describe the factors that influenced where one might find the species. While the 'niche' was not a novel word, even in an ecological context (see Gaffney 1975; Cox 1980), the use of the word by Grinnell (1904, 1917a,b) initiated what became the important current use and concept of the niche. In his discussion of the niche of the California thrasher, Grinnell included considerations involving the food of the species, the preferences of the bird for certain types of vegetation structure and other details that influenced where the species could be found. Grinnell, later (1928), defined the niche as the 'ultimate distributional unit, within which each species is held by its structural and instinctive limitations, these being subject only to exceedingly slow modification down through time.'

Ramensky (1924) and Gleason (1926) proposed an essentially similar view, referred to as the 'principle of species individuality' for plants (Whittaker, 1975).

The Grinnellian niche can be thought of as being defined by the attributes of the animal (or plant). Whittaker *et al.* (1973) felt that Grinnell's concept of the niche emphasised the habitat to the extent that the niche was synonymous with habitat ('the habitat or place niche'). While Whittaker *et al.* (1973) may be correct with respect to the current perception of Grinnell's work, a reading of Grinnell (and an inspection of the quotation above) indicates a concept of much greater breadth involving not only the habitat use, but the behaviour and physiology of the organism.

James *et al.* (1984) developed a critical review of the niche concept and its elaboration since the term was defined by Grinnell (1917a,b; 1924). They defined the Grinnellian niche as:

> . . . the range of environmental factors that are necessary and sufficient to allow a species to carry out its life history. Under normal conditions of

Table 5.1. *A comparison of the Eltonian niche and the Grinnellian niche of a species*

	Eltonian niche model	Grinnellian niche model
Objective	To understand communities in terms of resource division among co-existing species; to study the assembly of communities and their evolution (MacArthur and Levins 1967)	To understand population regulation in terms of the resources that limit the distribution and abundance of a single species throughout its geographical range
Focus	Emphasis is on direct or indirect, present or past interactions among co-occurring species; competitive exclusion (the elimination of overly similar species from communities); mechanisms by which species co-exist (Levins 1970). Also, an important consideration is the organisation of communities by interactions among co-occurring species (Gause, 1934; Hutchinson, 1957; Cody, 1974; Diamond, 1975). Interactions among species are important in the determination of species distribution, population regulation and evolution (Levins, 1968)	Emphasis is on the relationship between a species and attributes of the environment that impinge upon its life history. Interactions with other species are one of several factors that can be considered but usually are a secondary consideration
Relationship to evolution	Invokes natural selection in the evolution of communities and lineages (Cody, 1974). Associated constructs involve limitations in the degree of similarity between species (Roughgarden 1976) and evolution of regular differences among similar species (character displacement: Brown and Wilson, 1956)	Invokes natural selection for intraspecific adaptations in different parts of the species geographical range (Grinnell, 1904, 1924). Evolution in relation to membership in local communities is relatively less significant that evolution in relation to the geographic distribution of resources

| Methods and procedures | *Mathematical*: analysis of theoretical models usually derived from the Lotka–Volterra equations (see Eqs. 4.10 and 4.11). Niche overlap is sometimes equated with magnitudes of the competition coefficients (α and β parameters in Eqs. 4.10 through 4.17; see Levins, 1968)
Empirical: quantification of resources (e.g. food, habitat, etc.) and degree of sharing or overlap in these resources using a variety of observational, interpretive and inferential procedures | Consideration of species-specific requirements in terms of life history, physiology, behaviour and habitat in many locations. Determination of common features in the places where the species is found. Prediction of species density from resource and environmental descriptors |

Source: Table is slightly modified from that of James *et al.* (1984). In the original, the 'Eltonian niche model' was denoted the 'Hutchinsonian niche model' and is changed here (along with some of their text) to better reflect the present discussion and to acknowledge Elton's (1927) role in developing an alternative view of the niche.

reproduction and dispersal, the species is expected to occupy a geographical region that is directly congruent with the distribution of its niche, and the density of the species within its geographical range is expected to be correlated with the prevalence of these conditions. A study of the Grinnellian niche may seem overly broad because, except for some aspects of population biology, it is equivalent to the study of a single-species geographical ecology.

James *et al.* (1984) illustrate the Grinnellian niche using the wood thrush (*Hylocichla mustelina*) as an example. With the wood thrush, James and her colleagues found that the breeding range and the relative abundance of the species through its range co-varied with the vegetation structure of the eastern deciduous forest of eastern North America. The species did not appear to be related in its distribution with patterns of other, ecologically similar species of thrush across its range. This latter observation is important with respect to the issue of whether or not the species distribution is shaped by competitive interactions with other species: an aspect often associated with the Eltonian niche (see Table 5.1 and below).

The Eltonian niche

Elton (1927) formulated a concept of the niche that differed significantly from that of Grinnell (1917b). Elton emphasised the function of the species and defined the niche of the species as, '. . . the niche of an animal means its place in the biotic environment, its relations to food and enemies'.

Elton (1927) particularly focused on the relation of the species to its food and its predators, but he also made it clear that the niche included the species' place in the community with respect to shelter, competition and a range of other considerations. The Eltonian niche places such a strong emphasis on the function of the organism that it has been called the 'functional niche' (Clarke 1954) for this reason.

In the sense of Grinnell (1917a,b), two species having the same niche is unlikely because the niche of the species is described with considerable detail. Elton (1927) used a less precise concept of the niche and saw, for example, foxes in the Arctic that feed on guillemot eggs and also eat dead seals left by polar bears as being equivalent to hyenas in semi-arid Africa that eat ostrich eggs and the remains of lion kills (Griesemer 1992). Both Grinnell (1924) and Elton (1927) pointed out that two species in widely separated communities can be thought of as occupying the same niche, but such occurrences would be more common in the more broadly defined Eltonian niche.

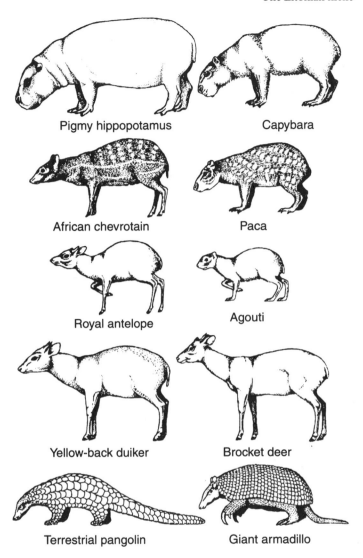

Figure 5.1. Morphological convergence among African (left) and Neotropical (right) rain forest animals. The convergence of form and function such as these was felt by Elton (1927) to indicate that they occupied similar niches. From Bourlière (1973).

In a phenomenon called 'convergence', unrelated species in different ecosystems evolve to the point of being strikingly similar (Fig. 5.1). Elton saw convergence as evidence of similar niches in geographically separated communities and reasoned that these similarities implied some sort of community constraint on the pattern of species niches (Giller 1984). If one could find analogous niches for species in different communities in widely disjunct places (Elton 1927), then one also could think of communities as having 'empty niches'.

The Eltonian niche concept and its theoretical descendants focus strongly on understanding the patterns and statistical distributions of rarity and abundances within communities in the context of a larger, unifying theory (Table 5.1). The emphasis is one of understanding the structure of communities expressed as numbers of individuals and species, patterns of rarity and commonness, and convergence of apparent roles across different taxa. The intent is the formation of a holistic theory of community structure that explains the patterns seen in all communities.

The development of a unifying community theory is a remarkable challenge and Elton (1927) posed the problem in a particularly clear and engaging fashion. Elton initially emphasised the trophic or feeding situation of a species as a principal determinant of the species niche. For many researchers this thermodynamically based view of the niche of the species is still a basis for developing an ecological theory.

The Eltonian niche and the competitive exclusion principle

Under the influence of the experimentalist G. F. Gause, the Eltonian niche was given a strong shift in emphasis from trophic interactions (which concentrate one's interest to the relations between predators and their prey, or herbivores and their food) to intratrophic interactions, particularly competitive interactions. The emphasis on internal factors that structure communities has remained central to the interest of researchers involved in competition studies and in other Eltonian niche-related issues (the limit of similarities among species, the degree to which different communities can be invaded by alien species owing to empty niches, etc.). Gause (1932) and his co-workers developed a set of experiments using test-tube populations of protozoans being fed by regular inoculations of yeast cells. These experiments were aimed at investigating the theoretical derivations of Lotka (1925) and Volterra (1926) for the interactions between populations of different species (Chapter 4). Gause worked with species of the protozoan *Paramecium* and conducted experiments on pairs of the three species *P. aurelia*, *P. bursaria* and *P. caudatum*. Experiments involving *P. aurelia* and *P. caudatum* resulted in *P. caudatum* losing population numbers (Fig. 5.2) over the course of the experiment. Gause's

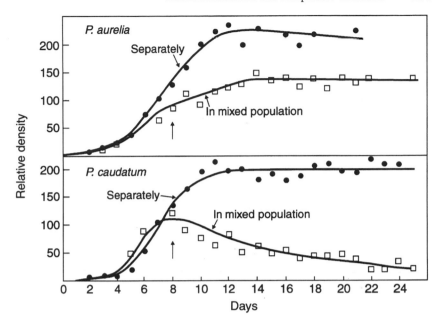

Figure 5.2. Competition experiments with *P. caudatum* and *P. aurelia* reported by Gause (1934). Closed circles indicate the population dynamics and approach to a carrying capacity when the respective species are grown separately. Open squares indicate the equivalent population dynamics when the two species are cultured together. Note the lowered carrying capacity of *P. aurelia* in the presence of *P. caudatum* and the tendency toward competitive exclusion of *P. caudatum* by *P. aurelia*.

experiments were not conducted long enough for the species that appeared to be losing the competition to actually become extinct (Arthur 1987), but the results were interpreted as the competitive exclusion of *P. caudatum* by *P. aurelia*. Experiments with *P. bursaria* interacting with either of the other two species resulted in both *P. bursaria* and the other species surviving until the end of the experiment.

Gause's experiments (Fig. 5.2) strongly resembled (and thus confirmed) the results predicted from mathematical analysis of dynamic equations for interacting populations (Lotka 1925; Volterra 1926). The agreement of formal mathematical descriptions of the growth, death and competition between species with experimental results had the promise of moving population and community biology to a more rigorous, more mathematically formal level. The experimental work of Gause (e.g. Fig. 5.2) was reported in two books, one in English (Gause 1934) and the other in French (Gause 1935).

Gause (1934) associated his results directly with Elton's (1927) ideas about the niche and wrote, '. . . as a result of competition two similar species scarcely ever occupy similar niches, but displace each other . . .' This idea became known as 'Gause's principle' or the 'competitive exclusion principle' (Hardin 1960). Gause (1934) based his idea that competition arose from the overlap of species niches on his observations of the species pair that demonstrated competitive exclusion (*P. aurelia* and *P. caudatum*). Both of these species appeared to feed in the upper part of the test tubes in which they were cultured. Either of these two species were able to survive in the absence of the other and in the presence of the species *P. bursaria*, which appeared to feed at the bottom of the test tubes.

Arthur (1987) summarises Gause's experiments and several subsequent experiments under laboratory conditions involving *Paramecium*, fruit flies (*Drosophila* spp.) and small beetles (*Callosobruchus* spp.). One finds that the details of how niche differentiation actually works is not always clear from these experiments. Nevertheless, competitive interactions, including competitive exclusion, appear to be a regular and repeatable feature in laboratory populations.

The importance of competition appears to be more debatable in natural populations. As will be discussed below, Schoener (1983) and Connell (1983) in simultaneous and independent surveys of the ecological literature found evidence for competitive interactions in a large percentage of the cases they examined. However, Arthur (1987) contends, 'There is probably not a single case in which we can conclusively attribute the disappearance of a species from, or its lack of successful invasion of, a natural environment to competition with another species'. This echoes Williamson's (1972) earlier observation, 'The evidence for stable competition in natural populations is extremely tenuous . . .'.

If, as Williamson (1972) and Arthur (1987) contend, direct demonstrations of stable competitive interactions and/or competitive exclusion (the hallmarks of Gause's interpretation of the Eltonian niche) are in short supply under field conditions, there remains a rich array of circumstantial evidence that has been used to support the concept that the structure of ecological communities is caused by competitive interactions. Much of this evidence is based on the interpretation of regular patterns in morphologies, abundances and spatial distributions of species in communities (Giller 1984). These include interpretation of differences among individuals (such as differences in morphology) and species (such as patterns of distribution and of abundances). In some cases, the presence of one species appears to cause the elimination of another. In other cases, species in the presence of one another differ in ways consistent with having different niches. In diverse assemblages of species, combinations of competitive interactions are interpreted as generating regular statistical features in the abundances of species, or in limits to the degree of similarity among some of the species. Williamson (1972) develops several categories of evidence of competition that are discussed as examples in the sections that follow.

Displacements and replacements

The displacement of one species by another can be seen as evidence for competitive interactions or competitive exclusion. The most striking examples are cases in which an introduced species invades a region and an ecologically similar resident species is eliminated with the systematic spread of the alien. The best documented cases of such occurrences are for relatively recent species invasions. These almost always are associated with human alterations of habitats over landscapes and regions: a situation that invariably clouds the interpretation of the responses in the populations of the species involved. Elton (1958) noted that successful invading species are likely to be found in disturbed, human-altered habitats: an observation that has been supported by most of the cases that have occurred since (Orians 1986).

An example of a 'displacement' cited by Williamson (1972) is the reduction in numbers of the native red squirrel (*Sciurus vulgaris*) in England and Wales associated with the invasion by the American grey squirrel (*S. carolinensis*) (Lloyd 1962). Williamson cautions that there are several potential difficulties in interpreting such dynamics as a result of competition. In the case of the two squirrels, his caution appears to be supported by subsequent studies.

Data documenting over 20 years (from 1960 to 1982) of geographic displacement of the red squirrel by the grey squirrel over eastern England (Reynolds 1985) show an apparent collapse of the population of red squirrels and an associated invasion by the grey squirrel (Fig. 5.3). However, detailed analysis of these data reveals that the likelihood of the red squirrel becoming extinct on a given 5×5 km grid square appears to be unrelated to the arrival of grey squirrels in that location. Some populations of red squirrels have become extinct as much as 18 years before the arrival of grey squirrels; others have remained viable as much as 16 years after the arrival of grey squirrels (Reynolds 1985). Several factors not related to competition are involved in the displacement. A pox virus

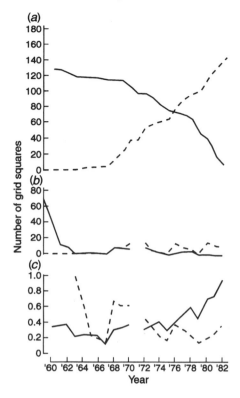

Figure 5.3. (*a*) Total number of 5 × 5 km grid squares in eastern England occupied by the red squirrel (solid line) and the grey squirrel (dashed line) in each year 1960–82. (*b*) Numbers of grid squares recorded for a species for the first time in each year. (*c*) Proportion of grid squares recorded for a species for the first time in each year. 1971 is eliminated from the calculations as this was a year of unusually intense surveying (see Reynolds (1985) for details and for year-by-year maps). Figure from Reynolds (1985).

disease is a sporadic and locally catastrophic factor in the decline of the red squirrel. Another complicating factor is that the two squirrels appear to have somewhat different habitat preferences and there have been systematic habitat changes for the region in the 22 years of the Reynolds' (1985) study. One factor that regularly clouds the interpretation of species displacements is the spatial and temporal heterogeneity of landscapes, the 'omnipresence of change' discussed in Chapter 2. The condition of stable, relatively homogeneous habitat that would make the interpretation of species displacements unequivocal does not occur.

'Replacements' are another category of evidence that can be interpreted as supporting the role of competition as a force in structuring communities (Williamson 1972). Replacements are cases in which species replace one another either geographically or with respect to using different habitats. An example that Williamson (1972) provides involves three species of Darwin's finches (*Geospiza* spp.) found on the Galapagos Islands. In his example (Fig. 5.4), *G. conirostris* is not found on islands occupied by the pair of species *G. fortis* and *G. scandens*. The feeding habitats of *G. fortis* and *G. scandens* are the most similar among the species of *Geospiza* to that of *G. conirostris*. This overlap is presumably the basis of a competitive interaction allowing the pair of species to replace *G. conirostris* over most of the islands.

In the cases of displacements and replacements, the changes in distributional patterns in time and space are interpreted in terms of competitive interactions. The evidence, in these cases, is circumstantial and its meaning can be debated, sometimes quite acrimoniously. The problem with circumstantial evidence is that it does not prove, in a rigorous sense, the importance or even the existence of competition.

Ecological differentiation and character displacement

There are other lines of circumstantial evidence possibly supporting but not necessarily proving the importance of competition as a present or past force in structuring ecological communities. From Gause's experimentation, there has been an expectation that co-existing species would differ in some ecologically significant way. An important corollary to the principle of competitive exclusion is the so-called 'principle of limiting similarity'. This is the idea that co-occurring species must necessarily be different in some way(s) that allows them to be in different niches and thus to co-exist. Of course, species usually differ morphologically to some degree, and a demonstration of such differences among species has been a part of most studies on the comparative morphology, behaviour or ecology of similar species (e.g., Fig. 5.5). It appears that, if a sufficient number of factors are considered, one can almost always find significant differences in the way that species fit into their environments. These differences can often be interpreted as reflecting different adaptations of the species. They can also be interpreted as a basis of niche differentiation. Often there are observable differences among species that are sufficiently similar to be judged to be potential competitors (similar habits, same genus, same food, similar use of space, etc.). However, there are regularities in the ways that similar species differ that imply a more structured limit to the similarity of species (Brown 1995).

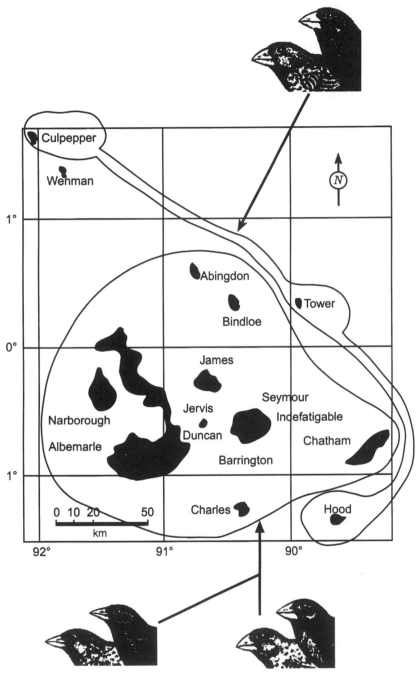

Figure 5.4.   Displacement in the distribution of *Geospiza conirostris* (illustrated at the top of the figure) on the Galapagos Islands by two species with similar feeding habits (*G. scandens* and *G. fortis*, respectively, on lower left and right of figure). *G. conirostris* does not occur on islands inhabitated by the other two similar species. Map from Williamson (1972), *Geospiza* illustrations from Lack (1953).

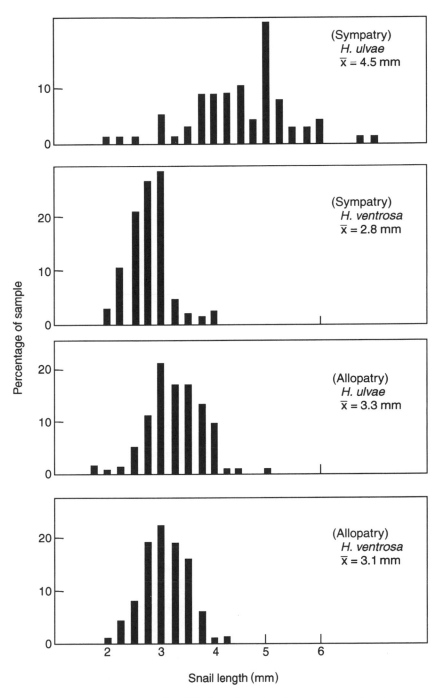

Figure 5.5. Frequency of shells of different lengths of two species of snails (*Hydrobia ulvae* and *H. ventrosa*) in a location where they co-exist (sympatry) and at locations where they do not co-exist (allopatry). From Fenchel (1975).

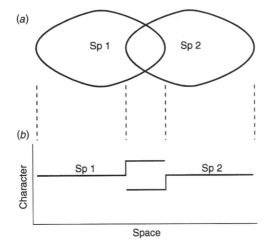

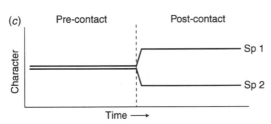

Figure 5.6. Concepts of character displacement. One would expect species to have differences or displacement of characters in areas in which their geographical distributions overlap. Similarly, one would expect species brought together at a location to demonstrate a subsequent displacement over time. (*a*) Geographical distributions. (*b*) Displacement in characters in response to overlaps in geographical distributions. (*c*) Character displacement over time for two species following contact of their ranges. From Grant (1972).

Character displacement occurs when two species have an overlap in their geographical distributions and are found to differ most greatly in some measurable feature or 'character' one from another in this zone (Fig. 5.6). Often the character's mean value changes with a ratio of about 1.3 between the members of an ordered sequence of species (Hutchinson 1959). An interpretation of such regular divergence in morphological characters of apparent ecological significance is that it implies differential use of resources and is a consequence of competition (see Pianka (1976) for a review of several such studies). Horn and May (1977) provide an analysis of the equivalent ratios of sizes of musical instruments, children's toys and frying pans that also appear to conform to the 1.3 rule for sizes.

While humorous, the observations of Horn and May are in keeping with Grant's (1972) commentary that,

> The principal conclusion to be drawn is that evidence for the ecological aspect of morphological character displacement is weak. Several examples of supposed character displacement have been shown to either be wrong or to be possibly correct but just as convincingly interpreted in other ways without recourse to character displacement. . . . Others illustrate the different but related principle of character release . . . Yet others are so complex that the unequivocal identification of character displacement is impossible, even though it may well have occurred . . .

One problem with evaluating circumstantial evidence for competition and the related phenomena that Gause (1934, 1935) and others (see Arthur 1987) have been able to demonstrate in the laboratory may involve the spatial complexity and environmental heterogeneity encountered under field conditions. Cases that seem on initial inspection to be clear examples of competitive exclusion, replacement or character displacement often become less clear-cut as more data are collected or when a greater range of factors is considered (Peters 1991).

Experimental evidence

The most obvious approach to avoid the interpretative difficulties based on circumstantial evidence is direct experimentation to understand the competitive interactions under field conditions. Such experiments would involve removing each of two or more potentially competing species in an area and observing an increase in the numbers of its competitor(s). Proper procedure would involve maintaining control populations. Because many populations fluctuate in numbers and vary in time and space, there can be a considerable number of replicates needed to control the variability.

In 1972, Williamson noted the rarity of field experiments and repeated an earlier plea of Varley (1957) for more experiments. Because of the interest in competition in structuring communities in the 1970s, there was a subsequent great increase in the number of field experiments involving competition. Arthur (1987), who had a strong interest in laboratory experiments on competition, tended to fault many of the field experiments on the grounds of experimental design and lack of experimental control. These problems are most easily explained by the substantial logistic difficulties in developing large-scale field experiments.

In 1983, Schoener and Connell independently reviewed the status of evidence for competition as a force working to structure the community by tabulating papers that claimed to investigate the outcomes of experiments on competition among two or more similar species. Connell (1983) reviewed papers in six journals (*Ecology, Ecological Monographs, Journal of Ecology, Journal of Animal Ecology, The American Naturalist* and *Oecologia*) published between 1974 and 1982. He identified evidence for competition in about 40% of 527 experiments. Schoener (1983) conducted a less restricted review and had somewhat different criteria for his enumeration (see Schoener 1985). He found about 90% of the experiments that he tabulated demonstrated competition.

In evaluating Schoener's (1983) and Connell's (1983) results, one should realise (as they point out) that an ensemble of experiments drawn from the literature is a biased sample. Experiments, under field conditions, are likely to emphasise species with populations that respond relatively rapidly. Because such experiments involve removing or manipulating populations, the species populations are likely to be easily captured and/or contained. The spatial heterogeneity of natural landscapes represents a source of experimental variation requiring replication. Therefore, competition experiments are likely to be conducted on species that do not range widely. Given the effort that is involved in conducting field experiments, investigators are likely to focus on species that will probably produce results, or at the least be 'interesting' in a scientific sense.

Subsequent reviews (e.g. Cornell and Lawton 1992) have tended to identify an overall trend for some communities to demonstrate experimental evidence for competition and for others not to do so. Connell's (1983) observation that competition experiments are not often replicated by independent observers (he only noted one case) still holds today.

Attempts to synthesise Elton's and Grinnell's niche concepts

Both Elton and Grinnell were similar in that they were attempting to formalise definitions of the way that an animal (or plant) fits into a community. It is also clear, in retrospect, that Elton and Grinnell defined and used the word niche in very different ways. The subsequent entanglement of the word 'niche' (and its already confused meaning) with 'principles' involving competitive interactions among species has not served to clarify the situation.

There have been several significant attempts to reduce the ambiguity in niche-related ideas. Hutchinson (1957, 1965) defined the niche of a species as a volume in a space with dimensions that were 'relevant' environmental variables. If the space had n such relevant dimensions, it is termed a n-dimensional space (or hyperspace). The species niche is a volume in such a hyperspace. This volume is bounded by some measure of the success or survival of the species (say for example that the fitness is greater than zero). The complete potential range of the species would be the 'fundamental niche' of the species. The set of conditions found in a given habitat was called the 'biotope'. In a particular location, the fundamental niche of the species can be restricted either because all the conditions under which a species might live do not occur, or because the species is excluded by competing species (Fig. 5.7). This restricted set of the conditions is called the 'realised niche'.

Hutchinson's (1957, 1965) geometrical or hyperspace interpretation of the species niche accorded well with the Grinnellian niche idea determining the factors that might control the distribution of a species. Since Hutchinson (1965) proposed that the overlap in the realised niches of two species related to the degree to which they might compete with one another, there is a direct link between the n-dimensional niche concept and the Eltonian niche and competition. The concept that the species niche could be thought of as an abstract volume in n-dimensional space (Hutchinson 1957, 1965) strongly suggests the use of a set of statistical methods collectively called multivariate statistics (Fujii 1969). Examples of the application of multivariate statistics as a niche analogue are presented below.

MacArthur (1968, 1970, 1972) defined the niches of species as their distributions with respect to a quantified resource-related variable (e.g. the size of seeds eaten by birds). For each species, the function describing its distribution is called a resource utilisation function (Arthur 1987; Schoener 1988). Several theoretical treatments of the niche (usually in the context of understanding abundance patterns generated by competitive interactions) have used resource utilisation functions as a basis for interpreting the relative rareness or commonness among an assemblage of related species or to interpret the relative absolute differences in species (Table 5.1). Often the variables to which these resource utilisation functions have been applied have been time, food and space (Brown 1995). Indeed, many of the studies have considered only one dimension, but there is no requirement that such niche studies have to be low-dimensional (Pianka 1981). Focal questions in these analyses involve relative locations of niches of species along the resource variable, the breadth of these niches and the overlaps among niches of different species (Fig. 5.7). Whittaker *et al.* (1973) produced an example of a two-dimensional

(a)

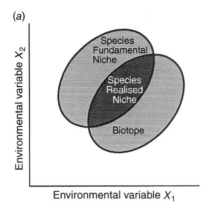

Environmental variable X_1

(b)

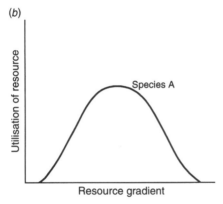

Resource gradient

Figure 5.7. (a) The niche hyperspace concept as developed by Hutchinson (1957). The entire range over which a species can survive with respect to a set of environmental variables is referred to as the fundamental niche. The realised niche is the overlap between the fundamental niche and the biotope (the set of conditions in a given situation). (b) A resource utilisation function is the use of a resource by a species with respect to a quantified resource gradient.

response surface (Fig. 5.8) in which one of the dimensions was a classic resource axis (prey length) and the other was a habitat dimension (height of the animal above the ground when feeding). They used this example to illustrate the separation of the habitat aspect of the niche (which they associated with Grinnell) with a resource utilisation dimension.

Quantification of the Grinnellian niche

Grinnell's (1917a,b) fundamental idea, and the concept reiterated by James *et al.* (1984), is that one can use a detailed description of the

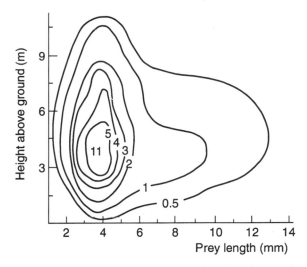

Figure 5.8. A niche response surface representing the capture of prey of different sizes taken at different heights above the ground by the blue-grey gnatcatcher (*Pilioptila caerulea*). The contour lines map the feeding frequencies (percentage of total diet with respect to these two niche axes. Data are collected for adult birds during the egg incubation period (July and August) in oak woodlands in California (data from Root (1967), used as an example in Whittaker *et al.* (1973)).

requirements of a species to understand its geographical distribution. The implementation of this concept provides an important basis for predicting the distributions and occurrence of species over large areas. Independent of subsequent developments of the niche concept, Grinnell's initial insights have proven valuable, particularly in the areas of wildlife management and conservation biology. This is likely to occur even more in the future as more powerful computer systems capable of developing dynamic maps of large areas are used (so-called Geographical Information Systems or GIS) with detailed, geographically extensive data sets.

The use of multivariate statistics to produce quantitative predictors of species occurrence is also an important development in this direction. An additional contributing factor is the increased availability of spatially extensive data sets from remote sensing or other sources. Examples of these aspects of the Grinnellian niche will be discussed in the sections that follow. The consequences of plants and animals distributing themselves on dynamically changing landscapes according to their niches will be discussed in Chapter 12.

If the niche is an ultimate geographic unit (Grinnell 1928), a logical way to determine the species niche is to collect a large number of observations on the conditions where a species is found and inspect this collection of observations for common patterns. If subsequent observation or experimentation indicates that such a determination of the niche can be used to predict the occurrence and non-occurrence of a species, then one would feel justified in claiming to have described the species niche at least in the manner that Grinnell (1917a,b, 1928) defined the term (James *et al.* 1984). Experimental tests are clearly preferred in such an evaluation, but it is impossible to perform experiments at the spatial scale of the geographical range of species in most cases. Predicting the response of an introduced species based on its performance and response to variables in their native range represents an, at least, quasi-experimental test.

The application of multivariate statistics in niche-related studies is in the spirit of formalising and quantifying the more intuitive procedure just described. Austin (1985) noted the strong similarities between the continuum concept in plant community ecology and the niche theory concept in animal ecology. The continuum concept is the concept that populations of plant species are distributed independently along important environmental gradients and, therefore, that one should expect plant communities to intergrade continuously along environmental gradients (Goodall 1963; McIntosh 1967; Whittaker 1975). A significant body of information on the distribution of plant species in response to measured features in the environment (a direct analogue to quantitative niche studies in animal ecology) has been developed. The methods often are discussed under the topic of 'direct gradient analysis' by plant ecologists (Waring and Major 1964; Whittaker 1967; Kessell 1979a) and have produced results that parallel those from studies of animals.

There has been some debate about which of several methods could best be used in niche quantification in animal community studies. An important topic in this debate concerns the best statistical procedure to be used to reduce the complexity of measurements of the abundance of multiple species in response to a large number of environmental variables (all potentially important). One method that has been used extensively is discriminant function analysis (Williams 1983). Discriminant function analysis is illustrated in a purposely over-simplified fashion in Fig. 5.9. In this figure, each point in the ellipses is an observation of the conditions under which individuals of two species are found with respect to two environmental variables, x_1 and x_2. The species have overlapping distributions along each of the variables but, owing to differences in the combinations of conditions under which they are found, do not actually co-occur

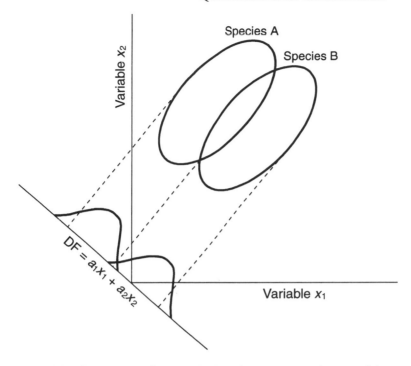

Figure 5.9. Discriminant function (DF) analysis as a quantification of the species niche. In this simplified diagram, the two groups are group 1, a cluster of points representing conditions with respect to environmental variables (x_1 and x_2) where a particular species occurs and group 2 is the analogous cluster of points for conditions in situations where the species is known not to occur. The two groups are different but show substantial overlap with respect to the two environmental axes. DF is a weighted combination of the two axes that maximally distinguishes the two groups (note the distribution on the DF axis). Modified from Smith *et al.* (1981a).

to any great degree in the same locations. The discriminant function (DF) is a weighted, linear index of the two variables that maximises the distance between the averages of the two species on the index. Given values of a new location with respect to variables x_1 and x_2, one can assess the likelihood that that location is the sort of place where one might find species A or species B.

In practice, discriminant function analysis has several limiting statistical assumptions involving the patterns of variability, the nature of the initial data set and the numbers of observations collected. The method has been used in a number of cases (Green 1971, 1974; Anderson and Shugart 1974; M'Closky 1976; Dueser and Shugart 1978, 1979; Baker and Ross

1981; Porter and Dueser 1982) to quantify the niches of a range of types of organism (from barnacles to mammals). Discriminant function analysis has been successful in cases in which differences in species are emphasised, or in cases in which the prediction of occurrence/non-occurrence of a species is important.

Other statistical methods such as principal components analysis (Rottenberry and Wiens 1980; Doncaster 1981; Burgman 1989) have proved valuable in cases in which the restrictive assumptions of discriminant function analysis have proscribed the method. In many cases, the differences in attributes of locations where different species are found (microhabitat use) appear to be sufficiently strong that the demonstration of dissimilarity in habitat use is clear regardless of method.

Initial studies using multivariate statistics were based on the use of multivariate procedures to express the patterns implied by the redefinition of the niche called Hutchinson's (1957, 1965) hyperspace niche concept (Fujii 1969; Green 1971; Shugart and Patten 1972; Litvak and Hansell 1990). It eventually became clear that the multivariate statistical methods were a logical extension of the niche concepts developed and elaborated by Grinnell (James et al. 1984). The relative success of discriminant function analysis and other methods (Miracle 1974; Johnson 1977a,b; Rottenberry and Wiens 1980; Doncaster 1981; Burgman 1989) in predicting the occurrence of species and the responses of species to novel habitats (Rice et al. 1981, 1986) attracted the interest of ecologists and wildlife managers involved in promoting or maintaining species diversity. Several major symposia (e.g. Capen 1981; Verner et al. 1986) have reviewed and provided case studies of methods based on multivariate statistics (and by extension the Grinnellian niche) to issues involved in managing wildlife, particularly non-game animals. As can be seen in the examples that follow, these applications cover a range of scales from local to continental. In the spirit of the Grinnell niche, they focus on determining the features that originate in the biology of a species and that control its geographical distribution.

Habitat selection in the ovenbird, *Seiurus aurocapillus*

The ovenbird, *Seiurus aurocapillus*, has been a focal organism for several studies on the relationship among the structure of the habitat, the availability of food in the habitat, and the size of the territory defended by birds in different locations (Zach and Falls 1976a,b,c, 1977, 1979). The birds forage by walking slowly and more-or-less continuously along the forest floor and picking invertebrates from the litter surface. They do not 'work' the substrate. They appear to be two-dimensional feeders.

Smith and Shugart (1987) studied the relationship between the micro-habitat (described by the nature of the ground cover, the density of shrubs and trees, etc.) and the presence of a male bird's territory for 23 ovenbird territories in a forested watershed in east Tennessee. The territories of the ovenbirds were mapped (using a method developed by Stenger and Falls (1959)) by collecting and mapping a minimum of 100 observations of each bird for each territory. Along with a determination of the vegetation structure (and its pattern of variation in the territories), the actual abundance of forest floor litter invertebrates was also sampled to obtain a direct measurement of the potential food supply for the ovenbirds.

There are several theories regarding the adaptive function of avian territorial behaviour. One of these is that the adaptive function of territories is to space pairs of birds over an area to assure that the food supply is adequate to successfully rear the young (Stenger 1958; Wilson 1975). Three hypotheses regarding the details of this theory are that:

1. The individuals directly monitor the prey abundance and adjust their territories accordingly (Armstrong 1965; Stimson 1973; Brown 1975; Gass et al. 1976; Salmonson and Balda 1977). This view is supported by observations that there are normally significant negative correlations between prey abundance and territory size, and by experimental results on changes in territory size when prey abundance is experimentally altered within the territory boundary (for a review see Franzblau and Collins (1980)).
2. The individuals defend as large an area as possible and the territory size is limited by intraspecific competition for habitat. Areas of higher quality would be defended more actively and would attract more competitors. The high cost of defending quality territories would produce smaller territories in desired areas (Hinde 1956; Lack 1966; Krebs 1971; Schoener 1971; Dunford 1977). Evidence for this hypothesis has been the correlation of territory size with indices of intraspecific competition (territory intrusion rate: Gibb 1956; Ewald and Carpenter 1978; Myers et al. 1979; population density of conspecifics: Krebs 1971).
3. The individuals may use the habitat structure for a basis of areas defended. Hilden (1965) suggested that birds may respond to proximate factors (habitat structure) that are related to prey abundance rather than monitoring food abundance as such (Howell, 1952; Stenger and Falls, 1959; Morse, 1976; Cody, 1968).

Smith and Shugart (1987) used discriminant function analysis (Morrison 1967) to describe the habitat preferences of the ovenbirds by determining which habitat variables best separated samples drawn from an area in the territory of an ovenbird and areas in the watershed not in an ovenbird's territory. The resulting discriminant function could be used to assign the likelihood that an area of unknown classification would be classed as ovenbird habitat or non-ovenbird habitat. Such a function conforms in a statistically formal manner to what Grinnell (1917a) referred to as the niche of a species (James *et al.* 1984).

One could use the discriminant function to map the expected pattern of occurrence of the bird over any area of known habitat features. The discriminant function value for a given area is strongly related to the food density of forest floor litter invertebrates. Indeed, if one compares the relationship between the size of ovenbirds' territories and the prey abundance that is actually measured or predicted by the habitat discriminant function, the predicted prey abundance is a better predictor of territory size than the actual prey abundance ($r^2 = 0.64$ for predicted prey abundance; $r^2 = 0.44$ for actual prey abundance; partial correlation analysis indicated that the relationship between territory size and actual prey abundance became not statistically significant when the effects of habitat structure were taken out). Very little of the variation in territory size in this species appears to be a result of interaction with neighbours (see Smith and Shugart (1987) for details).

The development of a multivariate statistical method of describing the microhabitat or niche preferences of the ovenbird allowed the prediction of changes in ovenbird habitat over time in response to ecological succession and timber harvest (Smith *et al.* 1981a,b; Smith 1986; see Fig. 5.10).

Analyses of animal habitat patterns using multivariate statistics as a basis have become an important feature in managing landscapes for the presence or absence of plants and animals (see Capen (1981) and Verner *et al.* (1986) for reviews and several case studies). In the earlier papers from the period when the applications of these methods were becoming established, one can see a continuing interest in the idea of using these multivariate statistics in the context of niche theory. There is also a development of a management interest in using several of these techniques to manage landscapes for a diverse array of species.

Managing bird diversity in riparian habitats

Rice *et al.* (1981, 1983a,b, 1986) worked for well over a decade determining the habitat preferences of birds found in riparian habitats in the

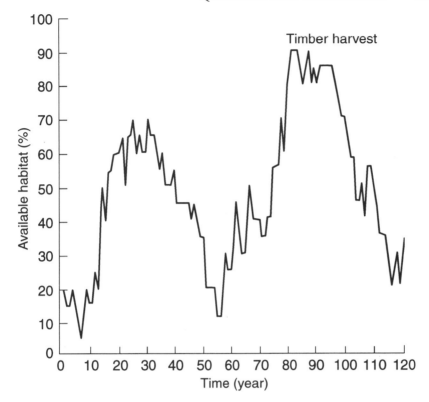

Figure 5.10. Prediction of habitat availability for ovenbirds following alterations in habitat caused by selective timber harvest simulated for 120 years. The habitat projection was made by incorporating a computer model that projects change in forest structure to the Grinnellian niche patterns of the ovenbird. From Smith *et al.* (1981b).

Colorado River area. Some of the support for the studies were provided by the US Bureau of Reclamation and involved studying the habitat use of birds on habitat created on sand islands made from the dredging of the Colorado River. These sand islands were devoid of vegetation but by pumping water from the river to small cottonwood (*Populus fremontii*) plants through a drip-irrigation system, a rapidly growing forest could be created. Since only trees that are planted and drip-irrigated become established in the initial stages of sand island succession, the structure of vegetation could be controlled to a great degree.

One of the initial interests of this research group (Rice *et al.* 1981, 1983a,b, 1986) was to manipulate the habitat on these artificially vegetated sand islands and create a habitat that would support a greater number of species per 100 ha than any native riparian habitat in Arizona.

This was accomplished in a few years using the species data set described below (R. D. Ohmart, personal communication).

Rice, Anderson and Ohmart (Anderson *et al.* 1983; Rice *et al.* 1983a,b, 1984) developed a data set by surveying birds present along 72 transects in representative habitats found in riparian sites in Arizona over several years. Each transect was 1.6 km in length. Based on discriminant function analysis, they were able to predict the presence or absence of all the bird species found in riparian habitats with 89.0% accuracy (90.8% of the time the discriminant function predicted a species to be present, it was correct; 88.0% of predictions of species absence were correct). Depending on season, the accuracy of prediction was between 87.1% and 91.3%. The investigators tested the ability of the set of discriminant functions that they had developed to predict the presence or absence of bird species in two test conditions (simple habitat structure: a salt cedar, *Tamarix chinensis*, stand; complex structure: a mature stand of cottonwood, *P. fremontii*; Goodding willow, *Salix gooddingii*; and honey mesquite, *Prosopis glandulosa*). They found that they could correctly predict presence or absence of species in these two cases with an accuracy of between 88% and 92%.

Aspects of these studies had important implications for the understanding of different habitat variables and the occurrence of bird species. The ability to interpret changes in habitat in terms of bird species presence for an entire assemblage of species has obvious management implications for the regulation of biotic diversity. Changes in the habitat structure of riparian zones will favour some species and disadvantage others. The eventual product of the study from a habitat manager's point-of-view was an interactive computer program capable of predicting which of the riparian bird species in Arizona would be favoured by a given habitat management practice.

Measuring the realised niches of tree species over a region

Austin *et al.* (1990) used a statistical technique (generalised linear models: Nelder and Wedderburn 1972; Dobson 1983; McCullagh and Nelder 1983) to estimate the ecological niches of five species of *Eucalyptus* trees over a study area ($\sim$40 000 km^2) in New South Wales and the Australian Capital Territory. The study was based on a data set of 6080 study plots (ranging from 0.04 to 0.25 ha) in which the presence or absence of the five species of interest along with that of 166 additional species were recorded. Other data on the environmental conditions at each plot were either directly measured (location, slope, rock type, aspect) or statistically associated (mean annual rainfall, mean annual temperature and amount of solar radiation) with the plot (see Austin et al. 1983, 1984).

Austin et al. (1990) mapped the niches of five major species of *Eucalyptus* (*E. rossii*, *E. muellerana*, *E. sieberi*, *E. pauciflora*, *E. maculata*). The niches were displayed as contours showing the likelihood of finding each species according to mean annual temperature and mean annual rainfall. Depending on the species, these niche contours were also developed for different aspects and different geology. An example for one of the species (*E. pauciflora*, Fig. 5.11) shows the species to be likely to occur in conditions of low temperature and relatively high moisture, as one might expect for this montane species. The species also demonstrates a second region of higher probability of occurrence at low rainfall and intermediate temperature levels.

The approaches and data sets developed for this study of tree niches have direct application in planning efficient surveys for conservation purposes. Margulis et al. (1987) developed a statistical model for the richness of *Eucalyptus* species as a function of mean annual temperature and mean annual rainfall and noted the use of this approach in planning nature reserves (Margulis and Stein 1989). The procedures developed by Austin and his colleagues have been featured in a recent book on cost-effective methods for developing biological surveys and identifying areas of high conservation value (Margulis and Austin 1991).

Patchiness in time and space: monitoring habitat change with satellites

According to the *Guinness Book of World Records*, an African weaver-bird (family: Ploceidae) called the red-billed quelea (*Quelea quelea*) is the most common bird on Earth. It breeds in colonies of as many as 10^6 pairs. These colonies may cover up to a few tens of hectares (Ward 1971; Wiens and Dyer 1977). Feeding on native annual grasses, often of the genera *Echinochloa*, *Panicum* and *Sorghum* (Ward 1965a), the species has exploded in density and in its magnitude as a pest animal with the increased production of cereal grains in Africa. While the magnitude of the destruction caused by the species may be over-estimated in some cases, the species is capable of destroying 10 to 20% of the production of large farms and the entire crop of subsistence farmers (Elliott 1979). The red-billed quelea has been described as the most numerous and destructive pest bird in the world (Ward 1973; Elliott 1989).

Quelea are nomadic birds that subsist over the dry season on a diminishing supply of dry seeds. With the onset of rains in the wet season

(a)

(b)

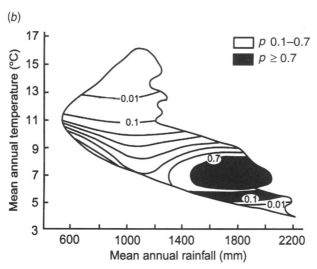

Figure 5.11. Environmental niche of the snow gum (*Eucalyptus pauciflora*). (*a*) The snow gum in typical montane habitat (photograph by H. H. Shugart). (*b*) The quantitative realised niche of the snow gum on granite in relation to mean annual rainfall and mean annual temperature. The area mapped in the space is the set of conditions found over the extent of the study area. The contours represent the probability (*p*) of finding the species. Along with the region of high probability of occurrence in cool, moist conditions, there is also a region of higher probability of occurrence at low rainfall and intermediate temperatures as well. From Austin *et al.* (1990).

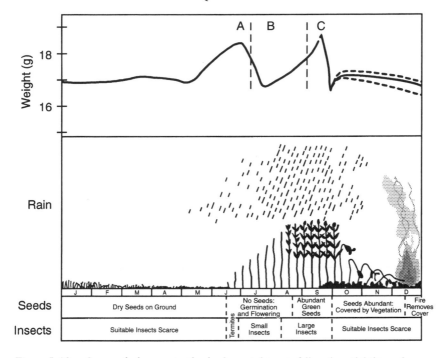

Figure 5.12. Seasonal changes in the body weight, rainfall and availability of grass seed and insects as food for the red–billed quelea (*Quelea quelea*) in an area with one wet season per year (e.g. Lake Chad). The initial weight increase (A) precedes the 'early rains migration' when red–billed quelea typically leave a given locality. The interval between the dashed lines (B) denotes the time the birds are absent from an area. The weight drops (C) as birds lay a clutch of eggs and initiate incubation. There is then a weight gain that varies with the availability of food supplies, the clutch size and other energy drains placed on the individuals rearing young. From Wiens and Johnston (1977) modified from Ward (1965b, 1971) and from personal communication from P. Ward with J. A. Wiens and R. F. Johnston (*fide* 1977).

(Fig. 5.12), the seeds that the species eats germinate and *Quelea* undergo a severe food shortage that lasts for 6 to 8 weeks. This shortage ends when grasses mature and fresh green seeds become available. At this point breeding colonies may be established in any area with suitable substrate for nesting (typically thorny *Acacia* spp.). The entire breeding cycle, from colony initiation to producing independent young requires 6 weeks: an exceptionally short interval for birds. In some cases, breeding areas may dry out prematurely and colonies are abandoned. In other cases, the rains may sustain a prolonged green period and several waves of breeding can occur in the same general area (Fig. 5.12).

The species is highly nomadic and responds to food shortages by moving (Fig. 5.12). The savanna ecosystems in which they live are a shifting mosaic of patches containing dry seeds, immature grasses, mature grasses with abundant green seeds and senescent grasses with no seed. By moving over distances of the order of 50 to 200 km, the birds move from one kind of patch to another (Jaeger *et al.* 1986). On the basis of observations at a fixed point, the arrival of huge flocks of quelea seems random.

Wallin *et al.* (1992) used monthly satellite images of Africa to predict the shifts in breeding habitat for the red-billed quelea over East Africa. Using monthly data compiled from the AVHRR (*A*dvanced *V*ery *H*igh *R*esolution *R*adiometer) instrument on-board a weather satellite, Wallin *et al.* (1992) calibrated a sample of known sites of *Quelea* breeding by registering data from ground-based and helicopter surveys of *Quelea* breeding colonies by C. C. H. Elliott (in Tanzania, Kenya and Somalia) and by F. Wilhelmi (in Somalia). The approach was to relate the vector of conditions associated with each point in East Africa to the presence or absence of a breeding colony at a given time.

The AVHRR are instruments on board the NOAA series of polar orbiting satellites. The sensor is designed to provide global daily coverage with a 16 km^2 resolution. One measure of the Earth's surface condition obtained from the satellite is an index called NDVI (*n*ormalised *d*ifference *v*egetation *i*ndex), which is computed as a corrected ratio of red (0.58–0.69 μm wavelengths) and infrared (0.725–1.1 μm wavelengths) (see Sellers 1985, 1987; Tucker and Sellers 1986). The application of the NDVI to detect patterns of the Earth's net primary productivity will be discussed later in Chapter 13.

Wallin *et al.* (1992) found that the series of changes in the NDVI index could be used to determine the start of the growing season over East Africa and that there was a strong relation between this and the date that a colony was initiated. Based on 159 samples, a regression between the computed start of the growing season and the initiation of quelea breeding accounted for 76% of the variance in the initiation data. Wallin *et al.* tested the ability of the information from the AVHRR sensor to detect the onset of breeding in different years and in different parts of Africa using a variety of observational and survey information for tests. The eventual result was a capability to map the temporal and spatial dynamics of quelea breeding habitat across East Africa (Fig. 5.13, colour plate).

Satellite mapping of the changes in niche-related features of the environment have been conducted for a wide range of species. For the AVHRR sensor that provides daily coverage globally at 16 km^2 resolution, these include a range of vegetation mapping studies (e.g. Goward *et al.* 1985; Justice *et al.* 1985; Tucker *et al.* 1985a, 1986; Malingreau 1986; D'Arrigo *et al.* 1987). Other AVHRR applications include the mapping of outbreaks of the desert locust, *Schistocerca gregaria* (Tucker *et al.* 1985b; Hielkema *et al.* 1986) and outbreak regions for the Rift Valley fever disease (Linthicum *et al.* 1987). Other satellite sensors have been used to model the habitats of grizzly bears (Craighead *et al.* 1982) and Australian marsupials (Saxon 1983).

Predicting the distributions of introduced species

Species introductions can be used to inspect how well the large area distributions of species in new situations can be predicted from niche considerations. The Australian Commonwealth Scientific and Industrial Research Organisation (CSIRO) has been a world innovator in well-studied releases of organisms for the purposes of biological control of introduced plants and animals. Their programmes have included intentional introductions of a moth to control *Opuntia* cactus, a disease to control introduced rabbits and a diverse assemblage of African dung beetles to bury the dung of the cattle (themselves non-native to Australia).

Introductions can also be used in an inverse fashion to test theories of distributions and to determine the likelihood of an introduced species continuing to expand. For example, Beerling *et al.* (1995) used a response surface approach somewhat analogous to that of Austin *et al.* (1990) discussed above to predict the distribution of *Fallopia japonica*, a plant native to southeast Asia and now a widespread alien over much of Europe. The species was introduced to Europe in the 1840s and it became widely cultivated as a garden plant. It has now spread through much of Europe as a naturalised alien.

Beerling *et al.* (1995) developed a climate response surface (a mapping of the likelihood of finding the species as a function of environmental variables) for *F. japonica* based on the recorded locations of the species in Europe using the *Atlas Florae Europaeae* (Jalas and Suominen 1979), which compiles the distributions of vascular plant species in 50×50 km^2 blocks. They fitted a climate response surface to the species distribution using three climatic variables (the mean temperature of the coldest month (°C), a growing-degree day or an annual temperature sum (summation of daily mean temperatures greater than 5°C) and an estimate of the ratio of annual to actual evapotranspiration). The resultant response function was then tested for its ability to reproduce the present distribution of *F japonica* in Europe and to predict the distribution of the species across its original range in southeast Asia. Woodward (1988) developed a similar analysis of the distributions of native European species now occurring as exotics in North America. Wilson *et al.* (1988) compared the niche relations in species introduced to New Zealand from Great Britain.

A group of Australian ecologists have developed an approach to mapping the climatic niche of species called the BIOCLIM program and have used this information in several cases involving species introductions. The BIOCLIM program uses a number of factors to map the climatic niches of different tree species.

1. Annual mean temperature (°C)
2. Coldest month minimum temperature (°C)
3. Hottest month maximum temperature (°C)
4. Annual temperature range (hottest month maximum temperature − coldest month minimum temperature) (°C)
5. Wettest quarter mean temperature (°C)
6. Driest quarter mean temperature (°C)
7. Mean annual precipitation (mm)
8. Wettest month mean precipitation (mm)
9. Driest month mean precipitation (mm)
10. Annual precipitation range (wettest month mean precipitation − driest month mean precipitation) (mm)
11. Wettest quarter mean precipitation (mm)
12. Driest quarter mean precipitation (mm).

Early applications of the BIOCLIM model involved studies of the climatic factors influencing the distributions of both plants and animals (Booth 1985; Nix 1986; Busby 1986). A principal application of the model has been to determine the suitability for transplantation of commercial tree species from Australia to Africa. This work has moved to testing of model results against field trials of a range of Australian *Eucalyptus* species mostly in Botswana, Malawi, Mozambique, South Africa and Zimbabwe (Booth *et al.* 1988; Booth 1991). Booth (1990) has also explored the use of the BIOCLIM approach to map the potential distributions of three major plantation tree species (*Eucalyptus grandis, E. tereticornis* and *Pinus radiata*) at the global level.

While examples of species whose distributions as exotic species differ significantly from that in their source ranges are plentiful, investigations of the controls of distributions of introduced species often show a

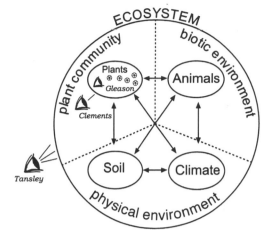

Figure 5.14. The point-of-view of major developers of ecosystem and community concepts, according to Allen and Hoekstra (1992). Gleason (1926) saw the plant community as a collection of individuals patterned by their unique interaction with the environment. Clements (1916) viewed the plant community as a highly integrated whole that was set in a particular physical environment. Tansley (1935) saw plants, animals, and their biotic and physical environment as components of the ecosystem. From Allen and Hoekstra (1992).

communality in the conditions between the distribution of the species as an exotic and the distribution of the species as a native.

Patterns of species abundance along environmental gradients: the continuum concept

The ecosystem was defined to resolve a controversy regarding the degree of higher organisation in ecological communities (Chapter 3). The original intent of the ecosystem concept by Tansley (1935) was to replace a community concept that allegorised the community as a super-organism with predictable appearance, internal interactions and development pattern (succession). Allen and Hoekstra (1992) felt that the ecosystem versus community controversy is a product of differing perspectives on the nature of vegetation (Fig. 5.14). According to Allen and Hoekstra, Gleason (1926) saw the plant community as a collection of individuals patterned by their unique interactions with the environment. Clements (1916) viewed the plant community as a highly integrated whole set in a particular physical environment. Tansley (1935) saw plants, animals and their biotic and physical environment as components of the ecosystem.

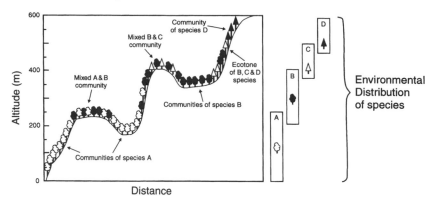

Figure 5.15. Patterns of variation in spatial distribution and with respect to the environmental changes associated with an altitudinal gradient, according to Austin and Smith (1989). The four species vary regularly in their distribution along the environmental conditions of the altitudinal gradient but produce a complex pattern of communities on the mountain landscape. Modified from Austin and Smith (1989).

Each of these scientists viewed the same system with a different perspective, and probably at different time and space scales. The debate over which of these views is 'correct' has abated over the years, but is without resolution. Currently, communities (as if they are natural objects) still are used by experimentalists and theoretical ecologists as experimental objects, as bases for experimental design and as consequences of theories. Applied ecologists and conservationists manage nature reserves and parks to preserve communities.

In vegetation science, an important alternative to the community that draws strongly from niche concepts (Grinnell (1917a), and particularly Gleason (1926)) is called the continuum (Cottam and McIntosh 1966). The essential idea of the continuum is that plant species vary in their response to environmental conditions and that the response of the species to environmental variation underlies the patterns in the vegetation. In a recent re-formulation of the continuum concept, Austin and Smith (1989) provide a useful example of the relation between the community and the continuum involving four hypothetical plant species distributed along an altitudinal gradient (Fig. 5.15). Species that are distributed continuously along an environmental gradient (altitude in Fig. 5.15) can form a range of assemblages that could be interpreted as communities (and ecotones between communities) when their environmental response is realised on a given landscape. Austin and Smith (1989) stress that the community (or associations) is a function of the landscape examined; the continuum applies to an abstract environmental space. In that the niche of a species is expressed as its distribution in an abstract environmental space, a continuum is the pattern of a collection of species niches.

For the distribution of species, there are three principal variations as to the pattern in a continuum (Austin 1985):

1. Gleason's (1926) original individualist concept with species abundance optima and the limits of species distributed independently along an environmental gradient
2. Continua could have major species regularly distributed along the gradient with optima more evenly spaced and minor species being independently distributed (Gauch and Whittaker 1972)
3. Strata (e.g. trees, shrubs, herbaceous plants) each with a continuum of regular or irregular species distributions and with either independent or correlated relations of the patterns between strata (Austin 1985; Austin and Smith 1989).

The evidence to distinguish between these three possibilities is not yet available (Austin and Smith 1989). The degree of spacing and the correlations in patterns in continua could arise from internal interactions among the species (discussed in the following chapter) causing regularities in dispersal or in similar responses of species to factors in the gradients causing aggregations.

In their re-formulation of the continuum concept, Austin and Smith (1989) compare several recent, and in some cases differing, views on the nature and response of plants to underlying environmental gradients. They separate the gradients into three types:

1. Resource gradients involving variation in 'resources' (e.g. light, water, mineral nutrients) consumed by plants and essential for plants to live
2. Direct gradients involving a factor with a direct physiological impact on the growth of plants but which is not consumed. Examples might be soil pH or air temperature
3. Indirect gradients, such as altitude, where the observed response to the gradient is actually a complex response to the regular changes in variables such as temperature and precipitation.

Austin and Smith's (1989) re-formulation of the continuum concept is based on considerations of the first two types of gradient (resource gradients and direct gradients). Presumably different physiological and ecological aspects of the species control the response to resource versus direct gradients. Indirect gradients are presumably a complex combination of underlying resource and direct gradients.

Concluding comments

Textbook writers usually have been very effective in defining the niche and habitat concepts in ways that communicate the gist of their meaning to the student. For example, '... *the habitat is the organism's "address", and the niche is its "profession"* ...' (Odum 1971). Research papers that deal with the niche begin by noting the confusion about the niche and then proceed to define the term for the immediate purposes of the paper, thus adding to the eschewed confusion. Why has an interest in the niche persisted despite an apparent confusion in what the 'niche' actually means? Why has this confusion been so resistant to efforts of well-respected ecologists to define more sharply the concept?

One partial answer is that the niche is a very appealing intuitive concept that was loosely associated with some very successful ideas in ecology when these ideas were young. Its disconnected origins incorporated both the factors controlling geographical distribution and the factors structuring communities. Grinnell's and Elton's work, the affiliation of the term 'niche' with competition by Gause, and the later incorporation into ideas about the patterns of communities with patterns in niches by Hutchinson and MacArthur all made the niche concept a fellow traveller in a company of exciting ideas. It is important to realise that the diverse views of the niche and the rich array of associated concepts contribute to an understanding of ecology at larger scales. Several of these interwoven developments – the factors that determine the distribution of species, the degree of equivalence of the roles of different organisms in different communities, the regularity of patterns with response to the conditions of the environment – are important in understanding the larger patterns and dynamics of ecosystems.

In the case of large-scale environmental change, one would like to have some insights that would allow the prediction of species that might be disadvantaged by the changes. The geographical aims of Grinnell's autecological and geographical concept of the species niche have potential to provide this capability. Indeed, the methods that are being explored using multivariate statistical analysis and large-scale data sets make these aims possible. Grinnell in 1927 recommended that the *American Ornithologist Union Check-List* (the recognised standard for species and subspecies of North American birds and their distributions) include the climatic limits of the ecological niche to provide a 'phylogenetically significant system for designating bird's ranges' (see James *et al.* 1984). In the present era of concern about global change and biodiversity change, one wishes that this lead had been followed and that 60+ years of such information was at our disposal today.

The Grinnellian niche concept places a focus on the attributes of a species and on how these attributes determine geographical distribution of a species. While Grinnell made it clear that it was important to consider interactions with other animals (both predators and competitors), these were not paramount factors to be considered, The Grinnellian niche is, in most of its aspects, an individual-organism or species–centric concept. The niche is an amalgam of the factors involved in understanding why a given organism might occur in a given location.

In the chapters that follow, Grinnellian niche concepts will emerge as important in several contexts. The mapping of potential animal habitat as a function of vegetation and the determination of shifts in species ranges associated with large-scale environmental change are concepts that can be associated with Grinnell's ideas about the species niche. One important aspect of the application of niche concepts is in the development of parameters for individual-based models of vegetation dynamics (Chapter 8). Another involves using the factors that appear to control the distribution of species under current conditions to predict the distribution of the species under altered environments (Chapter 11). The determination of which environmental factors determine the success and survival of a species and how these factors alter individual-organism performance is the quintessential Grinnellian niche question.

The Eltonian niche is a community-centric concept. Because one community might have a population that performed a function with no analogue to be found in another community, it is possible to have an empty niche in an Eltonian sense. This also suggests that the niches available all could be 'full' in some communities but perhaps not in others. The Grinnellian niche is defined on the sum of the conditions where the individuals of a species are (or could be) found and an empty Grinnellian niche is not well defined as a concept unless a species is designated. Both of these separate views of the niche have had an important influence on subsequent theoretical and conceptual developments in ecology.

The Eltonian niche and the rich theoretical treatments of competition (and other biotic interactions) as the structuring force for ecological communities are being reconsidered. Peters (1991) points out that the issue is not whether or not competition occurs, because it can be illustrated that it does; rather the issue is how competition can be used in predictive theories. Peters (1991) considers competition to be a concept that has failed to have been developed into theories worthy of the name. Levandowsky (1977) and Pielou (1977) have (among others) noted the lack of reality of the Lotka–Volterra equations. These criticisms have validity. There has been a considerable effort to reconcile patterns in the morphologies, distributions and abundances of species in communities with ideas as to how the communities might be structured. The result is a complex network of concepts that are internally consistent and close to the observational data, but which are also circular or tautological (Peters 1976, 1991). At the same time, because of the interest in understanding niche theory and community pattern, there has also been considerable progress in collecting a large observational data base on patterns of distribution and abundance of organisms. Further, there has been an increased appreciation for dynamic and interactive ecological systems developed in the process.

One regular observation in ecological communities is that if two species are considered with enough detail (or enough niche dimensions), one will almost invariably find some ecological difference in the species. There are two interpretations of observations of separation of species niches in communities. The first is that communities are (or were) highly biologically interactive and that competition has worked at some level (e.g., behaviour, evolution) to structure the communities so that there is little overlap. This would be the interpretation derived from considerations of competition and niche interactions. The second is that communities of animals (and plants) are relatively open; there is a large amount of available niche space for other species and, therefore, little overlap. The difference in the two views is whether or not the communities are so saturated with species that the addition of one species precludes the survival of another.

Cornell and Lawton (1992), while recognising that concepts about the nature of highly interactive communities are central in the development of community theory, note that evidence suggests that real communities lie on a continuum between biologically non-interactive and interactive, with the non-interactive cases being more in evidence. They note that, 'Finally, there seem to be numerous cases where there is open niche space ...', and later conclude that, '... many ecological communities should not be saturated ...' Wilson (1990) considers much the same issue (How the species of a community persist without competitive exclusion occurring) and finds that, for the flora of New Zealand, it is likely that environmental variation disequilibrates ecosystems before competition can manifest strong effects as a structuring force. Wilson states, '... it seems that all plant communities are always in a state of change in response to climate ...'. Zobel (1992) also considers factors that structure plant communities and notes, 'The diversity pattern of real plant communities does not suggest that the competitive exclusion of similar species can be the basis for explaining coexistence'.

It is inappropriate to suggest that the debate about the importance of competition has been resolved. It is interesting that the issue of environmental change has been injected into these discussions. Further, the degree to which communities are unsaturated, in the sense of Cornell and Lawton (1992), probably has a connection with the relative predictive success of methods using a Grinnellian niche construct as predictors of species occurrence.

Research on competition and the Eltonian niche has had a great effect on our appreciation of factors that either structure or might structure ecological communities. As was pointed out in Chapter 2, the patterns of ecosystem change over millennial time scales has been for communities to assemble and disassemble themselves. How these reconstituted communities can simultaneously be new constructions and display the rich pattern of seemingly highly co-evolved interactions among species remains a challenging problem. Understanding the underlying pattern of assemblages of plant species is particularly important in understanding the response of vegetation to environmental variables and will be discussed in Chapter 6.

6 · *Vegetation–environment relations*

The similarities in plant morphologies are often consistent in vegetation from equivalent environments – even when the plant species involved are unrelated. For example, in Mediterranean climates in southern Africa, Chile or Australia, the plants and vegetation resemble those in similar climates in France or Italy. The taxonomic affinities of many of the components of the vegetation in these areas are quite different. Vegetation composed of species that are relatively unrelated taxonomically can have striking similarities in the structure of the vegetation as a whole, in the morphology of the component individual plants and in the variation in the shapes and sizes of leaves (Fig. 6.1).

Do the morphological similarities in species in different vegetation also imply functional similarities in the ways the species influence the dynamics of the vegetation? This area of investigation is theoretically rich because it touches on the problem of the influence of biotic diversity on the responses of vegetation, and because it is concerned with the degree of self-organisation of natural ecosystems. A related theoretical issue involves determining the dynamic consequences of aggregating the species or individuals into different groupings in the process of formulating realistic models. Several applications of individual-based models (discussed in Chapter 8) involve the determination of the larger-scale consequences of individual biology on ecosystem structure and function. These applications will be the topic of Chapters 10 and 11, which deal with applications of such models to the problem of assessing the response of natural landscapes to environmental change. The issue of generality versus specificity inevitably arises when one attempts to synthesise the rich detail of ecological information about species (often description at relatively fine scale) toward abstractions aimed at depicting larger-scale responses of ecological systems.

Grime (1977, 1979a,b) developed a 'triangle' based on three primary

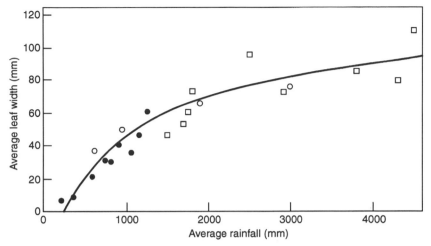

Figure 6.1. Average leaf width at low elevations as a function of average annual rainfall in tropical regions of Australia (O), Central America (□) and South America (●). The curve represents the relation $y = 32.7 \log (x/244.5)$. From Givnish (1986).

plant response strategies (Fig. 6.2). These strategies can be used to develop rules to predict the proportions of each strategies (and associated life forms) expected under a particular environmental regime. Grime recognised two types of external factor limiting the biomass of plants. The first was stress, involving the conditions that restrict plant productivity (shortage of light, water mineral nutrients, etc.). The second was disturbance, involving partial or total destruction of plant biomass (activities of herbivores, diseases, fire, frosts, etc.). These two external factors can operate independently, so that there are four possible combinations of high or low stress, and high or low disturbance.

Grime (1977) reasoned that the combined action of high stress and high disturbance created a condition from which the vegetation could not regenerate itself. Grime designated three primary strategies of plants. These represented extreme examples of the types of adaptation expected of plants in environments with different combinations of external controls. Low-stress and low-disturbance environments would ultimately favour species that were able to compete effectively against other species (competitor strategy); high-stress with low-disturbance environments should be dominated by plants of species able to tolerate the particular stress (stress–tolerator strategy); low-stress and high-disturbance environments should favour short-lived, fast-growing species (ruderal strategy). Grime related this scheme (Fig. 6.2) to the many instances in the literature

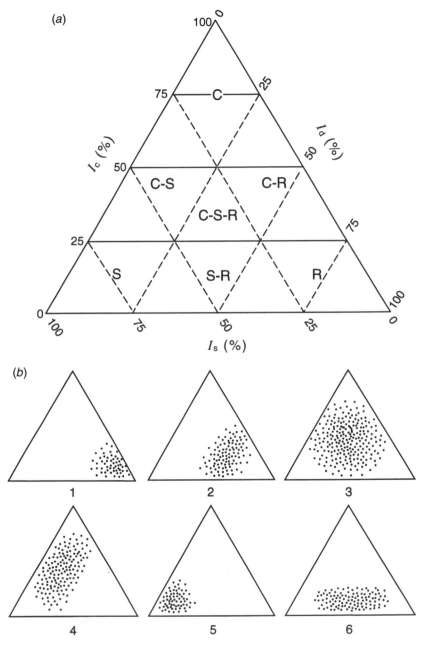

Figure 6.2. (*a*) Grime's (1977) model of competition, stress and disturbance in plant strategies. I_s is the relative importance of stress; I_d is the relative importance of disturbance; I_c is the relative importance of competition. (*b*) Diagrams showing the range of strategies that Grime (1977) associated with 1, annual herbs; 2, biennial herbs; 3, perennial herbs and ferns; 4, trees and shrubs; 5, lichens and 6, bryophytes.

of vegetation patterns responding to disturbance and site conditions. He also included a determination of the response of the three strategies to climatic change (Table 6.1).

The general problem developed by Grime (1977) is one of defining plant functional types (Smith and Huston 1989; Smith *et al.* 1997). Plant functional types connote species or groups of species that have similar responses to a suite of environmental conditions and can thus (perhaps) usefully be lumped into groups to interpret and predict ecosystem behaviour. Related or analogous terms are life-forms (Raunkiaer 1934), guilds (see Simberloff and Dayan 1991), plant forms (Box 1981), the aforementioned plant strategies (Grime 1974, 1977, 1979a,b; Tilman 1988), temperament (Oldeman and van Dijk 1991), etc. As will be discussed at the end of Chapter 9, one can also identify ecosystem functional types: aggregated components of ecosystems whose interactions with one another and the environment produce identifiable patterns in ecosystem structure and dynamics. Because ecosystems are abstractions (Chapter 3), both of these functional typologies necessarily are also abstractions. The distinction between plant and ecosystem functional types is largely in the emphasis on plant adaptations in the former and on ecosystem responses in the latter.

One of the essential aspects in Grime's concept that one can view plants as having three primary strategies is that of 'trade-offs': a single species cannot simultaneously be the best as a stress-tolerator, a competitor and a ruderal species. The idea of trade-offs, albeit with a different emphasis, is also expressed in Tilman's (1985) resource-ratio hypothesis (used to explain ecological succession), and other parts of his theoretical work on factors structuring plant communities (Tilman 1982, 1988). There are other, relatively higher-level ordering rules that could be used to pattern the fundamental properties of the vegetation that attend the environmental conditions at larger spatial scales than those considered by Grime, Tilman and others. These will be discussed below (e.g., Holdridge 1967; Box 1981; Woodward 1987b).

Historical roots of relating large-scale vegetation pattern to the environment

The changes in vegetation as one ascends a mountain are similar to those encountered as one moves toward higher latitudes. Writings about such associations of plants and their environments go back at least as far as Menestor in the fifth century BC (Chapter 3). Theophrastus (third century BC) understood that there were systematic changes in patterns of

Table 6.1. *Conditions of resource supply and attributes of importance in understanding response to climatic change for three primary plant strategies*

	Competitor strategy	Stress-tolerator strategy	Ruderal strategy
Resource condition	Resource continuously abundant but subject to local or progressive depletion	Resource continuously scarce	Resource temporarily abundant
Life forms	Herbs, shrubs and trees	Lichens, bryophytes, herbs, shrubs and trees	Herbs and bryophytes
Morphology	High–dense canopy of leaves; extensive lateral spread above and below ground	Extremely wide range of growth forms	Small stature, limited lateral spread
Life span	Long or relatively short	Long to very long	Very short
Longevity of leaves and roots	Relatively short	Long	Short
Leaf phenology	Well-defined peaks of leaf production coinciding with periods of maximum potential productivity	Evergreens, with various patterns of leaf production	Short phase of leaf production in period of high potential productivity
Reproduction	Established plants usually reproduce each year	Intermittent reproduction over a long history	Prolific reproduction early in life history
Proportion of annual production devoted to seeds	Relatively small	Small	Large
Perannation	Dormant buds and seeds	Stress-tolerant leaves and roots	Dormant seeds
Maximum potential relative growth rate	Rapid	Slow	Rapid

Photosynthesis and uptake of mineral nutrients	Strongly seasonal, coinciding with long continuous period of vegetative growth	Opportunistic, often uncoupled from vegetative growth	Opportunistic, coinciding with vegetative growth
Acclimation of photosynthesis, mineral nutrition and tissue hardiness to seasonal change in temperature, light and moisture supply	Weakly developed	Strongly developed	Weakly developed
Storage of photosynthate and nutrients	Most photosynthate and mineral nutrients are rapidly incorporated into vegetative structure but a proportion is stored and forms the capital for expansion of growth in the following growing season	Storage systems in leaves, stems and/or roots	Confined to seeds
Defence against herbivory	Often ineffective	Usually effective	Often ineffective
Litter decomposition	Rapid	Slow	Rapid
Associated regenerative strategies	Vegetative expansion, seasonal regeneration in vegetation gaps, numerous small, widely dispersed seeds or spores, persistent seed bank	Vegetative expansion, persistent juveniles, numerous small, widely dispersed seeds or spores	Seasonal regeneration in widely dispersed seeds or spores, persistent seed bank
Role in secondary successions in productive habitats	Relatively early	Late	Early

Source: From Grime (1993).

deciduousness and evergreenness with respect to climate (Hort 1916; Morton 1981). He discussed experiments in which plant species were transplanted to areas outside their natural range to determine if they would grow (or flower) and thereby explored plant–environment relationships directly. Theophrastus also observed the positive relationship between altitude and latitude with respect to their climates and vegetation.

Near the turn of the eighteenth century and with the investigation of newly discovered lands by explorers and naturalists, it was found that there was a high degree of convergence of form of plants from different geographical regions having comparable climates (Humboldt 1807). A considerable effort was made over the next century to understand, classify and map these patterns. By the middle of the nineteenth century, associations between the distributions of different types of vegetation and the climate had been drawn (e.g. Schouw 1823; de Candolle 1855). Explaining these patterns inspired the early plant geographers (e.g., Griesbach 1872; Drude 1890; Schimper 1898; and Warming 1909) to postulate relationships between climate and plant structure as the conceptual basis for mapping the global pattern of vegetation. Along with vegetation classifications based to some degree on features of environments, classifications of climates were also developed (e.g. Köppen and Geiger 1930; Thornthwaite 1931). Maps based on these climate classifications appear to have at least some of their boundaries determined by the locations of vegetation boundaries.

There is a variety of ways one can chose to map the vegetation in a given area (see Küchler (1967) for an extensive review). The issue of scale (Chapter 2) is particularly important in the development of vegetation maps. Printed maps with a high degree of spatial detail are by necessity only for a small area. To develop a map of the world vegetation, one must abstract the variability in vegetation pattern represented in the map to some degree. Otherwise, the map becomes so cluttered with notes, symbols and colours as to be unreadable. GIS provide the potential to use a computer to store and display information in map form at different resolutions and eliminate some of this 'map clutter' problem in displaying detail over large areas. Still, patterns of vegetation appear (and disappear) at different spatial resolutions.

For this reason, plant geographers have attempted to be formal in their development of systematic methods of abstracting the vegetation into maps. Embedded in these methods are important questions about the fundamental structure of vegetation at different scales. At what level of resolution is the structure of the vegetation the essential feature to map? At what level of resolution does one need to include the important species? Do the rules for appropriate resolutions change in different continents, in different regions?

These theoretical implications involved in making maps of vegetation were well understood by the early plant geographers, although there was an understandable variation in the solutions that they developed for these and similar questions (Smith *et al.* 1993).

The causes underlying the relationships between vegetation and climate are still being explored today at both the plant (Box 1981; Givnish 1986) and vegetation (Holdridge 1967; Woodward 1987b) level. The complexity of this task is considerable. For some factors there appear to be relatively straightforward physical and chemical explanations for the occurrence of certain features in the plants. For example, Woodward (1987b) notes that vegetation appears to be delimited by cardinal minimum temperatures that are related to ways that plants adjust biochemically and physically to low temperature (chilling or freezing).

Environmental variables are interwoven in relatively complex ways – such as the web of interactions in the plant use of water, intake of CO_2 for photosynthesis, and heat balance (Woodward 1987b). Plants at different parts of their life cycles are often differentially resistant to extreme environmental conditions. The Grinnellian niche of a seed is not necessarily the same as that of a seedling, which in turn may not be like that of the mature plant. Individual plants can often grow vigorously in areas in which they are unable to reproduce. Species often have 'sensitive' life stages that, for reasons that are not obvious, limit their ability to survive over large areas.

Time scale and adaptations to climate

Adaptations to climate and the overall response of plants to climatic conditions originate at several different scales of time resolution. In the case of hardening, cold stress may or may not be lethal to a plant depending on its recent history. In addition, adaptations at one scale may be overshadowed by a lack of adaptations at other scales.

At the biochemical level, one finds that different plants operate three different biochemical pathways for the process of photosynthesis: (a) the C_3 or Calvin Benson cycle; (b) the C_4 pathway; and (c) the crassulacean acid metabolism (CAM) pathway. Some species are facultative CAM plants (switching to C_3 when conditions are wet and to CAM when dry).

The plants using the C_4 pathway have an advantage in their ability to carry on photosynthesis under full light, high temperature and lower water availability when compared with those using the C_3 pathway. They also have a relatively limited life-form range. They are sensitive to mild frosts ($c. -5$ °C) in the early season. In some cases, species using the C_4 pathway are chilling sensitive and only grow at temperatures greater than $c. 10$ °C. Plants using the CAM pathway are not particularly efficient at photosynthesis compared with the other two kinds of plants but can survive in harsh, arid environments. There are differences in the proportion of plants employing these different biochemical pathways that are directly related to climate, with more plants using the C_4 pathway in the tropics and none using this pathway found in the tundra vegetation at high latitude.

Leaves of plants often vary in size and shape in response to climate variables. In capturing light to provide energy for photosynthesis, the leaves of plants can have a significant heat loading that is controlled by radiation of heat from the surface, by convection of heat from the surface, and by evaporative cooling. Succulent species have upper lethal temperatures of about 45–50 °C, but the temperatures tolerated by most species are lower. The leaves maintain a complex balance by opening stomata (small holes in the leaves) to let CO_2 needed for photosynthesis into the leaf and to evaporate water to cool the leaf. Trees and shrubs are predominately hypostomatous (stomata on the undersides of leaves). Herbaceous plants are both hypostomatous and amphistomatous (stomata on both leaf surfaces) with more hypostomatous species occurring in shady conditions and amphistomatous species being more frequent in open sites with ample water. Crop species are usually amphistomatous.

If leaves lose too much water or water is lost too fast, the leaf wilts. Wilting can change the leaf position and alter the heat loading to the leaf surface – allowing the leaf to avoid reaching high temperatures. In the interaction among radiative heat flow, convective heat flow and evaporation, the size of the leaf as well as its surface properties (shininess to reflect heat, hairiness to alter convective flow of heat and water vapour, etc.) can be important. The advantages of these features in different environments are associated with the similarity of leaves in equivalent climatic regimes.

At longer time scales, the different life stages of plants may have different responses to climatic conditions. Some plants can grow in environments in which they are unable to reproduce. Flowers may be sensitive to frost, or the growing season may be too short for the plant to have time to produce seed. Conditions needed for the seeds to be induced to germinate are in some cases related to temperatures (minima, maxima or summations of daily temperatures), exposure to light or moisture. Seedlings of some species seem unable to survive conditions that can be tolerated by the adult plants. In some cases, the smaller size of seedlings makes them more responsive (and, therefore, more sensitive to rapid variation in weather conditions) than the more massive, mature plants.

At even longer time scales, if variations in the climate are sufficiently pronounced, the plants growing in a given area may be ill-adapted for the new conditions and the development of the vegetation may be limited by the time required for the appropriate flora to migrate to the location. The species most responsive to climatic variation of the time scale of millennia may be those best adapted to long-distance dispersal: an adaptation not usually thought of as an adaptation to climate.

Traditionally, variations in the structure of plants and vegetation are primarily interpreted as adaptations to climatic or soil conditions (Walter and Breckle 1985). It is important to recognise that more or less similar structures can be associated with quite different environmental conditions. Even relatively simple indicators of leaf morphology, such as size or thickness, do not vary in concert with one another across environmental variables (Givnish 1986). To make matters more complex, environmental conditions themselves also tend to be interrelated (e.g. sunny sites are frequently dry, humidity and moisture are often correlated, etc.). Therefore, it can be difficult to isolate which factor is 'causing' a particular response in morphology from the vegetation. For these reasons, it can be difficult to interpret patterns of plant morphologies and patterns of environmental variables simply because the relations are potentially very complex. Further, the fundamental data tend to be observational rather than experimental.

Vegetation response to environmental gradients

One issue that has dominated plant ecology for much of the twentieth century has involved the response of the vegetation to systematic changes in environmental variables. One early realisation was that certain species were so tightly related to the environmental conditions of the areas in which they were found that their presence could be taken to indicate the environmental conditions. The idea was promoted rather extensively by F. E. Clements (1928) and forms a basis of a wide range of methods including such applications as evaluating conditions of pastures (by noting plant species indicative of over-grazing) to prospecting for minerals by using certain plants as indicators of soil conditions and, hence, the under-lying geology.

Often, Clements' views on the nature of the ecological community (Clements 1916) are contrasted with those of Ramensky (1924) and Gleason (1926), whose individualistic hypotheses were that 'each species is distributed in its own way, according to its own genetic, physiological, and life cycle characteristics and its way of relating to both physical environment and interactions with other species; hence no two species are alike in distribution' (Whittaker 1975). As was mentioned in Chapter 5, the concept is quite similar to those that were being discussed contemporaneously by both Elton and Grinnell with regard to the niches of animals. It is an indication of the breadth of Clements' concepts about vegetation pattern that Clements (1928) was involved in developing an individualistic theory of species-to-environment responses of plants (the 'indicator species'), and that he applied this theory to a wide range of practical problems.

There are several models of how groups of plant species might be distributed with respect to the environment. Austin (1985) identifies four alternative models (Fig. 6.3). In the first, the community-unit distribution along an environmental gradient (Fig. 6.3a), the community is composed of dominant species that through selection have become adapted to live with one another over some range of conditions. Competition among the dominant species in one community and the dominants of another com-munity creates sharp boundaries between different communities (Austin 1985). The second model, the individualistic continuum (Fig. 6.3b), has species responding to the environmental gradient, with each species response being idiosyncratic and independent of the other species. In this case, one would expect to have neither sharp boundaries nor strongly grouped species. In a third model, species that competed along the environmental gradient might be expected to partition the gradient in some more-or-less regular pattern (Fig. 6.3c). As a fourth model of plant distribution along an environmental gradient, if various of the species belong to different functional groups, then each functional group or stratum (e.g. canopy trees, subordinate trees, shrubs, etc.) might partition the gradient independently of the others (Fig. 6.3d). In this last case, if there are very many strata, say three or more, then the condition (shown in Fig. 6.3d) would be difficult to distinguish from the individualistic dis-tribution (Fig. 6.3b).

Austin (1985) pointed out that debate about whether the continuum (Fig. 6.3b) or the community (Fig. 6.3a) is the expected pattern of species

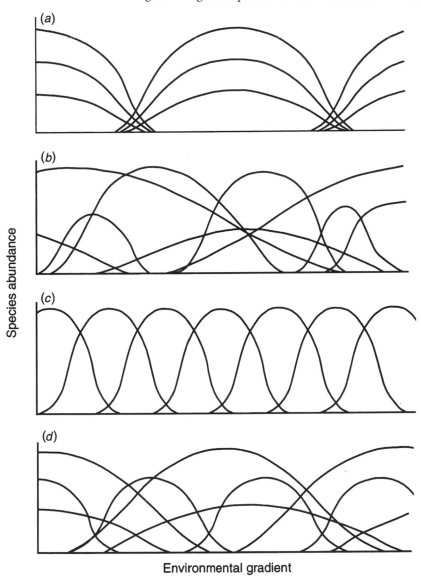

Figure 6.3. Four alternative models for the organisation of vegetation along an environmental gradient. (*a*) Community concept. (*b*) Individualistic continuum concept. (*c*) Resource-partitioned continuum. (*d*) Resource-partitioned continuum within strata, individualistic relationships between strata. From Austin (1985).

distributions (Daubenmire 1966; McIntosh 1967; Whittaker 1967; Dansereau 1968) is often focused on the issue of whether the data supporting the continuum concept were drawn from climax (equilibrium) systems. Discussions in earlier chapters (Chapter 2) identify some of the potential problems in assuming any vegetation is at equilibrium. Nevertheless, current consensus would appear to favour the continuum concept over the community concept (Austin 1985).

Global classifications of vegetation–environment relations

The categorisation of physiological, physiognomic or morphological characteristics of plants to specific environments has formed the basis of a number of global- or continental-scale plant classification systems (Smith *et al.* 1993). These approaches are based extensively on expected associations among climatic, edaphic and hydrologic features and broadly interpreted categories of existing vegetation complexes. The underlying relations can be either of a simple form, representing the direct relationship between environmental features and expected vegetation zones (Holdridge 1967; Budyko 1974; Stephenson 1990), or more complex rule-based models that attempt to incorporate in an explicit fashion more sophisticated, physiologically established range limits for types of plants (Box 1981; Woodward 1987b; Prentice *et al.* 1993; Neilson *et al.* 1992).

Both approaches (correlation between vegetation and the environment or rule-based relations between types of plant and the environment) are constrained in that they predict only the expected equilibrium or potential vegetation for a particular set of environmental variables. In both cases, there is an implication that the vegetation in a particular climate only has a single expected equilibrium condition. Transient responses (such as those expected in ecological succession) in response to changing environmental conditions cannot be evaluated directly by either method. In addition, vegetation classifications based on correlations with climate interpret existing interrelations between vegetation (or plants) and climate. They are unable to produce novel vegetation complexes not considered *a priori*. However, while these latter approaches are correlative, they frequently are encyclopaedic in the volume of information that they synthesise (e.g. Box 1981).

Early climate maps and most present-day global climate maps use information about the observed vegetation pattern to expand and interpret the limited climatic data. Maps of global vegetation are usually developed based on direct observations of the vegetation and topography. Global maps of potential vegetation based solely on climate stations are a recent development (Box 1981; Emanuel *et al.* 1985a,b; Prentice *et al.* 1993) and are strongly dependent on digital computers for data manipulation and display. These vegetation maps are based on the assumption that the same vegetation should be found in the same climate, even on different continents.

Relating the vegetation to the global environment

Historically, the interpretation of vegetation–climate relations has been a prominent feature of vegetation classification and mapping. The prediction of the vegetation expected in a given climate and the associated inferences about vegetation distributions amount to sweeping theories about the major factors that shape the global vegetation. Usually these theories involve hypotheses about the expected climax or equilibrium vegetation communities associated with specific environmental conditions (e.g. Clements 1916).

The degree to which large-scale boundaries of vegetation follow mapped patterns of climatic variables can be striking. This is frequently the case for temperature- or moisture-related variables (see Stephenson 1990). Several systems of vegetation–climate classifications are based on temperature and moisture indices as a necessary simplification for the mapping and interpretation of large-scale vegetation (Beard 1980).

For example, the Holdridge life zone system (Holdridge 1967; Holdridge *et al.* 1971) is a quantitative system that relates climatic and vegetation zones (called life zones) using indices of temperature and moisture available for plants. The life zone system has been used to evaluate regional and global vegetation patterns in several tropical countries in particular and it has been applied to predict the response of vegetation to global change (Emanuel *et al.* 1985a,b). The approach is based on relating the response of the vegetation to four interdependent environmental variables:

1. *Biotemperature*: the average of temperatures over time intervals with the substitution of zero for all intervals when the temperature is below 0°C or above 30°C (Holdridge 1967); the 0 and 30°C limits are felt to demarcate the range in which most plants typically are physiologically active
2. *Precipitation*: the amount of mean annual precipitation in millimetres
3. *Potential evapotranspiration*: calculated by multiplying the biotemperature

by 58.93, a constant derived by Holdridge. Result of the calculation is in millimetres

4. *Potential evapotranspiration ratio*: The ratio of average annual potential evapotranspiration to average total annual precipitation. A value of 1.00 indicates that the precipitated moisture equals potential evapotranspiration. Larger values imply an abundance of water; smaller values are associated with more arid climates.

Holdridge (1967) used these variables to develop a geometrical model of life zones (Fig. 6.4) that has been used to describe the vegetation of 12 Central and South American countries as well as elsewhere. The Holdridge life zone classification has a precise definition in terms of climate variables. Trained field workers can also apply the life zone classification on the basis of the physiognomy of the vegetation without direct reference to climatic data, or even in the absence of climatic data.

Figure 6.5a (colour plate) shows a Holdridge life zone map of Costa Rica (Smith *et al.* 1995) based on the Holdridge (1967) system that has been compared with favourable results to a map developed by Holdrige *et al.* (1971). Figure 6.5a was developed by using computer techniques to map the Holdridge (1967) life zones for Costa Rica using part of a global climate data set (Cramer and Leemans 1993) and elevational data from Costa Rica. There is good agreement between this map and that of Holdridge *et al.* (1971) maps, even on the relatively complex vegetation of Costa Rica and despite the fact that the maps were developed using different data sets and different mapping techniques. This replication is desirable in that the Holdridge life zone approach has been used rather extensively in assessments of the response of the global vegetation to global climate change (See Chapter 10). Two examples of applying the Holdridge model to climate change assessment are shown in Fig. 6.5b (for the effects of a uniform increase of +2.5 °C in the temperature and an increase of +10% in precipitation) and Fig. 6.5c (+3.6 °C in temperature and +10% precipitation). More examples of such applications are discussed in Chapters 10 and 11.

Relating plants and plant adaptations to the global environment

An alternative approach to developing a global vegetation classification is based on the determination of the distributions of the different components that constitute the vegetation. These plant- or life-form-based approaches differ from the Holdridge (1967) and related approaches by focusing explicitly on the plants, rather than the vegetation, as the

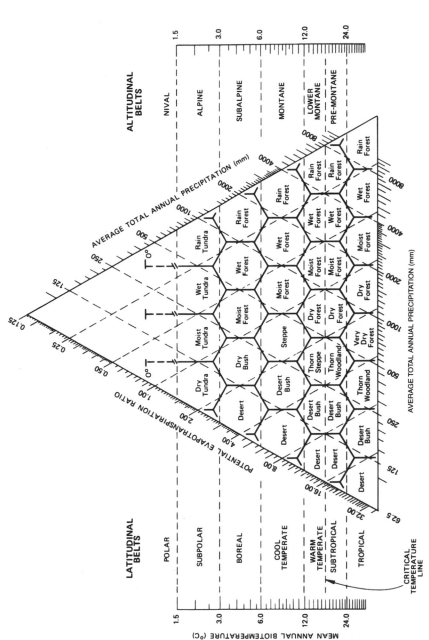

Figure 6.4. Diagram for the classification of world life zones based on climate variables developed by Holdridge (1967).

appropriate level to relate to the climatic conditions. The approaches are based on mapping the areas of the world where different types of plant can occur depending on the soil and climate. Overlaps of these distributions can be used to interpret or, in some cases, construct the expected vegetation from its elements and, thus, to infer the vegetation pattern. Vegetation classifications by Humboldt (1807), Schimper (1898), Rübel (1930) and Küchler (1967) all employ the physiognomy (general appearance) of the vegetation as a basis for their classifications and, to some degree, have an emphasis on the individual plants that make up the vegetation. To an extent, many of the classifications differ in their emphasis on features of the plants that are used to construct the classifications (Smith *et al.* 1993).

One advantage of considering the independent distributions of the plant life forms that constitute the vegetation as a basis for developing vegetation maps is that it is possible to predict novel vegetation associated with combinations of soils and climates that do not currently occur (Box 1981; Prentice *et al.* 1993). This gives the possibility of better understanding and retrospectively predicting the formation, dissolution and reformation of entire biomes noted in the Quaternary record (Huntley and Webb 1988; Prentice 1992; Chapter 2).

A classic and straightforward approach, strongly focused on the attributes of particular plants, is that of Raunkiaer (1934), who developed a functional categorisation of plants based largely on the relation of the 'perennating tissue' and the surface of the ground (Fig. 6.6). The perennating tissue is undifferentiated (meristematic) tissue that remains inactive during a cold or dry period and then grows to produce new plant structure with the return of a favourable time for growth. Familiar examples of perennating tissue are the buds that produce leaves and twigs in the springtime or at the onset of a wet season.

The position of perennating tissue relates directly to the plant's adaptation to climate. Plants have perennating tissue in more protected locations in harsh environments. The proportion of the different Raunkiaer life forms changes in a systematic way with climate (Fig. 6.7). The biological spectrum of life forms in local communities also responds over time to experimental manipulation of shelter (Grace 1987). Whitehead (1954, 1959) demonstrated this for alpine vegetation in Italy by building stone walls to increase protection and then observing changes in the abundances of the life forms making up the vegetation.

E. O. Box (1981) hypothesised that the links between plant form and plant function involved in a plant's water and energy balances were key in determining the best features to use for categorising plant life forms for

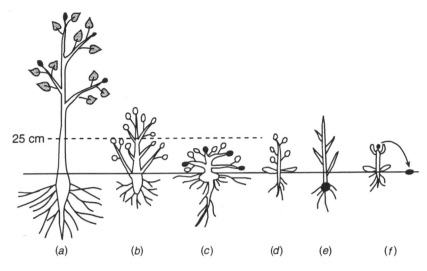

Figure 6.6. Plant life forms of Raunkiaer. Perennating tissues are shown in black, woody tissues are in grey, and deciduous tissues are unshaded. (*a*) Phanerophyte (tree or tall shrub) with buds more than 25 cm above the ground. (*b*) Chamaephyte, semishrub (suffrutescent low shrub) with some buds less than 25 cm above the ground. (*c*) Chamaephyte, subshrub, with buds less than 25 cm above the ground. (*d*) Hemicryptophyte, perennial herb with its buds at ground surface. (*e*) Geophyte, perennial herb with a bulb or other perennating organ below the ground surface. (*f*) Therophyte, annual plant surviving unfavourable periods only as a seed. From Whittaker (1975).

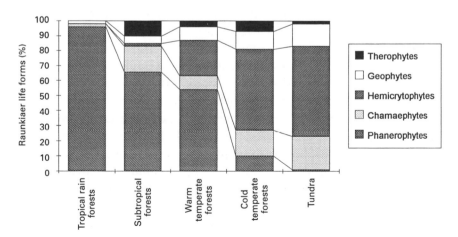

Figure 6.7. Percentages of Raunkiaer life forms in a series of relatively moist ecosystems occurring on a latitudinal gradient from the equatorial tropics to the Arctic. From Cain and Castro (1959), Whittaker (1960, 1975) and Whittaker and Niering (1965).

predicting global vegetation from environmental conditions. He postulated these main characters to be:

1. *Structural type*: primarily having to do with size, pattern of branching, woody, herbaceous or succulent
2. *Relative plant size*: referring to the height of a particular plant with respect to others of the same structural type; four designated types were tall (large), normal, short and dwarf
3. *Leaf type* (including photosynthetic stems and other photosynthetic organs): four recognised types were broad, narrow, graminoid (grass-like) and absent-leafed plants
4. *Relative leaf size*: the types were large, normal, small and very small
5. *Leaf structure*: referring to the 'hardness' of the leaves; categories were malacophyllous (herbaceous), coriaceous (leathery), sclerophyllous (both hard and stiff), succulent, ligneous (with woody photosynthetic parts), and pubescent (with a covering of fine hairs)
6. *Photosynthetic habit*: seasonality of photosynthetic activity; seven categories include evergreen plants, semievergreen plants (usually evergreen when conditions permit but losing their leaves during unfavourable times), raingreen plants, summergreen plants, suffrutescent plants (having perennial woody bases that produce shoots which are lost at the end of the growing period), marcescent plants (grasses and erect herbs) and ephemeral plants (capable of rapidly producing leaves or shoots from seeds or bulbs).

Based on different combinations of these attributes, the Box (1981) life-form classification produces descriptive, almost poetic, names for the different elements of the vegetation (e.g. tropical evergreen sclerophyll trees, xeric tuft-treelets, leaf-succulent evergreen shrubs).

Box associates each of the approximately 100 life forms (resulting from different combinations of the attributes listed above) with ranges of macroclimatic variables postulated to limit the distribution of each life form. These macroclimatic variables are TMAX (highest monthly mean temperature), TMIN (lowest monthly mean temperature), DTY (annual range of mean monthly temperatures), PRCP (annual precipitation), MI (annual moisture index, the ratio of total annual precipitation to annual evapotranspiration), PMAX (highest average monthly precipitation), PMIN (lowest average monthly precipitation) and PMTMAX (precipitation of the warmest month).

By comparing the ranges for each of the life forms with respect to each of these variables with the climate of a given location one can determine whether or not the life form should occur at that location. This determination is actually performed by a computer program called ECOSIEVE.

Given a list of which life forms could potentially occur at a given location, Box (1981) applies an additional set of criteria to determine which would be the dominant types (e.g. in moist situations, forest trees dominate small trees, which dominate large shrubs, then small shrubs, then forbs and ferns). The moisture index, MI, is also used to determine the total vegetative cover, the layering of the vegetation and other descriptive terms for the vegetation.

While elaborate, the system is straightforward in its application (Fig. 6.8). It produces descriptions of the vegetation that can be matched against actual vegetation to determine how well the classification actually works in a particular location (Table 6.2). With climatic and vegetation data from a large number of sites, one can generate and test the classification of vegetation, worldwide. Box (1981) used 1225 such sites in his original development of the model.

An important attribute of Box's approach to predicting the expected vegetation from climate data is that the methodology allows for the prediction of novel assemblages of vegetation associated with unusual configurations of climate conditions. Global maps based on Box's model are presented in Cramer and Leemans' (1993) paper in the context of assessing global-scale climate changes. Box's original work (1981) concludes with sensitivity analyses comprising varying global climatic conditions. If desired, one could also predict the expected vegetation with certain types of plant eliminated from a location.

Probably the principal disadvantage of the Box system is that the list of vegetation attributes that are predicted is fairly complex. This is also an advantage in that it allows for rich comparisons with actual vegetation (e.g. Table 6.2).

Prentice et al. (1993) extensively modified the Box approach to produce a biome model based on 13 plant types (seven tree types and six non-tree types). They included environmental limitations derived explicitly from physiological considerations, particularly those outlined in Woodward (1987b) and Woodward and Williams (1987). The expected ranges of each of the plant types with respect to the eight different controlling environmental variables is shown in Table 6.3. They developed extensive data sets for the terrestrial surface (excluding Antarctica) for each of the environmental variables at 0.5° latitude by 0.5° longitude cells (Leemans and Cramer 1991). In the case of temperature data, these data

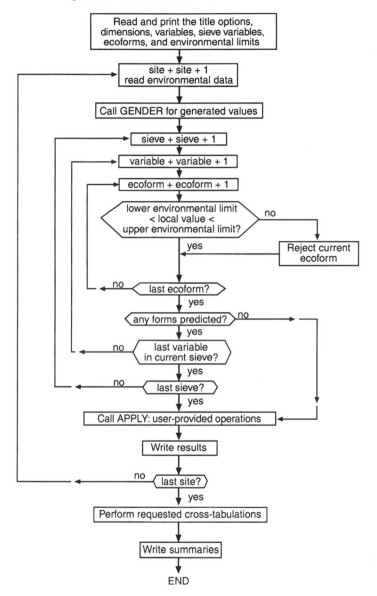

Figure 6.8. Sequence of procedures used in ECOSIEVE, Box's (1981) method of determining the life forms both present and dominant at any point on the terrestrial surface.

Table 6.2. Two examples of a validation test on the Box (1981) vegetation description. Box initially published 74 such test cases along with the results of 111 calibrations of the conditions limiting each life form at well-known sites. An asterisk (★) indicates potentially dominant life forms in closed canopy systems. Potential dominants in more open systems are indicated by plus (+) signs and the asterisks (★) are retained to indicate potentially larger, but widely spaced forms (e.g. trees in a savanna). The level of detail and the independence of the validation data strongly reinforce the robustness of the Box (1981) methodology

Location	Predicted vegetation	Actual vegetation
Fairbanks, Alaska		**General description: Continental, closed boreal spruce–hardwood forest, with scattered bogs and with lower, open forests on north slopes**
Trees	★Boreal–montane short needle trees + Boreal summergreen needle trees Broad–summergreen small trees	★ *Picea glauca, Larix alaskana, Betula resinifera, B. papyrifera, Populus baslsamifera*, etc.
Shrubs	Summergreen giant-shrub Xeric summergreen shrubs Broad–summergreen mesic shrubs Needle-leaved evergreen shrubs	*Arctostaphylos rubra, A. uva-ursi, Empetrum nigrum, Ledum decumbens, Ribes triste, Rosa acicularis, Salix alaxensis, S. bebbiana*, etc. *Vaccinium vitis-idaea, V. uliginosum*, etc., *Viburnum edule, Amelancier alnifolia*, etc.
Grasses, herbs	Short bunch-grasses Summergreen forbs Xeric cushion–herbs Xeric thallophytes	Microthermal tundra and meadow herbs (e.g. *Carex, Scirpus, Eriophorum, Ranunculus, Oxytropis, Epilobium* spp., *Castilleja pallida, Artemesia frigida*)

Table 6.2 (*cont.*)

Location	Predicted vegetation	Actual vegetation
Makurdi, Nigeria		**General description: raingreen wooded savanna (wet-savanna zone)**
Trees, shrubs	Tropical xeric needle trees Tropical evergreen sclerophyll trees Palmiform tuft-trees Tropical evergreen microphyll trees + Xeric raingreen trees + Broad–raingreen small trees Raingreen thorn–shrub Palmiform tuft-treelets Tropical broad–evergreen shrubs Xeric evergreen tuft-treelets Evergreen giant-shrub Xeric rosette-shrub	★*Daniella oliveri*, ★*Lophira alata*, ★*L. lanceolata*, ★*Parinari polyandra*, *Terminalia* spp., *Afzelia africana*, *Bridelia ferruginea*, *Buttrospermum parkii*, *Cussonia barteri*, *Hymenocardia acida*, *Vitex doniana*, *Annona*, *Detarium senegalense*, *Gardenia*, *Lannea*, *Prosopis africana*, etc.
Grasses	Tall cane-graminoids + Short bunch-grasses	★*Hyparrhenia* spp., ★*Andropogon* spp., ★*Pennisetum* spp., *Panicum*, *Schizachyrium*, *Tephrosia*, *Eriosema*, etc.
Forbs etc.	Raingreen forbs Succulent forbs Broad–raingreen vines Broad–wintergreen epiphytes Xeric halophytes Bush stem succulents	*Eulophia* spp., *Indigofera*, *Curculigo pilosa*, *Amorphophallus* spp., *Kaempfera*, *Costus spectabilis*

Table 6.3. *Environmental constraints for each plant type in the BIOME model of Prentice* et al. *(1993)*

Plant type	T_c min.	T_c max.	GDD min.	GDD$_0$ min.	T_w min.	Moisture α coefficient min.	max.	D
Trees								
Tropical evergreen	15.5					0.80		1
Tropical raingreen	15.5					0.45	0.95	1
Warm–temperate evergreen	5.0					0.65		2
Temperate summergreen	−15.5	15.5	1200			0.65		3
Cool–temperate conifer	−19.0	5.0	900			0.65		3
Boreal evergreen conifer	−35.0	−2.0	350			0.75		3
Boreal summergreen		5.0	350			0.65		3
Non-trees								
Sclerophyll/succulent	5.0					0.28		4
Warm grass/shrub					22	0.18		5
Cool grass/shrub			500			0.33		6
Cold grass/shrub				100		0.33		6
Hot desert shrub								7
Cold desert shrub				100				8

Note: T_c is the mean temperature for the coldest month, GDD are the growing degree days (the summation of daily temperature departures greater than a 5 °C base), GDD$_0$ are the growing degree days (the summation of daily temperature departures greater than a 0 °C base), T_w is the mean temperature for the warmest month, α is the Priestly–Taylor coefficient of annual moisture availability and D is the dominance class for each of the plant types in the model.

sets are corrected for elevation. Digital soil maps of soil texture from Zobler (1986) at 0.5° latitude by 0.5° longitude resolution were used to compute the moisture-holding capacity of the soil and to derive measures of moisture availability (and soil storage of water).

In this approach to developing a global-scale model, there are relatively few plant life forms and many of these life forms have environmental ranges such that they do not co-occur. Therefore, one can enumerate the number of combinations of the life forms. In the case of the Prentice *et al.* (1993) model, there are 15 combinations or ecosystem complexes (Table 6.4). Prentice *et al.* compared the resultant map of world biomes with a map developed by Olson *et al.* (1983) for actual global vegetation and found substantial agreement between the two maps.

Table 6.4. *Combinations of plant types from Prentice et al. (1993) that produce the biomes that are mapped by the BIOME model of global vegetation. Also shown are comparable categories from a world vegetation map from Olson et al. (1983) that were used to test the accuracy of the BIOME model*

Combination of plant types	Biome name	Olson *et al.* (1983) Ecosystem complex name(s)
Tropical evergreen trees	Tropical rain forest	Evergreen equatorial forest
Tropical evergreen trees+ tropical raingreen trees	Tropical seasonal forest	Tropical seasonal forest
Tropical raingreen trees	Tropical dry forest/savanna	Tropical dry forest and woodland, tropical savanna and woodland
Warm temperate evergreen trees	Broad-leaved evergreen/warm mixed forest	Broad-leaved evergreen or partly deciduous forest, warm conifer forest, partly evergreen broad-leaved or subtropical conifer forest, south–temperate broad-leaved and/or conifer forest, broad-leaved south temperate forest, deciduous warm woods with conifers, tropical montane complexes
Temperate summergreen trees+ cool–temperate conifer trees+ boreal summergreen trees	Temperate deciduous forest	Deciduous (summergreen) forest
Temperate summergreen trees+ cool–temperate conifer trees+ boreal evergreen conifer trees+ boreal summergreen trees	Cool mixed forest	Cool hardwood–conifer forest
Cool–temperate conifer trees+ boreal evergreen conifer trees+ boreal summergreen trees	Cool conifer forest	Cool conifer forest

Boreal evergreen conifer trees+ boreal summergreen trees	Taiga	Main taiga, northern or maritime taiga, southern continental taiga
Cold–temperate conifer trees+ boreal summergreen trees	Cold mixed forest	Not mapped by Olson et al.
Boreal summergreen trees	Cold deciduous forest	Included in taiga by Olson et al.
Sclerophyll/succulent	Xerophytic wood/shrub	Semi-arid woodland or low forest, mediterranean types, succulent and thorn woods and shrub, dry or highland tree or shrub
Warm grass/shrub	Warm grass/shrub	Warm or hot shrub/grassland
Cool grass/shrub+ cold grass/shrub	Cool grass/shrub	Cool grassland/shrub, Siberian parklands, Tibetan meadows, cool irrigated dryland
Cold grass/shrub	Tundra	Wooded tundra and treeline, tundra, cold irrigated dryland
Hot desert shrub	Hot desert	Warm to hot desert and semidesert, sand desert, warm–hot irrigated dryland
Cool desert shrub	Semidesert	Cool semidesert shrub
No type occurring	Ice/polar desert	Polar or rock desert, ice

Adaptation of plants to the environment

Most global vegetation mapping systems (e.g. Holdridge 1967; Box 1981; Prentice *et al.* 1993) use relations between temperature and moisture as a basis for interpreting what sorts of vegetation (or plants) should occur in a particular environment. In a sense, the same methods mentioned in Chapter 5 in the context of the Grinnellian niche are applied at different scales in vegetation mapping. However, in exploring climate–vegetation relationships, assemblages of plants with particular attributes rather than single species are evaluated with respect to environmental conditions. The fundamental roots of some of the relationships between plant types and the environment is in the physiological ecology of terrestrial plants.

Low temperatures

An example of a climatic factor that has a profound effect on the physiognomy (structural appearance) of the vegetation in different parts of the world is the minimum temperature and the relation of minimum temperatures to different physiological adaptations to chilling. Table 6.5 lists cardinal minimum temperatures and the expected vegetation associated with these temperatures. These minimum temperatures are related to different mechanisms for adapting to low temperatures. For example, the freezing point of pure water (0 °C) and the point at which supercooled water will freeze (−40 °C) are major divisions in the list of cardinal minimum temperatures. At temperatures dropping below 0 °C, water in plants is likely to freeze, with associated damage to the plant tissues. At temperatures of −40 °C, supercooled liquid water (water below the freezing point but still in liquid form) will spontaneously form ice. Associated with the 0 and −40 °C temperatures are different adaptations (such as using supercooling to prevent freezing at low temperatures) to allow plants to tolerate lower temperatures. The cardinal temperatures tabulated in Woodward (1987b) have been used more or less directly by Woodward (1987b), Prentice *et al.* (1993) and Neilson *et al.* (1992) to predict global vegetation patterns.

Plants can change in their ability to tolerate stressful environmental conditions. Through the process known as hardening, a plant can tolerate a lower temperature by being previously exposed to progressively lower temperatures over a period of time. If two plants of the same species, one protected in a greenhouse and the other exposed to outside conditions for increasing intervals, are exposed to a low temperature, then it is the

Table 6.5. *Cardinal minimum temperatures and expected dominant physiognomy*

Temperature range (°C)	Phenomenon	Comment(s)	Expected physiognomy
Over 15	Temperature not limiting	Most productive types of vegetation are broad-leaved evergreen (Cannell 1982) under these conditions, under inadequate rainfall expect drought-deciduous vegetation (Axelrod 1966; Doley 1981)	Broad-leaved evergreen when rainfall is adequate
−1 to 15	Chilling	Change in physiology involving plant membranes has a cost in energy required and an associated loss in physiological efficiency (Lyons 1973; Quinn and Williams 1978)	Broad-leaved evergreen when rainfall is adequate
−15 to 0	Freezing and supercooling	Ability to withstand these temperatures often associated with tolerance of frost drought and summer drought (Levitt 1980); there are structural changes in the cell membranes, increased in cryoprotectants, increased thickness in cell walls and in leaves (Paleg and Aspinall 1981)	Broad-leaved evergreen
−40 to −15	Freezing and supercooling	Dormant buds, xylem and cambium of deciduous trees can have survival limits between −15 to −40 (Sakai and Weiser 1973; George *et al.* 1974); while frost-tolerant needle-leaved trees may make photosynthetic gains in the winter (Schulze 1982), the high photosynthesis rates of deciduous vegetation confers an advantage	Broad-leaved deciduous
Less than −40	Freezing and supercooling	Extensive adaptations in chloroplasts (Senser and Beck 1977), cell membranes (Ziegler and Kandler 1980) and cell osmotic properties (Larcher *et al.* 1973)	Evergreen and deciduous needle-leaved (coniferous)

Source: From Woodward (1987b).

greenhouse plant that will show the damage. Hardening involves an adjustment or an acclimation to low temperature.

Optimality of plant form and function

One view of the difficulties in interpreting the patterns of plant adaptations may be found in the position of Harper (1982) in describing what he called 'eureka ecology'. This is the idea that observed similarities in the plants in the same environment are interpreted as examples of parallel adaptations to the environment, while observed differences in the plants are interpreted as cases of niche differentiation allowing the species to minimise competition and divide resources. In general, questions involving evolution and adaptation are difficult to approach experimentally because of the time scales involved.

Recently, the interpretation of supposed adaptations in plants and animals has been made in terms of the degree a given adaptation appears to be optimal with respect to a particular environment. This approach is appealing because it allows the inclusion of a more mechanistic interpretation of the potential causes of adaptations. These analyses are particularly valuable in enriching the traditional explanations of plant responses to the environment with modern understandings of plant physiology and biophysics.

Parkhurst and Loucks (1972), in a paper inspecting the optimal size of leaves in different environments, outlined the basic ideas in this approach, which they called 'the principle of optimal design', as 'Natural selection leads to organisms having a combination of form and function optimal for growth and reproduction in the environments in which they live'.

Gould and Lewontin (1979) voiced concerns similar to those of Harper (1982) and further criticised studies based on assuming the optimality of parts of plants and animals as adaptive solutions to functional problems. Givnish (1986) responded to these criticisms in some detail and noted that, 'The main role of an optimality argument should be as a heuristic device, giving a quantitative means for testing particular hypotheses of how variation in a given trait contributes to plant competitive ability'.

Rather subtle differences in the interpretation of the function of plant organs can lead to different interpretations of appropriate morphologies. For example, in two classic papers, Parkhurst and Loucks (1972) and Givnish and Vermelj (1976) attempted to derive the optimal shape of a leaf under the assumption that leaf-shape adaptation to environmental conditions could be considered the consequence of an optimisation of the carbon, water and heat budgets of the leaves. Each study grappled with including effects of leaf size in the heat-balance equations, the water-flux equations (which are strongly tied to the heat equations through the evaporation terms) and resistance-based water flux equations.

Parkhurst and Loucks (1972) assumed that natural selection favours plants with leaves that maximise the efficiency of water-use by the plant. Givnish and Vermelj (1976) made an alternative assumption that leaves should optimise the payoff of having leaves of a given size versus the costs of maintaining leaves of a given size. Either assumption seems reasonable. Each study produced a diagram of optimal leaf size and environment. These diagrams differed in their predictions of leaf size in low-light, high-moisture conditions (Fig. 6.9*a,b*). It seems clear from the comparison of these two studies that relatively subtle differences in assumptions about what is being optimised in plants can have considerable differences in certain environments. One might see this outcome as a short-coming of one or other of the studies. An alternate view is to see the two studies as working out the critical observations needed to separate which of two plausible theories about the function of leaves is most likely to be operative.

This example case is illustrative of the types of consideration that may arise from applying classic optimisation techniques to develop a theoretical basis for relating form to function. It is not always clear what natural selection is optimising. There are several obvious choices ranging from optimising (or maximising) survival, fitness, efficacious use of energy, water or nutrients, occupancy of space, etc. It is quite possible that two similar species of plant functioning in the same environments may, because of their past evolutionary and biogeographical histories, be optimised through natural selection toward different goals. Nevertheless, it is clear that studies of optimal form with respect to the functioning of a plant in a particular environment provide considerable insight into what ranges of adaptations are possible and the manner in which particular plants perform in particular environments.

Life history-based classifications of species roles

Box (1981) notes that the interest in relating vegetation structure to climate and other environmental factors that was a hallmark of early plant ecologists and plant geographers gave way in the past decade or two to the sharpened interest in plant demography. The rich insights of the early

(a)

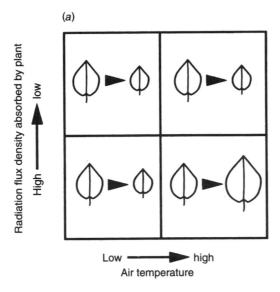

(b)

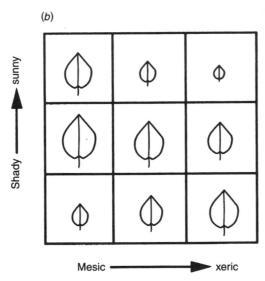

Figure 6.9. The optimal sizes of leaves in different environment conditions derived by assuming that leaves are in some sense optimised to the environment. (*a*) Parkhurst and Loucks' (1972) prediction of changes in leaf shape under natural selection. Leaf-shape changes are based on the assumption that leaves tend to optimise their water-use efficiency. (*b*) Givnish and Vermelj's (1976) prediction of leaf shape based on the assumption that leaves tend to optimise the difference between gains associated with leaf size minus costs (e.g., increased respiration, need for structure to support large leaves, etc.) associated with leaf size. While the axes of the two diagrams are different, they clearly illustrate that reasonable assumptions about the optimisation goal can produce rather different expected outcomes.

ecologists should not be overlooked in developing functional types for large-scale interrelating of vegetation and climate. One of the challenges in developing plant functional types is interweaving the physiognomic adaptations emphasised in the early systems with population and community dynamics.

Structural classifications (Raunkiaer 1934; Hallé 1974; Hallé and Oldeman 1975; Webb *et al.* 1970; Walker *et al.* 1981) of the life forms that constitute a vegetation have been discussed to some degree above. There are also several categorisations of plants based on population and life-history attributes of the species. These include categorisations of *r*, *K* and adversity strategies (MacArthur and Wilson 1967; Southwood 1977; Greenslade 1983) based on the reproductive rates of species. An '*r*' species would have a high reproductive rate and be able to demonstrate explosive population growth to take advantage of favourable conditions. A '*K*' species would have a set of biological attributes allowing it to prosper in a crowded community that was near the maximum (*K*) that a given environment could hold. An adversity strategy species would have attributes that would ensure its long-term presence even in the face of a harsh environment.

Early and late successional species (Budowski 1965, 1970; Whittaker 1975; Bazzaz 1979; Finegan 1984) would be identified by having a suite of attributes that might allow success at different times in an ecological succession. Other categorisations are exploitative and conservative species with respect to limiting resources (Bormann and Likens 1979a); ruderal, stress-tolerant and competitive species (Grime 1977, 1979a,b); and gap and non-gap species (Hartshorn 1978; Shugart 1984, Brokaw 1985a,b). In all of these approaches, the intent of the categorisations is to simplify the complexity of the different life histories and attributes of the plants that constitute the vegetation in a manner that provides insight into the appearance and the possible dynamics of the vegetation.

The vital attributes classification combines a life-history-based classification of species (age or life stage when an individual can reproduce, longevity) with the categorisation of the response of the species to disturbance events (Noble and Slatyer 1980). Vital attributes (Noble and Slatyer 1980) are those attributes of a species that are vital to its role in a vegetation replacement sequence. Noble and Slatyer defined three categories relating to (a) method of arrival and persistence of the species at the site during and after a disturbance, (b) ability to establish and grow to maturity in the developing community, (c) time taken for the species to reach critical life stages. The characteristics include such features as longevity, persistence of seed bank and the ability to establish following disturbance or in the presence of established vegetation. Species are then classified into 'types' based on the combination of attributes exhibited. Although there may be differing mechanisms responsible for a particular vital attribute, the classification relates only to the outcomes of those mechanisms. Qualitative predictions of a species' presence or absence as a function of time following disturbance can then be made and the consequences of different disturbance regimes explored.

The vital attributes approach does not explicitly consider the response of species to environmental factors but relates establishment and growth under conditions associated with disturbance and the presence of vegetation. The scheme is intended to deal primarily with successional dynamics on a site in the absence of changes in the underlying environmental conditions (e.g., climate). However, these restrictions are not inherent to the approach and the vital attributes scheme has been modified to simulate vegetation dynamics in both time and space (Noble *et al.* 1988). The vital attributes approach can be used to construct computer models that not only indicate the likely composition of the vegetation but also provide an indication of the successional dynamics of the vegetation under a particular disturbance regime. This approach will be discussed in more detail in Chapter 9.

The vital attributes approach and other functional type approaches to predicting vegetation response to change are based on the constraints imposed on plants by different environmental conditions, such as resource availability or disturbance regime. After classifying environmental conditions, plant strategies based on responses to those environmental conditions and the associated life-history characteristics, either explicitly or implicitly considering the consequences of resource allocation, are used to devised the life-history classification.

There is to some degree a circularity of defining the environmental conditions in terms of the plant response. For example in the concept of the K species (MacArthur and Wilson 1967; Whittaker and Goodman 1979; Greenslade 1983), the carrying capacity of the environment (K) is defined in terms of the numbers or biomass of a particular species.

Also, the characterisation of the species (i.e., their strategies) is dependent on their role within the present vegetation. The categories of early versus late successional plants are actually a description of the role of the species in the temporal dynamics of a community. To categorise a species as an early successional species does not necessarily provide insight as to why the species responds in time the way it does. Therefore, one is unable to predict how the same species may respond in a different context, either under different environmental conditions or in the presence of a novel assemblage of other species. The role of a species in the community or ecosystem is a function of the prevailing environmental conditions (e.g. climate, geology) and the suite of species present; it is context sensitive. There is a need for an approach for defining key characteristics and processes of plants that determine their response to environmental conditions, competitive interactions and ultimately their distribution in time and space.

Concluding comments

The observation that the vegetation appears to respond to climatic conditions is an early observation and one that is essential to the regional and global-scale interpretation of terrestrial vegetation. The approaches to mapping global or continental vegetation pattern in response to climate are of two basic sorts. First, one can relate the patterns in the vegetation to climate by the development of appropriate theory or by statistical correlation (Holdridge 1967). The disadvantage of using this method to predict the Earth's vegetation response under conditions of an altered climate lies in the possibility of relatively novel ecosystems existing under changed climatic conditions (a point that was discussed in Chapter 2).

Second, one can relate the climate to the types of plant that compose a given community and use rules for the manner in which the types of plant in a given climatic setting should interact to produce the vegetation pattern (Box 1981; Prentice *et al.* 1993). At the base of these approaches is a fundamental notion of the relative adaptability of particular plant morphologies to environmental conditions. Initial plant component-based approaches to interpreting the vegetation and environment relation over large areas have been expanded to include aspects of the plant population biology as well as morphology in recent years.

In Chapters 13 and 14, the use of the models of vegetation–climate interactions discussed in this chapter will be discussed as applications to provide initial assessments of the response of the global vegetation to climatic change. The challenge in applying these methods to determine change in vegetation is the lack of transient or successional responses in the approaches.

7 · *The mosaic theory of natural landscapes*

When viewed at an altitude of one or a few kilometres, most terrestrial ecosystems are noticeably patchy. The grain of the patchiness may be at the scale of the individual crowns of trees in a mature forest landscape. It may correspond to the patches that have been burned at different times in the case of fire-prone landscapes, or it may arise from a number of other causes including landslides, wind damage or other disturbances. Whatever generates the observed heterogeneity, prediction of the structure of a terrestrial landscape is strongly based on understanding the patterns and dynamic changes of the patches that make up the landscape mosaic.

Theoretical views regarding the dynamics of landscape mosaics have been developed in forests to a significant extent. Regardless of other sources of spatial heterogeneity, a forest canopy is a mosaic of tree crowns. When the individual trees that are the elements of this mosaic die and open a gap in the canopy, several ecological responses are initiated in the small area below the canopy opening. These responses eventually lead to the repair of the forest canopy and over the course of time are re-initiated by the death of the new dominant replacement tree. The implications of the cyclical nature of small-scale forest dynamics were clearly elucidated by Watt in 1925 and later, in a classic paper, in 1947. While Watt was clearly a pioneer with respect to these concepts among English-speaking ecologists, the ideas have also been seemingly independently discovered or significantly re-introduced by others including Aubréville (1933, 1938), Leibundgut (1959, 1978), Sukachev (1968b), Oldeman (1978) and Remmert (1985, 1991).

The mosaic concept of vegetation dynamics

It is rarely clear where some of the ideas about the more fundamental nature of landscapes ultimately arise. One would expect observant hunters or gatherers to capitalise on a knowledge of the parts of a hetero-geneous landscape that are most productive. Oldeman (1991) points out that forest dwellers in various parts of the world have a rich vocabulary describing processes and patterns regarding tree falls, patches in forests and the like. He notes *chablis* and *volis* in French, *rytä* in Finnish, and *traa* and *loo* in Dutch, as such terms in the respective European languages. Oldeman speculates that the variety of grape used to produce chablis wine may have originated from a variety of *Vitis vinifera* found in small forest openings, or *chablis* in older French idiom.

The cyclical nature of the local dynamics of forests has been under-stood by European foresters for the past two centuries and formed a basis for the management of forest landscapes. Oldeman (1991) illustrates this point by producing a 200-year-old drawing from Cramers (1766) showing the conversion of a broad-leaved German forest into a conifer-ous forest along a transect with small patches of conifers among the canopy openings in the hardwood forest. Whitmore (1982) also acknowl-edges the development of a practical understanding of forest dynamics by foresters that anticipated the founding of ecology as a science. It is appropriate that the individual who introduced the mosaic dynamic concept to ecology, A. S. Watt, was himself a forester (Grieg-Smith 1982).

Watt (1925) in his doctoral research had noted the cyclical replacement of beech (*Fagus sylvatica*) in mature forests in southern England (Fig. 7.1). In 1947, in his presidential address to the British Ecological Society, he extended these concepts to a range of other systems (bogs, sparse grass-lands, alpine communities, heathlands). Watt's (1947) address presented several examples of two types.

First, he developed cases in which the seemingly helter-skelter pattern of a vegetation could be ordered to elucidate an underlying process(es) causing the pattern. The development of the process interpretation required inference by the observer to deduce the appropriate ordering of the patterns in the vegetation. For example, the heterogeneity in open Breckland grasslands (Fig. 7.1*a*), in heathlands (Fig. 7.1*b*) and beech forests (Fig. 7.1*c*) all can be inferred to originate from similar dynamics at the spatial scale similar to the dominant plants. The small-scale dynamics involved the colonising, growing, maturing and senescing of individual plants at a particular location.

Second, Watt discussed cases in which the local processes seen in the first type of example were naturally synchronised to some degree. This caused the vegetation pattern to move in a regular and predictable way over the landscape. This second type of example seems particularly

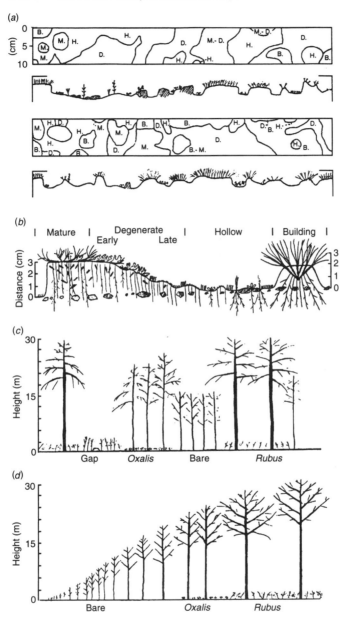

Figure 7.1. Examples from Watt (1947) of patterns in plant communities. (*a*) A sparse grassland at Breckland dominated by *Festuca ovina*. Two map transects 160 cm by 10 cm are shown in two views: from above and in section. (*b*) Diagrammatic representation of phases of the Breckland grassland with a reordered pattern of changes over time progressing from left to right. (*c*) A transect through an old forest dominated by beech (*Fagus sylvatica*). (*d*) The phases of development in a beech forest ordered by stage of development.

compelling in that one does not need to infer the order of the pattern as a preamble to understanding the generative processes. In Watt's (1947) original paper, the examples were drawn from alpine situations (Fig. 7.2) in relatively adverse locations. Several of Watt's alpine examples involved the same species: notably shrubs such as heather (*Calluna vulgaris*) and low-statured herbs and grasses. Subsequent to Watt's classic examples, there have been descriptions of vegetation of larger stature demonstrating this same tendency for spatial arrangement in a regular pattern ('self-organising') in a manner to suggest the underlying processes controlling the dynamics of the system. Two of these, the 'tiger-bush' and the 'fir wave', will be discussed in the sections that follow.

Tiger bush or *brousse tigrée*

Following the Second World War, extensive aerial surveys for purposes of mapping and land inventory revealed the presence of vegetation patterns in the form of parallel stripes of alternating vegetated and non-vegetated zones in certain arid and semiarid areas (Cornet *et al.* 1992). The pattern of occurrence of these patterns, which resemble the stripes of a tiger from the air and hence the name tiger bush, or *brousse tigrée*, appears to be worldwide. The phenomenon has been reported in Africa (Hemming 1965; White 1970, 1971; Boudet 1972), North America (Montaña *et al.* 1990; Cornet *et al.* 1992; Montaña 1992; see Fig. 7.3), and Australia (Slatyer 1961; Pickup 1985; Tongway and Ludwig 1990). The systems occur under conditions of:

1. Arid or semiarid climates with few but high intensity rains
2. Gentle regular slopes (0.25 to 1.0%)
3. Soils with low permeability producing sheet runoff (Cornet *et al.* 1992).

The stripes have been observed to migrate slowly upslope in several studies.

For example, in the Mapimí Biosphere Reserve in the Chihuahuan Desert of Mexico, the stripes have a width of between 20 and 70 m and are 100 to 300 m in length. Within each stripe there are five relatively distinctive zones (Fig. 7.4). The stripes themselves represent a feedback between the vegetation, soil and climate. The presence of vegetation increases the soil organic matter content and this further increases the water permeability in the upper 30 cm of soil below the vegetation stripe (Cornet *et al.* 1992). Delhoume (1988) and Cornet *et al.* (1992) have analysed the moisture depths in the five zones delineated in Fig. 7.4.

(a)

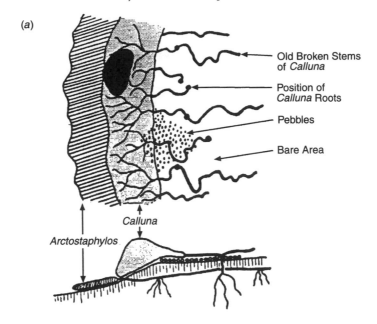

Old Broken Stems
of *Calluna*

Position of
Calluna Roots

Pebbles

Bare Area

Calluna

Arctostaphylos

(b)

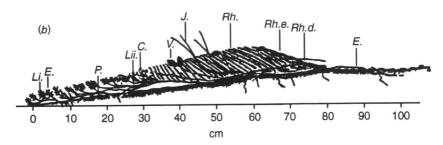

Figure 7.2. Examples of self-organised, moving patterns in vegetation documented by Watt (1947) as examples of the product of patterns in vegetation responding to underlying processes. The communities move across landscapes from right to left as diagrammed. All of the systems are found in alpine situations in Scotland. Unlike the example in Fig. 7.1, these patterns are directly observable, not a product of inferring the time sequence of the pieces making up a landscape mosaic. (*a*) Pattern of *Calluna* and *Arctostaphylos* in double strips found in highly exposed locations in the Cairngorms. The *Arctostaphylos* invades bare soil and the *Calluna* moves over and suppresses the adjacent *Arctostaphylos*. (*b*) More complex but similar system found at higher elevations in the Cairngorms. Li., Lii. and P. are mosses; E. is *Empetrum hermaphroditum*, C. is *Cladonia rangiferina*, V. is *Vaccinium myrtillus*, J. is *Juncus trifidus*, Rh. is *Rhacomitrium lanuginosum*, Rh.e. is eroded *R. lanuginosum*, Rh.d. is dead *R. lanuginosum*.

(a)

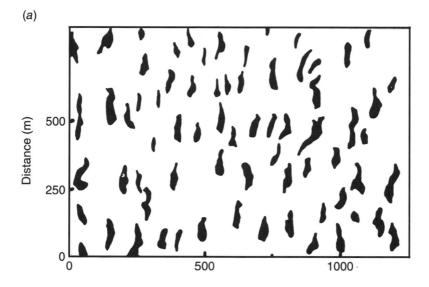

(b)

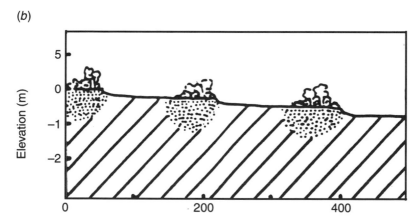

Figure 7.3. Schematic diagram of two-phase mosaic commonly found in the Bolsón of Mapimí, southern Chihuahuan Desert, Mexico. (*a*) Aerial view showing alternation of vegetation arcs (in black) and bare areas. (*b*) Idealised cross-section of the landscape showing the distribution of the vegetation. Stippled areas below the vegetation indicate areas to which moisture is distributed after a rain. From Montaña (1992).

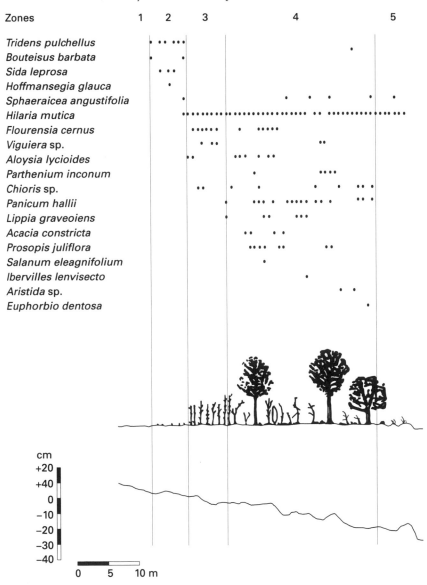

Figure 7.4. Section across a vegetation stripe in the two-phase mosaic commonly found in the Bolsón of Mapimí, southern Chihuahuan Desert, Mexico. Species presence and absence are indicated in the table above the schematic diagram. The stripes can be divided into five zones that differ in structure and composition. The zones are: 1. bare area; 2. deposition area with pioneer vegetation; 3. screen of *Flourensia cernua*; 4. dense bushy area; 5. downhill bare area. The stripe moves in an upslope direction (from right to left). From Cornet *et al.* (1992).

Cornet *et al.* (1992) found that the bare zone had water storage amounting to 10 to 27% of the rainfall, the pioneer zone 60 to 150%, the upslope vegetation area (zone 3 in Fig. 7.4) had 170 to 530% and the downslope vegetation area (zone 4 in Fig 7.4) had 160 to 210%. The percentages higher than 100% reflect the increased contribution of runoff water to the stock of water in the vegetated stripes. The two major woody species forming the vegetation stripes (in Mexico, *Prosopis glandulosa* and *Flourensia cernua*) both have increased regeneration success on the upslope side of the stripe (Montaña *et al.* 1990). According to Montaña (1992), the upslope side of the stripe is colonised during years of relatively higher rainfall, and mortality on the downslope side of the stripe is accentuated in relatively dry years. Thus, the movement pattern is variable in time and space and generally reflects the climatic variability.

The tiger stripe pattern of vegetation in semiarid and arid zones is similar to the organised patterns seen in Watt's (1947) discussions of pattern and process with regard to alpine vegetation growing on rather harsh sites in the Cairngorms in Scotland (Fig. 7.2). The occurrence of these regular patterns in different locations (Africa, the Middle East, Australia, Mexico) and involving different species of plants speaks to the generality of the phenomena. These systems are in some senses special cases in which the feedbacks between the environment and the plant processes produce regular moving patterns. (A computer model of the dynamics of this type of vegetation in Africa will be discussed in Chapter 12.) The areas covered by such patterns can be extensive. For example, the mosaic striped shrublands (Fig. 7.3) cover 32% of a 172 000 ha area mapped by Montaña (1988) in Mapimí, southern Chihuahuan Desert, Mexico.

Fir waves

High-altitude conifer forests can exhibit a striking pattern described as 'wave regeneration' (Sprugel 1976). The spatial arrangement features with crescent-shaped bands or parallel stripes of dead trees on mountainsides. These features appear in the high-altitude forest of Japan, as noted by Oshima *et al.* (1958), 'In the dark green of the gentle southwest slope of Mt Shimagaree with subalpine coniferous forest, several whitish stripes horizontally running in parallel with each other can be seen in a distant view so distinctly that the mountain has been named "mountain with dead tree strips" according to its curious physiognomy'. Sprugel (1976) studied the analogous phenomena in the New England region of the

(a)

(b)

Prevailing wind direction

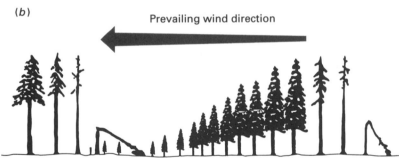

Figure 7.5. Wave-regeneration in the *Abies balsamea* forest of the north-eastern USA. (*a*) General appearance of the wave mortality. (*b*) Diagrammatic section through a regeneration wave. From Sprugel (1976).

USA and found that the stripes of dead trees represent a moving wave of synchronised mortality followed by regeneration of the balsam fir *Abies balsamea* (Fig. 7.5).

Sprugel (1976) surveyed transects through the stripes of dead trees at 10 m intervals and determined the ages of trees and structure of the forest in each location. The age structure and the physical structure of the forest indicated a travelling wave of mortality followed by regeneration and growth. The time needed for the wave to travel through the length of a cycle appeared to be of the order of 55 years. The waves were found to travel in the direction of the prevailing winds and were oriented perpendicularly to these winds. Velocities of movement of the waves were between about 1 and 3 m year^{-1} and there was a strong negative correlation between these velocities and the slope. Fast-moving waves were either on level ground or were moving downhill. Sprugel found that wind, the formation of rime-ice on the exposed trees on the edge of the wave, winter desiccation along the edge of the wave resulting from increased exposure, cooling from summer winds along the edge of the wave (with associated reductions in photosynthesis rates) and the increasing age of the trees as they approach the moving edge were all possible factors maintaining the wave patterns. Sprugel notes that these factors could work in concert or combination and are not mutually exclusive.

Beyond the interest in the occurrence of a pattern in the dynamics and structure of forests in different locations, the striking aspect of the fir wave phenomena is the strong resemblance to the pattern of gap regeneration proposed by Watt (1925, 1947). Watt's conceptualisation of the process (Fig. 7.1d) is the very image of the fir wave recorded by Sprugel (1976) for balsam fir forest (Fig. 7.5b). The fir wave may be a special case in which the internal processes and the pattern of important environmental factors make an underlying pattern/process interaction more observable. The importance of the fir wave and the tiger bush phenomena (along with the fascinating aspect of vegetation generating regular pattern) is that both phenomena reinforce the interpretation of Watt (1947) of the underlying cyclical processes of growth, death and rebuilding as the generator of vegetation pattern.

Roles of species on mosaic landscapes

One can categorise plant species into simple 'roles' with respect to the manner in which their life-history attributes affect their likelihood of occupancy of small areas (e.g., Grime 1974, 1977, 1979a,b) making up a larger landscape (see Fig. 6.2, p. 146). If, as was discussed in Chapter 6, 'strategies' used by species emerge from an interplay of form and function at the individual plant level, they are part of a larger spectrum of system rules for form–function, pattern–process interactions. At appropriate levels, these rules can be expressed as theories. In practice, such theories are constrained (either explicitly, or implicitly by the domain of their successful applications) to particular space and time scales. One theory can 'work' in producing successful predictions at one scale, while another theory, with different fundamental assumptions, can also 'work', albeit at a different scale. This latter feature is consistent with a hierarchical concept of vegetation pattern.

Shade-tolerant versus shade-intolerant species

An important dichotomy between types of plant in ecological succession in forests and other closed canopy systems involves the way the plants use light. Figure 7.6 illustrates typical responses of net photosynthesis of plants to changes in the light level. Bazzaz (1979) illustrated the tendency for plants found in early succession to have high levels of carbon fixation at high light levels and lower levels of carbon fixation at low light intensities relative to late successional plants (Fig. 7.6a). This pattern, manifested at the leaf tissue level, has been implicated as a mechanism that gives plants different tolerances to shading and, thus, can be used to explain patterns in the successional changes of ecological systems (e.g. Bazzaz 1979; Bazzaz and Pickett 1980; Kozlowski et al. 1991).

Horn (1971) examined trees with equivalent tissue level responses to changes in light levels but with different geometries with respect to leaf layering and produced the whole plant net photosynthesis curves illustrated in Fig. 7.6b. These curves demonstrate the same pattern summarised by Bazzaz (Fig. 7.6a). Plants differ in their ability to utilise light because of their geometry (Horn 1971) as well as their physiology (Bazzaz 1979). It seems likely that combinations of physiology and geometry might amplify the effect of a seeming trade-off in the ability of a plant to use light at different light levels. One might expect a plant with physiological responses that allow it to perform relatively better under low light levels to also have a geometry that supports similar performance under lower light regimes.

Strugglers versus gamblers

Oldeman and van Dijk (1991) developed a scheme for identifying 'tree temperaments' among tropical trees that was based on combinations of three size (or life-stage) categories and two responses to light levels: 'gamblers' having many of the attributes associated with early successional plants in the sense of Bazzaz (1979), such as fast growth rates and intolerance to shading, 'strugglers' having slower growth rates and relatively greater shade tolerance. These two strategies with respect to light were combined with three life stages of the tree (seedling, pole and mature) to

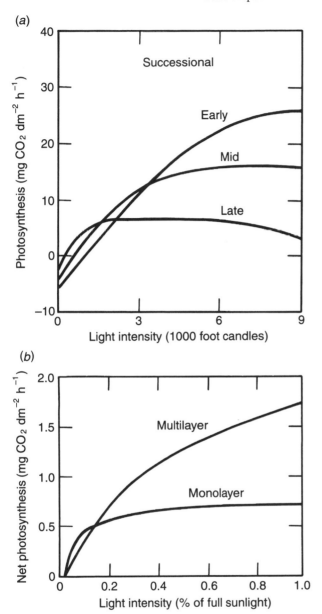

Figure 7.6. Typical responses of net photosynthesis in plants as a function of light level. Note that at the tissue level and at the canopy level, there is a reversal of photosynthetic efficiency at high versus low light levels. (a) Idealised light saturation curves for early, mid and late successional plants (from Bazzaz 1979). (b) The effect of light on net photosynthesis in multilayered and monolayered trees (from Horn 1971). 1 foot candle $= 1.076\,lx$.

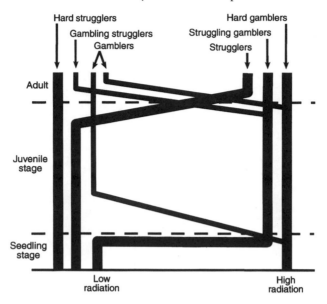

Figure 7.7. Six tree temperaments according to Oldeman and van Dijk (1991). Hard gamblers and hard strugglers remain constant in shade tolerance through their entire lives. Other 'temperaments' change in their shade tolerance at some stage in the life cycle. Strugglers correspond to classic 'shade-tolerant trees' or 'climax species'. Hard gamblers are the classic 'light-demanding trees' or 'pioneer species'. The relative frequency of a temperament in tropical rain forest is indicated by the thickness of the bar.

produce a categorisation of tree strategies depending on whether the tree behaved as a 'gambler' when small, a 'struggler' when somewhat larger, etc. (Fig. 7.7). While Oldeman and van Dijk (1991) did not explicitly draw on Horn's multilayer–monolayer dichotomy of tree geometry (Fig. 7.6b) to ascribe response curves for the strategies they developed, the functional consequences are form is evident in the development of their tropical tree temperaments.

Gap dynamics in the forest mosaic

Ideas about the mosaic dynamics of landscapes have developed in forests to a significant extent. A forest canopy is a mosaic of tree crowns. When the individual trees that are the elements of this mosaic die and open a gap in the canopy, a number of responses are initiated in the small area below the canopy opening. These responses eventually lead to the repair of the forest canopy. The implications of the cyclical nature of small-scale forest dynamics were clearly elucidated by Watt in 1925 and later in 1947. While Watt was clearly a pioneer with respect to these concepts among English-speaking ecologists, the ideas have also been seemingly independently discovered or significantly re-introduced by others, including Aubréville (1933, 1938), Leibundgut (1959, 1978), Sukachev (1968b), Oldeman (1978) and Remmert (1985, 1991).

In a special publication honouring A. S. Watt (Newman 1982), Whitmore (1982) asserts,

> that forests of the world are fundamentally similar, despite great differences in structural complexity and floristic richness, because processes of forest succession and many of the autecological properties of tree species, worked out long ago in the north temperate region, are cosmopolitan. There is a basic similarity of patterns in space and time because the same processes are at work.

Whitmore (1982) then provides evidence for this view by discussing forests from several parts of the world including Watt's (1925) beech forests in southern England, mature forests in the north temperate zone in general (Jones 1945), *Nothofagus* forests in Chile (Veblin 1979; Veblin *et al.* 1979, 1981) and several tropical rain forests. Whitmore reiterated this point in 1989, 'I believe that the existence of a forest cycle . . . fits our present knowledge about forest dynamics'. As one might expect, there are other ecologists (e.g. Lieberman and Lieberman 1989; Runkle 1989), who see the response of forests to canopy gaps as being more complex than Whitmore (1982, 1989) claims. Some of this difference of opinion seems to arise in the preferences for generality versus specificity in explanations.

For forests, the fundamental concept is that, at a rather small scale (say 1/10 to 1/20 ha), the long-term behaviour of a forest is quasi-periodic (Fig. 7.8). Starting when a small plot of land in a mature forest is dominated by a single large tree, the large tree shades the ground and reduces the survival of smaller trees and seedlings below. There may be a few smaller trees that are tolerant of shading that may survive under the large tree but even these trees eventually will die as well. The large tree dominates the resources (light, water and nutrients) that are available at the site and blocks other trees from growing at the location. When this tree dies, the forest floor (where there had previously been little chance of a young tree's survival) becomes a nursery for small seedlings and saplings. There is adequate light and other resources and hundreds of small trees survive and begin to grow toward the canopy. As these trees grow, they begin to compete with one another and some of the trees lose to more vigorously

Figure 7.8. Forest dynamics in the vicinity of the fall of a large tree and its associated destruction (the *chablis* in the sense of Hallé *et al.* 1978).

growing competitors. Eventually one tree manages to win the race to be the local canopy tree and begins to eliminate the others. This represents the closure of the cycle with a large tree again dominating the site.

The basis of this view of small-scale forest dynamics is an appreciation of the architecture of forest canopies and of the implications of this architecture for forest processes. Trees can grow to sufficient size to alter locally their own microenvironment and that of subordinate trees (Whitmore 1982; Brokaw 1985a,b; Silvertown and Smith 1988; Oldeman 1991). The magnitude of this effect can depend on the species, shapes and sizes of trees. The environment, in turn, has a profound influence on the success of different species, shapes and sizes of trees. Therefore, there can be a feedback from the canopy tree to the local microenvironment and subsequently to the seedling and sapling regeneration that may become the next canopy (Shugart 1984, 1987).

The environmental alteration from a canopy tree is most easily observed for the forest light environment (Hartshorn 1978). The nature of the leaf area profile and canopy geometry are dominant factors in the amount and pattern of the light at the forest floor (Anderson 1964; Cowan 1968, 1971). Other tree–environment interactions are also potentially important and plants can alter their local environment with respect to other variables, for example throughfall (Zinke 1962; Helvey and Patric 1965), soil moisture (Shear and Stewart 1934; Swift et al. 1979), soil nutrient availability (Zinke 1962; Chapin et al. 1986) and soil thermal regime (Dyrness 1982; Chapin et al. 1987; Bonan 1989a,b; Bonan and Shugart 1989). Shugart (1984), Oldeman (1991) and Remmert (1991) provide discussions of the ecological consequences of the mosaic nature of ecosystems.

To describe the gap-phase dynamics of a forest it is initially useful to concentrate on the hypothetical stages or phases of the gap cycle. Different authors with somewhat different objectives have created a nomenclature for these stages (for example, see Table 5.1 in Oldeman 1991, p. 168). In the spirit of not creating a new nomenclature in cases in which an already useful one exists, the Oldeman (1991) nomenclature will be used in the discussion that follows (see Fig. 7.9). Oldeman's phases include the tendency for the forest in the location of a newly formed canopy gap to increase initially in phytomass, to reach a period of relative equilibrium and then to decrease in biomass with senescence of dominant trees. This reflects changes in the pattern of production noted by Watt (1947) for forests as well as for a variety of other ecosystems.

Of course, the distinctions in stages (or phases) of the forest cycle are

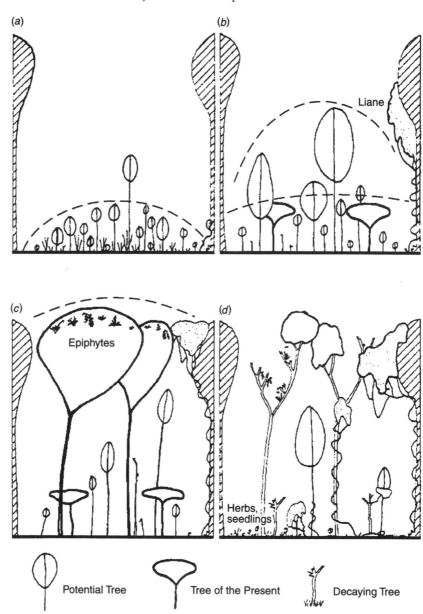

Figure 7.9. Simplified, idealised architectural pattern of the development phases of an 'eco-unit' associated with a tree fall. (*a*) Innovation phase with a lens-shaped structure of activated propagules. (*b*) Aggradation phase with closure of the canopy. (*c*) Biostatic phase with mature vigorous canopy trees. (*d*) Degradation phase with dominant trees becoming senescent. Slightly modified from Oldeman (1991).

arbitrary divisions of what is a more or less continuous process (albeit one marked by what, in some cases, are relatively distinct events such as the sudden fall of a large tree, the formation of a closed canopy or the germination of seeds formally dormant). Whitmore (1982) criticises the antecedents of the Oldeman nomenclature (Hallé *et al.* 1978; Oldeman 1978) as being too complex, and it should be understood that the phases that follow are intended as a conceptual model. The initial discussions will be restricted to cases at the scale of the area affected by the death of a large canopy tree. Multiple tree death events and the dynamics of forest mosaics will then be treated in the following section.

The death of a large tree in a forest with a relatively intact canopy can create a relatively abrupt change in the conditions on the forest floor beneath the former dominant tree. Oldeman (1991) calls this the 'zero-event'. The area associated with the death of a large tree is referred to as the gap (Watt 1947; Shugart and West 1980; Whitmore 1982, 1989). It is generally recognised that the gap can be quite heterogeneous (Brandani *et al.* 1988). Hallé *et al.* (1978) refer to the fall of a tree and its attendant destruction as the *chablis*, in part to sharpen the focus on one sort of heterogeneity. When a large tree falls the area beneath the crown of the former canopy-dominanted area is subject to a large change in the light level at the forest floor (see also Barton *et al.* 1989). If the tree is uprooted there may also be an exposure of mineral soil, the creation of a heap of soil in the root mat and a pit left by the removed root ball. On the area adjacent to this zone, the fall of the tree may have relatively little effect. The trunk of the fallen tree lies on the ground creating a microsite with different conditions to the rest of the area. Unless the falling tree has happened to strike a relatively large tree and toppled it, the light levels on this adjacent site are relatively unchanged. In the area in which the crown of the tree has fallen, there can be a considerable destruction of the other trees formerly growing in that location, as well as an exposure of mineral soil and increased levels of sunlight.

Another source of variation in the conditions associated with the generation of a gap in the forest canopy is the size of the gap that is created (Fig. 7.10). Depending on gap size, there can be complex gradients within the gap in which the gap-filling processes are primarily by vegetative means, from immigrated seeds, or from the seed bank (Oldeman 1978, 1991; Bormann and Likens 1979a; Gillon 1984). The details of these patterns can be expected to vary with the biology of the species in the particular forest (Putz 1983; Putz and Appanah 1987; Oldeman 1991) and with variations in the physical environment, for example owing to the interaction between

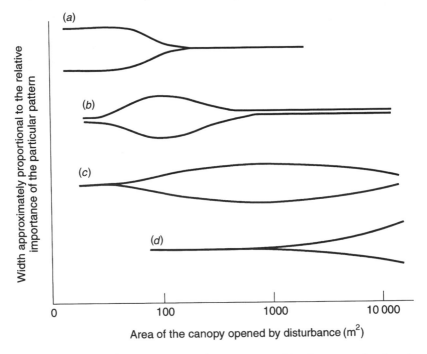

Figure 7.10. Gap-filling patterns in the New Hampshire forests as related to the size of canopy gaps. The width of each section is proportional to the importance of each process at the different spatial scales. (*a*) Canopy closure by lateral branch growth of canopy trees or height growth of subcanopy trees. (*b*) Height growth of suppressed juveniles of seed or sprout origin (*Fagus grandifolia, Acer saccharum* and *Tsuga canadensis*). (*c*) Height growth of new juveniles of seed origin (*Acer rubrum, Betula alleghaniensis, Fraxinus americana, Liriodendron tulipifera, Quercus rubra* and *Prunus serotina*). (*d*) Height growth by new juveniles of seed origin (*Betula populifolia, Betula papyrifera, Populus tremuloides, Populus grandidentata*). From Burrows (1990); modified from Bormann and Likens (1979a).

canopy height and sun angle (Canham 1988). One would expect the heterogeneity on the north and south sides of gaps to increase with latitude in response to the changes in mean sun angle. The relatively high sun angles in the tropics would tend to allow smaller gaps to let direct sunlight to the forest floor than can occur in temperate and, particularly, boreal situations.

In the case of tropical forests, Denslow (1980) noted a spectrum of environmental changes in the forest floor as a function of the size of the gap created by a canopy-tree's death. She classified forest trees into categories according to their ability to use gaps for regeneration. In her categorisation, large-gap specialists typically had highly shade-tolerant seedlings that produced seeds that germinated only in very high-temperature and high-light conditions. She categorised small-gap specialists as those trees that had seeds that germinated in shade and required the presence of a gap overhead to grow. Understory specialists were also found in the tropical rain forest that did not seem to require gaps at all for either regeneration or growth. Denslow (1980) pointed out that there was actually a continuous gradation across these general categories and that, in essence, the categories were only to be thought of as examples of typical strategies.

Following the death of a large tree, there follows what Oldeman (1991) calls the 'innovation phase'. There typically is a great proliferation of the seedlings of woody plants and, in many cases, a diverse array of herbaceous species as well. The sources of heterogeneity in microsites in gaps can have a considerable effect on the germination and seedling success. The alterations in the environmental conditions also can alter the survival of individual trees already in the forest understory in and near the gap (as a result of side-lighting). These factors and the relative proportions of microsite conditions are influenced by a wide range of factors including the size of the tree that falls, whether it dies standing and rots slowly to the ground, breaks in the wind at the mid-trunk or is torn out roots and all by its fall. The weather conditions following the gap-creating event can influence the eventual suitability of the gap and its microsites for establishment of different species. The complexity of microconditions associated with the fall of a large tree working in concert with the regeneration requirements of the species (Grubb, 1977) can be a source of great potential diversification in tropical (Hartshorn 1978; Denslow 1980; Brokaw 1985a,b; Swain and Hall 1988) and temperate forests (Forcier 1975; Whittaker and Levin 1977; Shugart 1984; Koop and Hilgren 1987).

The diverse mixture of microsites, exogenous and internal variations in the environment and chance variation as to whether or not a given species has propagules in a given gap all combine to make a gap in a forest a rich substrate for the expression of differences in regeneration features of species (Grubb 1977). The actual responses are likely to be idiosyncratic to the particular forest systems, but the elevated germination, establishment and death of trees in the initial history of a particular canopy gap are a unifying feature for many forests (Whitmore 1982). The 'stages' in the gap-phase cycle of forests can be thought of as having different levels of predictability (Shugart *et al.* 1986) and the innovation phase is most unpredictable.

When the canopy closes (the beginning of the 'aggradation phase' of

Oldeman (1991)), the trees present in the gap begin to grow and compete with one another for light, and the overall biomass begins to accumulate (Bormann and Likens 1979a,b). The canopy may be quite variable in height at the time of canopy closure (Fig. 7.9) with a lower layer of slower growing and more shade-tolerant trees in some cases. As discussed earlier, the details depend on the forests and the particular species involved. The dominant process in the area of the gap tends to be forest growth and thinning as the dominant trees increase in height and leaf area. Given a knowledge of the biology of the species involved, most foresters and ecologists can predict the future of the forest composition in the gap with a fair degree of confidence for a considerable number of years.

Eventually, the forest structure builds to a point in time where the accumulation of plant biomass is in a state of relative balance with processes that reduce the amount of biomass in the area of the former gap. This 'biostatic phase' (Oldeman 1991), sometimes referred to as the mature phase (Baumphase), is the stage in which one can harvest saw-logs (Fig. 7.9). After a time the tree(s) dominating the forest in the location of the former canopy gap begins to age and becomes increasingly vulnerable to disease, parasites and other sources of mortality. In this, the 'degradation phase' (Oldeman 1991), the condition of the dominant trees predisposes them to death. The death of a large tree is a complex consequence of the state of the trees and the conditions to which they are exposed (Franklin *et al.* 1987). Aging trees have increased demands on the photosynthate that they produce to meet a variety of needs associated with their large size (maintaining root tissue, producing chemical compounds for defence against insect pests, etc.). Depending of the allocation patterns in a particular species, there can be a predictable pattern of mortality. For example, species that, when faced with a shortage of photosynthate with respect to demand, allocate less to herbivore defence compounds become more vulnerable to insect attack. Species that allocate less to roots become subject to root rots and topple in winds when their weakened root systems cannot support them. The complexity of such patterns in photosynthate allocation and in mortality is great, but the pattern of mortality can be quite predictable in many species of tree.

Mosaic-scale biomass dynamics in forests

If one considers the dynamics of living material (biomass) on a single-canopy-sized piece of a forest over multiple generations of a gap generation and filling cycle, the expected changes in biomass are for a

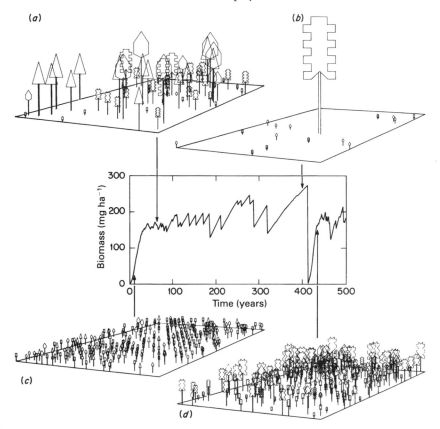

Figure 7.11. Biomass dynamics expected over several generations for a small plot of land with gap-phase replacement. The curve increases as the gap fills and eventually becomes dominated by a single large tree. The abrupt drop in living biomass is associated with the death of this canopy dominant. This curve is based on output from a computer model simulating the expected biomass dynamics of a small plot of forest land (see Chapters 8 and 9). From Shugart (1984).

quasi-cyclical or saw-toothed curve (Fig. 7.11). Bormann and Likens (1979a,b) considered the dynamic response of a watershed in New Hampshire following a clear-cutting to be a summation of several such saw-toothed curves initially synchronised by the clear-cutting application at time zero. The distances between the 'teeth' in the saw-toothed, small scale biomass curve are determined by how long a particular tree lives and how much time is required for a new tree to grow to dominate a canopy gap. Thus, the curves for several small plots of land eventually become desynchronised from chance differences in the timing of the death of a

(a)

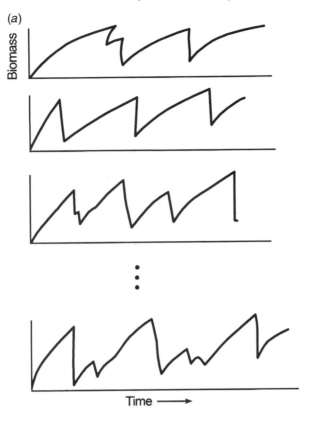

(b)

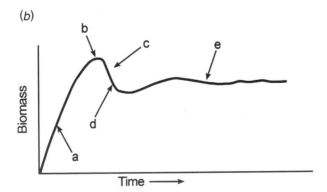

Figure 7.12. Biomass dynamics for an idealised landscape. The response is from a relatively large, homogeneous area composed of small patches with gap-phase biomass dynamics (Fig. 7.11). (*a*) The individual dynamics of the mosaic patches that are summed to produce the landscape biomass dynamics. (*b*) The landscape

particular tree. The summation of several of these curves for local biomass dynamics can be used to predict the biomass of the entire mosaic.

The larger-scale biomass dynamics (Fig. 7.12) is a simple statistical consequence of summing the dynamics of the parts of the mosaic. If there has been a synchronising event, such as a clear-cutting, one would expect the mosaic biomass curve to rise as all of the parts are simultaneously covered with growing trees (a in Fig. 7.12b). Eventually, some patches have trees of sufficient size to dominate the local area and there is a point in the forest development when the local drops in biomass are balanced by the continued growth of large trees at other locations and the mosaic biomass curve levels out (b in Fig. 7.12b). If the trees over the area have relatively similar longevities, there is also a subsequent period when several (perhaps the majority) of the pieces that constitute the forest mosaic all have deaths of the canopy-dominant trees (c in Fig. 7.12b). Over time, the local biomass dynamics become desynchronised and the biomass curve varies about some level (e in Fig. 7.12b). The variation in biomass is produced by averaging of the variations in the saw-toothed curves (e.g. Fig. 7.11) that are summed to produce the biomass response (see Bormann and Likens 1979a,b; Shugart and West 1981; Shugart 1984). The condition represented by e in Figure 7.12b can be taken as the mature forest.

What is the structure of a mature forest system? Whittaker (1953) reviewed the Watt (1947) pattern and process concept to redefine the climax concept of Clements (1916). Similar ideas were developed by Bormann and Likens (1979a,b) in their 'shifting-mosaic steady-state concept of the ecosystem' as well as Shugart (1984) with the 'quasi-equilibrium' landscape. In these views, the structure of a mature forest (at the scale of several hectares) is as a heterogeneous mixture of patches in different phases or stages of gap-phase replacement (Figs. 7.1c,d, 7.8 and 7.9). The mature forest should have patches with all stages of gap-phase dynamics and the proportions of each should reflect the proportional duration of the different gap-replacement stages.

biomass dynamics curve has four sections as indicated. a. Increasing landscape biomass curve rises as all of the patches are simultaneously covered with growing trees. b. Local drops in biomass are balanced by the continued growth of large trees at other locations. The landscape biomass curve levels out. c. If the trees have relatively similar longevities, there is a period when several (perhaps the majority) of the patches that make up the forest mosaic all have deaths of the canopy-dominant trees. d. The local biomass dynamics become desynchronised. e. Finally, the landscape biomass curve varies about some level.

The occurrence of such patterns has been documented for several different mature forest systems. The presence of shade-intolerant trees in mature undisturbed forest in patches is one observation consistent with the mosaic dynamics of mature forests (tropical rain forests: Aubréville 1938; Jones 1955–6; Whitmore 1974; Knight 1975; Hartshorn 1978; temperate forests: Jones 1945; Raup 1964; White 1979; Oliver 1981; Peterken and Jones 1987). (See Whitmore (1982) for review and discussion.) The scale of the mosaics in many natural forests is somewhat larger than one would expect from gap filling of single tree gaps (Rackham 1992), indicating the importance of phenomena that cause multiple tree replacements. Also, the relatively long records (*c.* 40 years in most cases) that are available for forests indicate a tendency for the forest composition to fluctuate, with species showing periods of relatively weak recruitment of individuals to replace large trees and strong recruitment in other periods (see Jones (1945) and Rackham (1992) for temperate forests, Swain and Hall (1988) for tropical forest examples).

Whether or not actual landscapes respond like the idealised patterns shown in Fig. 7.12 is problematical. The responses that are depicted are theoretical and are based on assumptions that the landscape dynamic can be obtained by averaging the individual dynamics expected for the pieces of the landscape mosaic. In Chapter 2, it was pointed out that at increasing spatial scale, one would expect different processes to become more important. This sort of spatial-scale effect is certainly in evidence in the different response expected for gaps of different size discussed above (also see Fig. 7.10). In a cleared landscape, the migration of seeds from mature trees may vary with the size of the clearing, or the assumption of a homogeneous condition may not be appropriate. The degree to which these sorts of consideration alter the biomass response of a landscape is not well documented.

In many respects, the mosaic dynamics of landscapes are a theoretical idealisation that is best used to interpret (rather than to predict) ecosystems pattern. Many of the details of the mosaic dynamics theory are developed and tested using computer models of forest dynamics. For this reason, the expected responses of mature forest landscapes and the biomass dynamics of forest systems will be discussed in more detail in Chapter 9.

Mosaic dynamics in ecosystems other than forests

The dynamic responses of non-forest systems may, at least in some cases, also be determined from fundamental responses of mosaic elements summed to produce a larger or scale landscape pattern. The mosaic dynamics that have been discussed for forests may be expected to occur in a wide range of systems dominated by sessile organisms in a manner analogous to trees in forests. Such systems include coral reefs (Connell 1978; Huston 1979; Pearson 1981; Colgan 1983) and other marine communities (Karlson 1978; Sousa 1979; Paine and Levin 1981; Taylor and Littler 1982; Dethier 1984; Sousa 1984). In these marine systems, a mosaic theory much like that for forests has been developed.

In terrestrial systems, Watt introduced his ideas about pattern and process with a variety of examples (Figs. 7.1 and 7.2). Christensen (1985) describes the mosaic dynamics for a wide range of heathlands, with patches burned by wildfires forming the mosaic elements. Remmert (1991) discusses the importance of mosaic dynamics in forests but also provides examples of dynamic terrestrial mosaics in Mopane (*Colophospermum mopane*) woodlands and grasslands in Africa, in thorn shrublands and grasslands in Argentina, in Mongolian grasslands, and in meadows in Iceland. In many of Remmert's examples, the agents generating the mosaic patterns are animals. Mueller-Dombois (1991) has documented several cases of synchronous die-backs followed by regeneration of seedlings producing a mosaic pattern in forests in the Pacific area.

Concluding comments

Spatial heterogeneity seems one of the essential aspects of terrestrial ecosystems. The observation of patchiness in terrestrial systems is indubitably from antiquity. Ecologists have made considerable recent progress in understanding the manner in which underlying processes generate such patchiness. In a wide range of terrestrial ecosystems, it is possible to order the patterns of patches of a landscape mosaic into ordered cycles that appear to reveal the operation of underlying processes. There are systems (tiger bush, fir waves) that owing to the particularities of the vegetation and the environment actually form organised arrangements of these patterns. While these seemingly self-organised systems are not particularly common, they do occur in a wide variety of locations and with a range of vegetation types.

The mosaic dynamics of terrestrial ecosystems are particularly well developed as a theoretical concept in forest ecology. Some of this development is the result of the progress made in practical forestry over the past two centuries. The size of mature trees and the damage done by their fall is also at a scale that is naturally observed by humans. In forests, the local influence of a large tree on its associated microenvironment is sufficient to produce a considerable impact on the environment when the tree dies. Tree birth, growth and death cycles in the gaps left in the canopy of a forest after a large tree falls are processes that can produce a mosaic character in a forest independent of external factors – at least in theory. In reality, the tendency for forests to generate a canopy-tree-scale mosaic interacts with external factors that advantage or disadvantage trees to degrees that vary with different stages of their life cycle and the particular species.

The mosaic dynamics of forests seem to be applicable to a range of other terrestrial and marine ecosystems, albeit at different spatial scales. The duration of the processes that generate and maintain dynamic mosaics is such that there are relatively few studies that actually follow the processes through a cycle. In the case of forests, this arises from the great longevity of trees, but even in systems with relatively more short-lived plants there are few detailed multiple-generation studies. The expected dynamics of different ecosystems behaving as dynamic mosaics in a changing environment will be treated in Chapter 9.

Part 3
Ecosystem models

8 · Individual-based models

Models are simplifications of more complex realities. It can be surprising the degree to which successful models of a wide range of ecological interactions assume unimportant aspects that one rationally might think otherwise. The central issue is often, 'At what scale is it appropriate to leave a given process out of an ecosystem model?'

A principal simplification for the mathematical models used in ecology traditionally has involved assumptions about homogeneity (Chapter 4). For example, models of population growth typically use the numbers of individuals in a population as a state variable and ignore the statistics of the ages and sexes in the population.* Intuition tells us that the ages and sexes in a population should have important consequences on birth and death rates. Models of element cycles ignore the spatial heterogeneity that seems a dominant feature of terrestrial ecosystems (Chapter 10). Food web models represent transfers of material as relative constant flows from one homogeneous compartment to another despite the non-homogeneous distributions of prey and predators.

Science progresses by simplifying the systems under consideration. In experimental science, one attempts to simplify by controlling, to as great a degree as possible, all extraneous factors with respect to a given experimental objective. Similarly, the formulation of a dynamic model is embedded with assumptions involving which factors can be left out of one's accounting of system dynamics. It may be true in a poetic sense that, 'one cannot touch a flower without trembling a star', but when computing the motions of stars and planets, the effects of touched flowers do not loom large: it is the disregarding of the effects of flowers on stars that

---

* A notable exception to this generality is the work of Von Foerster (1959) who developed a model of cellular population dynamics with age structure. During the 1960s and 1970s, this model was generalised by several scientists (e.g., Sinko and Streifer 1967; Gurtin and MacCamy 1979). The model is now being used for policy making with regard to human population control in China (Song and Yu 1981; Song *et al.* 1985; Song 1988).

Table 8.1. *Six criteria influencing the usefulness of analytical and simulation approaches to individual-based modelling. As each criterion becomes more important, one is forced toward a simulation approach*

Criteria	Analytical approach		Simulation approach
Spatial environment	Homogeneous	→	Heterogeneous
Demographic stochasticity	Unimportant	→	Important
Rare events	Unimportant	→	Important
Biological and/or environmental discontinuities	Unimportant	→	Important
Number of individuals	Large	→	Small
Biological complexity	Simple	→	Complex

Source: From Gross *et al.* (1992).

allows progress in astronomy. The consideration of the reciprocal (the effects of the motions of stars on flowers and other things) is the grist of astrology. Appropriate abstraction is critical to progress in science.

Some of the simplifications found in models arise from representations that are appropriate at particular spatial and temporal scale: discussed in Chapter 2 under the topic of scale. Appropriate scales of phenomenon arise frequently in the formulation of ecological models. At certain temporal scales, particular phenomena change so slowly that they can be considered constant. If one is averaging over a large spatial area, then phenomena that seem relatively discrete in their occurrence at smaller scales appear continuous. For example, a wildfire is a discrete event at the scale of a few hectares. But at the global scale, there is a continuous presence of wildfire that perhaps varies in magnitude with seasonal or climatic conditions.

Another consideration that often motivates model simplification involves whether given formulations can be solved and manipulated analytically. Models that can be analysed with mathematical formality are usually relatively simple in their structure. As models are elaborated to include spatial variation, the deaths and births of individuals, rare events and other complexities, the degree to which the formulations are subject to formal analysis is decreased (Table 8.1).

Development of individual-based models in ecology

Since the early 1970s, a broad range of ecologists (animal behaviourists, population biologists, community and ecosystems ecologists) have independently created a diverse array of individual-based models (Huston *et al.* 1988). This model development has arisen independently for different classes of organism. The catalyst for this development was undoubtedly the increased availability and power of digital computers – the necessary tools to solve most of these models. The earliest such models were developed by population ecologists (Holling 1961,1964; Rohlf and Davenport 1969) interested in incorporating animal behaviour into population models, and foresters (see discussion below) interested in relating tree growth to forest yield. For example, Holling (1964) analysed the shape of the foreleg of praying mantids (*Hierodula crassa*) to determine the size items that could be grasped and incorporated these into models of predation. This approach has been taken in studies of the consequences of behaviour and morphology for a range of fish, insects and birds (see Huston *et al.* (1988) for review).

Increasingly, models that simulate the dynamics of ecological systems by accounting for changes in each of a large number of individuals in the system have been developed and applied in population and ecosystems ecology (Huston *et al.* 1988; DeAngelis and Gross 1992). Huston *et al.* (1988) point out that one advantage of such models is that two implicit assumptions associated with the more traditional state variable approach used in ecological modelling populations are not necessary. These are the assumptions that (a) the unique features of individuals are sufficiently unimportant to the degree that individuals are assumed to be identical, and (b) the population is 'perfectly mixed' so that there are no local spatial interactions of any important magnitude. Most ecologists are interested in variation in individuals (a basis for the theory of evolution and a frequently measured aspect of plants and animals) and appreciate spatial variation as being quite important.

These assumptions seem particularly inappropriate for trees, which are sessile and which vary greatly in size over their life span. This may be one of the reasons that tree-based forest models are among the earliest and most widely elaborated of this genre of model. Because of their applications in issues concerned with terrestrial ecosystems change, individual-based forest models will be discussed in some detail below and in the chapters that follow. Before reviewing tree-based forest models, it is appropriate to discuss the application of individual-based models, in general.

Individual-based models of plant and animal populations

The dynamics of populations when represented by individual-based models can differ significantly from an equivalent state-variable representation of the same system (Huston *et al.* 1988; Lomnicki 1988; DeAngelis and Gross 1992). In an early example, Hilborn (1975) in a model-based study found that, for zooplankton feeding on phytoplankton, the rates of predation were such that the populations should go to a local extinction (owing to overgrazing). The survival of the two populations in a large body of water such as a lake was dependent on colonisation of open patches by phytoplankton diffusing in from other areas. These phytoplankton could then multiply in the local absence of predatory zooplanktors. Upon the diffusion into the local water volume of zooplankton, a local extinction would eventually result. The resultant phytoplankton bloom, followed by predator bloom, followed by extinction cycle is something of a fluid analogue of the mosaic dynamics discussed in Chapter 7.

An important implication of Hilborn's work was that if phytoplankton–zooplankton populations actually worked exactly as his model indicated and if one were able to perfectly measure the model parameterz (feeding rates, growth rates, etc.) in laboratory studies then the application of this 'perfect' model with an exact parametrization to the dynamics of a lake would lead to the incorrect conclusion that the populations should become extinct (see also Skellam 1973). In general, in models of individuals dispersed in space (Hilborn 1975; Caswell 1978; Cain 1985), the time to extinction of interacting populations is longer than the non-spatial equivalent models (Huston *et al.* 1988).

A significant feature in population models in which the variations in sizes of individuals are considered explicitly is the striking differences in population dynamics with respect to that expected from a population of average individuals. Often populations of individuals with different variances in population sizes (but the same average size) can be strikingly different (Huston *et al.* 1988). This occurs in populations structured by competitive interactions (Fig. 8.1), cannibalism (Fig. 8.2) or predator–prey interactions (Fig. 8.3).

Variation in spatial position also can have a pronounced effect on the dynamics of systems. This is clearly the case for plant competition for light (Diggle 1976; Aikman and Watkinson 1980; Pacala and Silander 1985; Huston and DeAngelis 1987), but the same effects can occur in competition for below-ground resources in plants (Weiner 1986; Huston and DeAngelis 1987).

The effects of spatial pattern have been well known from practical experience in agronomy and forestry for some time. In many agricultural (and forestry) experiments, the determination of spacing of plants has a profound effect on the thinning response and the size of the individuals in a crop.

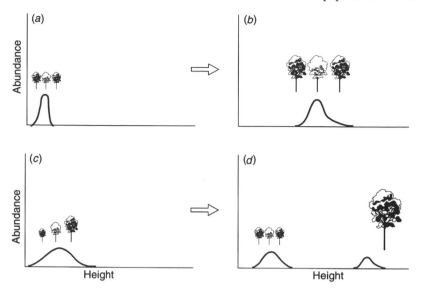

Figure 8.1. The effect of initial variation in the height distribution of plants on the consequences of competition for light. Initial height distribution with low variance (*a*) leads to a population of stunted individuals of similar size (*b*). Initial height distribution with the same average height as (*a*) but with a greater variance (*c*) leads to a bimodal population (*d*) with a few large individuals and several suppressed individuals. Note that individuals of average height are common in (*b*) but do not occur in (*d*). From Huston *et al.* (1988).

In behavioural studies, results analogous to those for plants have been obtained for the effects of the interactions of territory size (Tanemura and Hasegawa 1980; Huston *et al.* 1988; DeAngelis and Gross 1992). Individual-based models have been used to represent a range of behavioural topics as diverse as bird flocking behaviour (DeAngelis and Gross 1992), insect dispersal patterns (Myers 1976; Myers and Campbell 1976) and human resource-sharing patterns (Weinstein *et al.* 1983).

Individual-tree models of forests

Among the earliest individual-based models in ecology were models of forests based on the growth of the individual trees that make up the forest stand. These models were developed by quantitatively oriented foresters and were focused from their inception toward practical issues in production forestry. That trees were an initial focus for individual-based models springs naturally from the ecology and biology of trees.

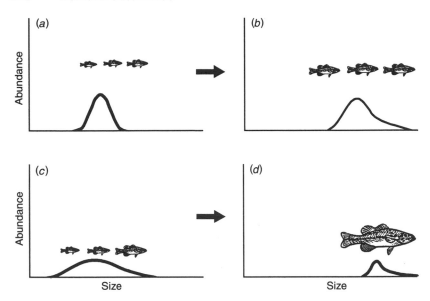

Figure 8.2. The effect of initial size variation on the outcome of cannibalistic interactions among young largemouth bass in aquaria. When the initial variance in the population is low (*a*), none of the young fish are able to eat the others and all the individuals survive growing at a relatively slow rate (*b*). When the initial variance in size is high (*c*), the largest individuals can eat the smaller and the population eventually contains fewer, but larger, individuals (*d*). From Huston *et al.* (1988) based on results from DeAngelis *et al.* (1979).

Trees are difficult organisms to study for several reasons. The typically high mortality rates of small trees imply a need to study many individuals to estimate death rates. Also, the scaling-up of what is known of the physiology of tree tissues to the level of the whole plant has proved to be a daunting problem. At present, it is difficult to derive mechanistic predictions for either annual or decadal tree growth expected under a particular environmental condition even for major commercial tree species growing in plantations (presumably the most straightforward case). In addition, trees have the potential to live for very long periods of time (multiple centuries to millennia). These features conspire to make trees (and forests) difficult experimental subjects and to elevate the importance of inferring the processes that produce the observed patterns in forests.

Harper (1977, p. 600) summarised the challenge of understanding forest ecology with regard to tree demography in two sentences bridging adjacent paragraphs,

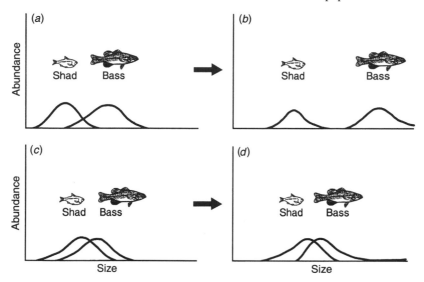

Figure 8.3. The effect of relative sizes of predators and prey on trophic interactions. When there is a large initial size difference between predators (largemouth bass) and prey (shad) sizes (*a*), the bass are able to feed on the shad and grow rapidly to a large size (*b*). When the predator–prey size difference is small (*c*), most of the bass are unable to consume shad (which are almost as large as the bass). The bass grow slowly and more shad survive (*d*). From Huston *et al.* (1988) based on results from Adams and DeAngelis (1987).

> In a sense a tree has to master all trades – to be successful in a variety of life stages and to meet the hazards of each layer of the vegetation that it penetrates.
>
> The study of trees is a study of short cuts; the long life and large size of trees makes many of the conventional methods of plant biology impossible or unrealistic.

The first statement implies that the understanding of tree demographics may ultimately require a synthesis at some level seen in much of terrestrial-population ecology. The second statement mentions the 'short cuts' that are more or less mandated by the biology and ecology of trees. One of the principal 'short cuts' in forest biology is to use the statistical distributions of tree sizes to attempt to infer what population or growth processes are controlling forest dynamics. Tree size is often quantified by tree height or diameter.

An emphasis on sizes of individual trees in the understanding of forest change has been traditional in forestry and forest ecology for over 200 years. This, associated with an interest in the changes in sizes of trees with changes in numbers (from birth and death processes), leads naturally to an interest in simulating the changes in individual trees as a basis for under-standing forest dynamics. Some of the first individual-based models used in ecology were for trees.

A related development involved the prediction of the changes in the average sizes of trees in a given location (Clutter 1963; Sullivan and Clutter 1972). The approach in this case is to attempt to determine the change in biomass (or the productivity) of a forest as a product of changes in numbers in response to thinning as the trees grow larger and the changes in average tree size as a function of growth and crowding. For reasons mentioned in the section above, the prediction of the dynamics of the average individual in a population of spatially interactive individuals is a non-trivial consideration. In particular, the dynamics of the average is related to a non-linear effect of the variance in the sizes of individuals in a population (Fig. 8.1). Statistical approaches to predicting forest yield from growth of an average tree (Clutter 1963; Sullivan and Clutter 1972), even in the case of approaches that attempt to account for changes in the vari-ance of sizes (e.g., Suzuki and Umemura 1967a,b, 1974), have been most successful for forests of one species being grown in plantations of similar age. Even in this case, the prediction of the average size of trees in a planta-tion over time is a non-trivial problem.

Early individual-based forest models: the contribution of forestry

Computer models that simulate the dynamics of a forest by following the fates of each individual tree in a forest stand were developed initially in the mid-1960s. The earliest such model was developed by Newnham (1964) and this was followed by similar developments in several Schools of Forestry (Lee 1967; Mitchell 1969; Lin 1970; Bella 1971; Arney 1974; Hatch 1971). The models predicted change using a digital computer to dynamically change a map of the sizes and positions of each tree in a forest (Table 8.2).

These initial models were as complex and as detailed as their descen-dants. The models have become increasingly used as the computer power available to ecologists has increased (Munro 1974; Shugart and West 1980). The essential approach was to represent the three-dimensional spatial interactions among individual trees. This dynamic mapping technique was applied for even-aged plantations but applications in more complex forests soon followed. After these first modelling efforts, individual-based tree models tended to feature simplifications of these initial models.

Individual-based models in forestry were focused on solving several

Table 8.2. *Forest dynamics models. Different types of forest model emphasise different features of the individual tree biology. Following Munro (1974) and Shugart and West (1980), the models are categorised as to whether they explicitly consider the forest age or size structure (age structure), whether they consider more than one species of tree (diversity), and whether they consider the geometry of interactions among individual trees (geometry). Examples are chosen from earlier publications*

Age structure	Diversity	Geometry	Description	Examples
Even-aged	Mono-species	Spatial (2- and 3-dimensional canopy geometry)	Models of plantations (often of genus *Pinus*) used to determine spacing, growth and yield of forests	Hegyi (1974)
Even-aged	Mono-species	Non-spatial	Models of plantations. Typical models relate a statistically based difference (or differential) equation for the growth of an average tree to statistically based estimators of numbers of trees. The numbers and size models are interdependent	Sullivan and Clutter (1972)
Even-aged	Mixed-species	Non-spatial	Estimates of yield tables for mixed aged stands based on statistical relations between expected biomass (or size of trees) and stand density. Strong emphasis on stand thinning or other mortality sources	Solomon (1974)
Mixed-aged (mixed-size)	Mono-species	Spatial (2- and 3-dimensional canopy geometry)	Strong emphasis on the canopy geometry of competition and on the growth rates of individual trees. Used on a range of commercial coniferous species of trees that tend to be clear-cut and naturally regenerate	Mitchell (1975)

Table 8.2 (*cont.*)

Age structure	Diversity	Geometry	Description	Examples
Mixed-aged (mixed-sized)	Mono-species	Non-spatial	Simulate the change in the mean and variance in the diameter or height distributions over time. Often partial differential equations or size-structured differential equations	Suzuki and Umemura (1974)
Mixed-aged (mixed-sized)	Mixed-species	Spatial (initially 3-dimensional canopy geometry)	Highly detailed models of 'natural' forests with 3-dimensional canopy geometry, seed rain computed for each tree, etc.	Ek and Monserud (1974a, b)
Mixed-aged	Mixed-species	Non-spatial	Models of tree replacement. Often formulated as Markov models. Often used in landscape-scale applications	Horn (1975a, b)

Source: Modified from Shugart (1984).

general problems involving stand yield tables. Modern forestry uses stand yield tables as a scientific basis for making decisions about the expected productivity of forest stands under different conditions. (Yield tables use the site index (the expected height growth of trees at a given age) to calibrate the expected changes in tree sizes and densities based on study plots remeasured over decades or centuries.) Such information is valuable for planning harvest and thinning schedules.

Stand yield tables are calibrated by using data on the performance (height and diameter growth) of individual trees and forest change for trees on study plots with different densities of trees per unit area and under different conditions. Some of the European yield tables are based on continuous studies of such plots for as long as 200 years. The large detailed data sets required to develop yield tables represent a weakness in the approach. If one is interested in the growth and yield of plantations of an exotic, such as radiata pine (*Pinus radiata*), a species that is being planted in dozens of tropical and subtropical countries as a source of softwood, then how does one develop a yield table in the absence of a long data calibration set? Similar predicaments arise in evaluating altered forestry practices such as forest fertilisation, in projecting responses of forests affected by air pollution and in predicting yield in forests with genetically improved species. Indeed, any condition that alters the performance of individual trees from the conditions for which a yield table is calibrated is a potential source of difficulty.

The early individual-tree-based simulators were focused at taking what was known from yield tables and other data sets and developing a more flexible, quantitative methodology for prediction. Some of the earliest attempts to apply such models were very successful and produced results of surprising detail. For example, Hegyi's (1974) model of growth of Jack pine (*Pinus banksiana*) based on Arney's (1974) model of Douglas fir (*Pseudotsuga menziesii*) is capable of predicting the size of individual trees on sites of different quality. The diameters of the trees are predicted at height intervals along the trunk (Fig. 8.4) and compare favourably with direct measurements. Models developed by Mitchell (1969, 1975) and Arney (1974) produce similar detail of the diameter and crown size at regular intervals for individual trees.

Gap models

An important subcategory of individual-organism-based tree models that has been widely used in ecological (as opposed to traditional forestry

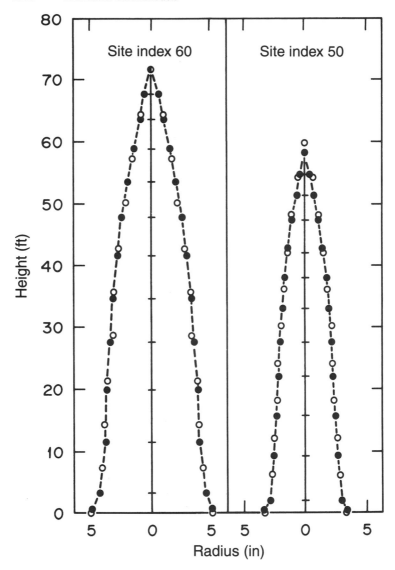

Figure 8.4. Stem profiles (1 in = 2.5 cm) of 60-year-old Jack pine trees from sites with site indices of 15 m and 12 m (60 ft and 50 ft) compared with simulation results from Hegyi's (1974) model. Actual sample tree (○), simulated tree (●).

applications) is the so-called 'gap' model (Shugart and West 1980). The first such model was the JABOWA model (Botkin *et al.* 1972; Botkin, 1993) developed for forest in New England. Several gap models for a wide range of forests (and, eventually, grasslands and savannas) were subsequently developed (Shugart et al. 1992b; Urban and Shugart 1992) for a wide range of locations (Fig. 8.5). The earlier versions were developed with an interest in using models for inter-ecosystems comparison (Shugart 1984) and adhered relatively closely to the JABOWA formulation. More recent models have diverged particularly in the addition of spatial competition among trees, a range of modifications to allow simulation of a wider range of environmental conditions and additional biological and/or physiological mechanisms.

General structure of gap models

As is the case with many of the earlier individual-based models used in forestry, gap models simulate the establishment, diameter growth and mortality of each tree in a given area (Fig. 8.6). Calculations are on a weekly to annual time step. At least initially, gap models were developed for plots of a fixed size. Many of the models focus on a size unit (c. 0.1 ha) approximately that of a forest canopy gap (Shugart and West 1980).

Gap models feature relatively simple protocols for estimating the model parameters (Botkin et al. 1972; Shugart 1984). For many of the more common temperate and boreal forest trees, there is a considerable body of information on the performance of individual trees (growth rates, establishment requirements, height–diameter relations) that can be used directly in estimating the parameters of such models. The models have simple rules for interactions among individuals (e.g. shading, competition for limiting resources, etc.) and equally simple rules for birth, death and growth of individuals. The simplicity of the functional relations in the models has positive and negative consequences. The positive aspects are largely involved in the ease of estimating model parameters for a large number of species, the negative aspects with a desire for more physiologically or empirically 'correct' functions.

Many of the more recent gap models have functional relationships that are different from those used in the earlier gap models (e.g. JABOWA (Botkin et al. 1972) and FORET (Shugart and West 1977)). Gap models differ in their inclusion of processes that may be important in the dynamics of particular sites being simulated (e.g. hurricane disturbance, flooding, formation of permafrost, etc.) but share a common set of characteristics.

Each individual plant is simulated as an independent entity with respect to the processes of establishment, growth and mortality. This feature is common to most individual-tree-based forest models and

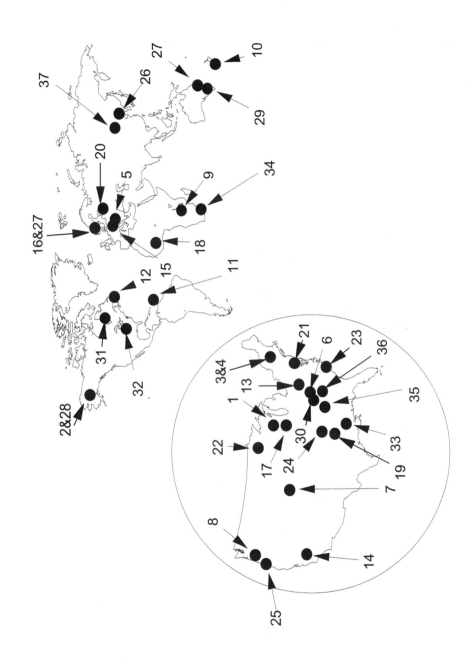

Figure 8.5. Individual–based gap models: numbers correspond to the numbers on the map

		Citation	Ecological system simulated
1	FORTNITE	Aber and Melillo (1982a)	Wisconsin mixed wood forest
2	LOKI	Bonan (1989a)	North American spruce–fir forest
3	JABOWA	Botkin et al. (1972)	Northern hardwood forest
4	JABOWA II	Botkin (1993)	Northern hardwood forest
5	FORCLIM	Bugmann (1994)	Swiss montane forest
6	FORANAK	Busing and Clebsch (1987)	Appalachian spruce/fir forest
7	STEPPE	Coffin and Lauenroth (1989a,b)	North American short-grass prairie
8	CLIMACS	Dale and Hemstrom (1984)	Pacific northwest coniferous forest
9	MIOMBO	Desanker and Prentice (1994)	Miombo woodland
10	FORENZ	Develice (1988)	New Zealand forest
11	FORICO	Doyle (1981)	Puerto Rican montane rain forest
12	SMAFS	El Bayoumi et al. (1984)	Eastern Canadian mixed wood forest
13	OVALIS	Harrison and Shugart (1990)	Appalachian oak–hickory forest
14	SILVA	Kercher and Axelrod (1984)	Mixed conifer forest
15	FORECE	Kienast and Kuhn (1989)	Central European forest
16	FORSKA	Leemans and Prentice (1987)	Scandinavian forest
17	EXE	Martin (1992)	Temperate/boreal forest transition
18		Menaut et al. (1990)	West African humid savanna
19	FORAR	Mielke et al. (1978)	Arkansas pine–oak forest
20	SJABO	Oja (1983)	Estonian conifer forest
21	SORTIE	Pacala et al. (1993)	Northern hardwood forest
22	LINKAGES	Pastor and Post (1986)	Temperate/boreal forest transition
23	FORFLO	Pearlstine et al. (1985)	Southern USA floodplain forest
24	SWAMP	Phipps (1979)	Arkansas floodplain forest

Figure 8.5 (cont.)

		Citation	Ecological system simulated
25	SUCSIM	Reed (1980)	Douglas fir forests
26	KOPIDE	Shao (1991), Shao et al. (1991)	Chinese mixed pine–deciduous forest
27	KIAMBRAM	Shugart et al. (1980a)	Australian subtropical rain forest
28	BOFORS	Shugart et al. (1992a)	Boreal forest of Eurasia and North America
29	BRIND	Shugart and Noble (1981)	Australian *Eucalyptus* forest
30	FORET	Shugart and West (1977)	Southern Appalachian deciduous forest
31		Sirois et al. (1994)	Boreal forest/tundra transition
32	FORENA	Solomon (1986b)	Forests of eastern North America
33	FORMIS	Tharp (1978)	Mississippi River floodplain forest
34	OUTENIQUA	Van Daalen and Shugart (1989)	South African temperate rain forest
35	FORCAT	Waldrop et al. (1986)	Southern oak–hickory forest
36	FORNUT	Weinstein et al. (1982)	Southern Appalachian deciduous forest
37		Yan and Zhao (1995)	Chinese mixed pine–deciduous forest

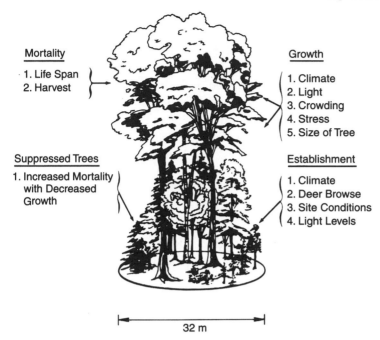

Mortality

1. Life Span
2. Harvest

Suppressed Trees

1. Increased Mortality
   with Decreased
   Growth

Growth

1. Climate
2. Light
3. Crowding
4. Stress
5. Size of Tree

Establishment

1. Climate
2. Deer Browse
3. Site Conditions
4. Light Levels

32 m

Figure 8.6. Schematic diagram of the function of a typical gap model.

provides sufficient information to allow computation of species- and size-specific demographic effects.

The model structure emphasises two features important to a dynamic description of vegetation pattern: (a) the response of the individual plant to the prevailing environmental conditions, and (b) how the individual modifies those environmental conditions. The model is hierarchical in that the higher-level patterns observed (i.e. population, community and ecosystem) are the integration of plant responses to the environmental constraints defined at the level of the individual.

Growth

In gap models, the growth of an individual is calculated using a function that is species specific and predicts, under optimal conditions, an expected diameter increment given a tree's current diameter. This optimum increment is then modified by the environmental response functions and the realised increment is added to the tree.

The central assumption in formulating the diameter increment equation is that growth in trees is the consequence of two opposite processes.

The positive part of the rate of volume accumulation is assumed under optimal conditions to increase as a linear function of the leaf area of the tree. The magnitude of this positive growth rate depends on the net photosynthetic rate of the tree per unit of leaves. Contrary to this positive rate is a negative rate that is associated with the energetic cost of maintaining a given volume of living tissue. Presumably, as a tree increases in size, these negative costs increase until they begin to approach the positive growth in magnitude, and the rate of growth of the tree slows and stops.

Botkin et al. (1972) included these effects in the initial JABOWA model using the following derivation. The volume of a tree is roughly that of a cone with the appropriate diameter and height:

$$V \propto D^2 H \tag{8.1}$$

where: V is a tree's volume, D is a tree's diameter, and H is a tree's height.

A function that approximates the relation between diameter and height (Ker and Smith 1955) is:

$$H = 137 + b_2 D - b_3 D^2 \tag{8.2}$$

The 137 in the equation refers to the fact that tree diameters are usually measured at 'breast height' or 137 cm. (This is abbreviated to DBH.) Equation 8.2 is for a parabola. If the maximum height of a tree is obtained when it reaches its maximum diameter ($dH/dD = 0$, when $H = H_{max}$, and $D = D_{max}$), then

$$b_2 = 2 \left(\frac{H_{max} - 137}{D_{max}} \right) \tag{8.3}$$

and

$$b_3 = \left(\frac{H_{max} - 137}{D^2_{max}} \right) \tag{8.4}$$

Assume that the growth of volume of a tree has a response much like that in the logistic equation (Eq. 4.9, p. 72), so that the change in volume increases with the photosynthetic production of the leaves, but that this rate of increase is slowed as the volume increases (reflecting the costs of respiring tissue and other losses in the plant). This can be expressed as

$$\frac{d[D^2 H]}{dt} = rL_a \left(1 - \frac{DH}{D_{max} H_{max}} \right) \tag{8.5}$$

where L_a is the leaf area and r is a rate parameter.

If one assumes (following Whittaker and Marks 1975) that:

$$L_a \approx cD^2 \tag{8.6}$$

then one can substitute and perform the indicated operations on Eq. 8.5 to find:

$$\frac{d[D^2H]}{dt} = rL_a\left(1 - \frac{DH}{D_{max}H_{max}}\right)$$

$$\frac{d[D^2(137 + b_2D - b_3D^2)]}{dt} = (rcD^2)\left(1 - \frac{DH}{D_{max}H_{max}}\right)$$

$$\frac{d[137D^2 + b_2D^3 - b_3D^4]}{dt} = (rcD^2)\left(1 - \frac{DH}{D_{max}H_{max}}\right)$$

$$\frac{dD}{dt}(274D + 3b_2D^2 - 4b_3D^3) = rcD^2\left(1 - \frac{DH}{D_{max}H_{max}}\right)$$

$$\frac{dD}{dt} = \frac{GD\left(1 - \dfrac{DH}{D_{max}H_{max}}\right)}{(274 + 3b_2D - 4b_3D^2)} \tag{8.7}$$

where G equals cr from earlier equations.

Clearly, this growth equation is designed to give an approximation of the growth rate and pattern of a species from a minimal amount of information. Several gap models are based on alternative growth equations (as well as other equations in the models). Some of these alternatives continue the appealing feature of ease of parameter estimation but have improved functions in terms of better representing tree growth.

The simplest variation in growth rate equations for gap models is the modification of Phipps (1977), who used measured tree-ring increments to assign growth rate (at three different levels) equations for the various species in the SWAMP model of successional dynamics in floodplain forests of the White River in Arkansas. Reed (1980) in his SUCSIM model used an empirically derived allometric function for leaf area in the place of Eq. 8.6 and based the photosynthesis rate in Eq. 8.5 on leaf biomass of the tree raised to a power.

Moore (1989) developed an alternative formulation, currently used in the FORCLIM model (Bugmann 1994). Moore noted that the initial formulation used in the JABOWA model (Eq. 8.7) implied that the losses associated with large size in trees increased at a rate greater than the increase in the tree's volume. He developed an alternative formulation based on a simplification of a carbon metabolism-based representation of the constructive versus destructive metabolism (*sensu* Korzukhin and Antonovski 1992) of the individual tree. In this formulation (Moore 1989; Bugmann 1994), one considers the change in the volume of the tree:

$$\frac{\Delta V}{\Delta t} = rL_a - mV \tag{8.8}$$

where r is a photosynthesis rate and m is a respiration rate.

Using the relationships already outlined (Eqs. 8.1, 8.2 and 8.6), Moore (1989) derived the diameter increment equation as:

$$\frac{\Delta D}{\Delta t} = GD \left(\frac{1 - \dfrac{H}{H_{max}}}{274 + 3b_2 - 4b_3} \right) \tag{8.9}$$

This derivation is probably a better representation of the initial growth equation used in most gap models in terms of the assumptions stated in the initial formulation (Botkin *et al.* 1972).

Another important modification of the original JABOWA formulation is that of Leemans and Prentice (1987), who used a different equation for the height-to-diameter relation (Eq. 8.2) of the form:

$$H = b_1 + (H_{max} - b_1) \left(1 - e^{\frac{-sD}{H_{max} - b_1}} \right) \tag{8.10}$$

where b_1 is the height at which the tree is measured (137 cm) and s denotes the initial increase of height with diameter (Huang *et al.* 1992).

Leemans and Prentice (1987) derived an equation for tree volume increment that included the effects of vertical canopy geometry:

$$\frac{\Delta(D^2 H)}{\Delta t} = \int_B^H S_L(\gamma P_z - \delta z) dz \tag{8.11}$$

where S_L is the vertical density of leaf area; γ is a species-specific growth scaling constant; P_z is the proportion of the maximum possible annual net assimilation achieved by the leaves at depth z in the canopy; δ is a species-specific cost factor; H is the tree height; and B is the height of the clear tree bole ($H - B$ is the depth of the tree crown).

This formulation of the growth equation has the advantage of incorporating much of the canopy geometry directly into the formulation. The rules for determining the geometrical features of the tree (e.g. the 'B'

parameter) incorporate the conversion of sapwood into heartwood using the pipe model of tree performance of Shinozaki *et al.* (1964).

Pacala *et al.* (1993) argue that gap models have evolved to a point at which the parameters in the models need to be tied to direct empirical observations. Their model developed toward this goal (SORTIE) uses several regression relations to improve the response of an individual-based model intended as an extension of the existing gap models. Growth is driven by an index of local light availability (G_{LI} of Canham (1988)) that intergrates the annual light condition (a complex function of changing sun angles, ratios of diffuse and direct radiation, and the canopy pattern). The Pacala *et al.* (1993) growth equation is:

$$\Delta r = r \left[\frac{P_1 G_{LI}}{\left(\frac{P_1}{P_2} + G_{LI} \right)} \right] + \alpha \tag{8.12}$$

where r is the radius of a tree; Δr is the annual increment (tree ring) for a tree; P_1 and P_2 are fitted regression parameters; G_{LI} is the 'general light index' (Canham 1988); α is a normally distributed variable with a mean of zero and a variance of C (predicted $\Delta r)^D$ where C and D are estimated constants.

Similarly the height of a tree is estimated by a regression equation

$$H = A \left(1 - e^{-\frac{S}{A} D} \right) + \alpha \tag{8.13}$$

where A is the asymptotic height (comparable to H_{max}); S is the slope of the height increase at zero diameter; and α is a normally distributed variable as in Eq. 8.12.

As Bugmann *et al.* (1995) point out, the differences in growth rate formulations in gap models represent different views as to how to represent the growth process in trees. There has been relatively little intercomparison of model performance using different growth equations but since the parameter demands of the different formulations are similar this seems a logical exercise to be undertaken.

Spatial scaling

The horizontal position of each individual is influenced by (and influences) the growth of all other individuals on the plot. The earlier models were constructed under an assumption of horizontal homogeneity (particularly of the light environment at a given level in the forest canopy) inside the simulated plot. The size of the simulated plot is critical in these cases (Shugart and West 1979). The spatial scale at which the earlier models operate is for an area corresponding to the zone of influence of a single individual of maximum size. This allows for an individual growing on the plot to achieve maximum size while at the same time allowing for the death of a large individual to influence significantly the light environment on the plot (Shugart and West 1979).

Environmental constraints and resource competition among trees

All gap models simulate individual tree response to light availability at height intervals on the plot. Even in the early gap models, the vertical structure of the canopy was modelled explicitly. The sizes of individuals (height and leaf area predicted by regressions on each tree's diameter) are used to construct a vertical leaf-area profile. Initially, the leaf area above a given tree was computed as the sum of leaf areas of all trees taller than the focal tree (Botkin *et al.* 1972). Using a light extinction equation (the Beer–Lambert equation), extinction of light through the forest canopy can be computed as:

$$I_L = I_0 e^{-k L_L} \tag{8.14}$$

where I_L is the light at some level; L; I_0 is the light at the top of the canopy; L_L is the leaf area above level, L; and k is a constant.

Equation 8.14 can be used to determine the light available to each tree. Several authors recognised that the assumption that each tree was shaded by the taller trees had the net effect of assuming that a given tree had all its leaf area-related shading as if the leaves were distributed homogeneously at the top of the tree. Leemans and Prentice (1987) went to considerable effort to develop a growth equation that distributed the leaf-area effects of each tree through its canopy depth (see Eq. 8.11) and Pacala *et al.* (1993) incorporate canopy effects directly in their growth equation (Eq. 8.12). Smith and Urban (1988) developed a spatially explicit model (ZELIG) that could be derived from any gap model parametrisation that computes the three-dimensional canopy interactions among trees and, in this sense, is convergent on the earlier forestry models (see Chapter 13).

Other resources are incorporated to varying degrees in different versions of the model; these other constraints include soil moisture, fertility (often available nitrogen, specifically), temperature, as well as disturbances such as fires, hurricanes, floods and windthrow. In most of the models, the environmental responses are modelled via a 'constrained potential' paradigm. In this, a tree has a maximum potential behaviour under optimal conditions (i.e. maximum diameter increment, survivorship or establishment rate). This optimum is then reduced according to the environmental context of the plot (e.g. shading, drought) to yield the realised behaviour under ambient conditions. The curves describing the response of species to environmental resources tend to be generic curves that scale between 0.0 and 1.0, and species are often categorised into a small number of functional types.

The limitation effects of resource shortages in the models were initially computed in a multiplicative fashion (Botkin *et al.* 1972; Shugart 1984; Botkin 1993). This would mean that the growth of a tree when light was 90% of the optimal and water availability was 80% of optimal would be 72% ($0.90 \times 0.80 = 0.72$). Later models may use other approaches, notably assigning the growth reduction to that of the most limiting factor (Pastor and Post 1986).

Competition in the model depends on the relative performance of different trees under the environmental conditions on the model plot. These environmental conditions may be influenced by the trees themselves (e.g. a tree's leaf area influences light available beneath it), or may be modelled as extrinsic and not influenced by the trees (e.g. temperature).

Competition may operate in two modes. Competition for light is asymmetric and exploitative: a tree at a given height absorbs light and reduces the resource available to trees at lower positions in the canopy. In most gap models, competition for below-ground resources (water and nutrients) is symmetric. Each tree experiences a resource level common to the plot. This approach to modelling the competition process has the result that competitive ability depends strictly on the context of the modelled gap.

The tree that has the best performance, relative only to other trees on the plot, is the most successful. Competitive success depends on the environmental conditions on the plot, which species are present and the relative sizes of the trees. Each of these varies through time in the model.

Pastor and Post (1986) used an individual-tree simulation model to examine the effects of climate, soil moisture and nitrogen availability on sites with different soils. They tallied the factor (temperature, light, soil moisture or nitrogen availability) that was most limiting to growth on a tree-by-tree basis over successional time. They found that the most limiting factor varied as a function of forest successional stage and as a function of the position of a tree in the canopy. During the first 50 years of succession when tree heights were below 10 m, nitrogen was seen to be the most limiting factor. For trees of 10 m or less height in the stands in years 100 to 150 of the succession, light was the most growth-limiting factor, but for taller trees (10 to 20 m in height), nitrogen was the most growth limiting. By years 200 to 250 of the simulation, the demand for nitrogen by the tallest trees (20 to 30 m tall) in conjunction with the amount of nitrogen sequestered in living biomass made nitrogen the most limiting factor for most trees, regardless of layer. Also, in years 200 through 250, soil moisture was the most important factor for many trees and light was important as a factor only in the smallest trees.

The overall pattern of most-limiting factors (Fig. 8.7) was one of a

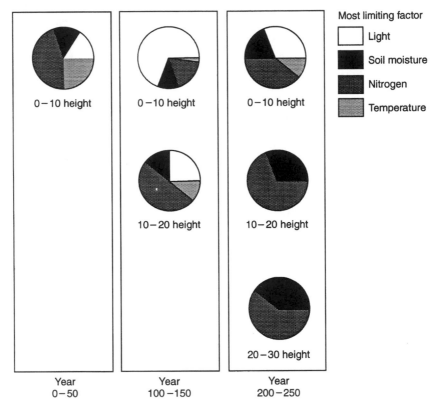

Figure 8.7. The factors most limiting to the growth of a tree as a function of height (m) and time in a simulated succession (starting with bare soil in year zero) produced with an individual-based model of northern hardwood forest in the USA (based on Pastor and Post (1986)). The sizes of the sections of the circles reflect the numbers of trees in the sample that are most limited by the factor indicated. From Shugart and Urban (1989).

change in the most limiting factor as a complex function of time, species–environment interactions, size and age of trees. A dominant tree over the time periods shown would probably be nitrogen-limited through its entire life; a subordinate tree (that might replace the dominant in the next generation) would have run a gauntlet of being limited by a variety of factors. Trees of the same species but of different heights might be limited by a multiplicity of factors in a pattern that changed over successional time.

Mortality

The death of individuals is modelled as a stochastic process. Most gap models have two components of mortality: age related and stress induced. The age-related component applies equally to all individuals of a species and depends on the maximum expected longevity for the species; this age is typically of the order of 300 years, yielding an annual survivorship rate of around 1–2%. The probability of mortality of each individual tree is determined as a stochastic function. Each tree is assumed to have an intrinsic mortality rate such that, under normal conditions, 1% of the individuals in a cohort could be expected to live long enough (a parameter called AGEMX) to attain their maximum height and diameter. Thus, the probability of mortality is:

$$P_m = 1 - e^{\frac{-4.605}{\text{AGEMX}}} \tag{8.15}$$

where P_m is the intrinsic probability of mortality.

AGEMX is the age at which one would expect 1% of a cohort to be alive and, by assumption, the age at which one might expect an individual to reach its maximum size.

Trees that are growing slowly have an increased probability of death. In the earlier gap models, trees with a minimum diameter increment of only 1.00 mm per year are subjected to additional mortality, $P_s = 0.368$, where P_s is the probability of survival of a tree suppressed (owing to shade or other factors). This has the effect of allowing only 1% of a suppressed cohort to survive 10 years. In more recent gap models, stress is defined with respect to a minimal diameter increment (typically 10% of optimum growth for a given size of tree). Individuals failing to meet this minimal condition are subjected to an elevated mortality rate. Pacala *et al.*(1993) elaborate the controls on tree mortality as a growth-related function with

$$M_{(g)} = e - U(\bar{r}_s)^V \tag{8.16}$$

where $M_{(g)}$ is the mortality rate for a given growth rate; U and V are constants; and $\bar{r}_5$ is the arithmetic mean of the last 5 years of growth.

The mortality of plants, in general, and certainly of trees, is a relatively poorly known aspect in plant ecology and the formulations in many of the gap models reflect the state of knowledge. The tendency in most of the existing gap models has been to use relatively straightforward descriptions of mortality in the anticipation of additional information.

Establishment

Several authors (van der Pijl 1972; Whitmore 1975; Grubb 1977; Bazzaz and Pickett 1980) have discussed species attributes that are important in differentiating the regeneration success of various trees. The complexity of the regeneration process in trees and its stochastic nature makes it nearly impossible to predict the success of an individual tree seedling. This problem is compounded when spatial effects are also considered (discussed later in Chapter 12). Malanson and Armstrong (1996) and Malanson (1996) have applied a spatially explicit gap model to investigate the effects of dispersal probability on forest dynamics. They found that relatively rare instances of a given species of tree establishing from a distance away could have profound influences on the forest dynamics and diversity. Unfortunately, such events are also very difficult to observe, much less quantify, under field conditions. Most gap models are designed to treat regeneration in trees from a pragmatic view that the factors influencing the establishment of seedlings can be usefully grouped in broad classes (Kozlowski 1971a,b; van der Pijl 1972; Grubb 1977; Denslow 1980). Tree establishment and regeneration are largely stochastic, with maximum potential establishment rates constrained by the same environmental factors that modify tree growth. Each simulation year, a pool of potential recruits is filtered through the environmental context of the plot, and a few new individuals are established.

Tests of gap models

Mankin *et al.* (1977) and Shugart (1984) divide model testing into two basic types of procedure (verification and validation) and see model application as a measure of a model's usefulness. In verifying a model, the model is tested on whether it can be made consistent with some set of observations. In validation procedures, a model is tested on its agreement with a set of observations independent of those observations used to structure the model and estimate its parameters.

When testing a model, it is important that it can simulate the pattern of the system under the constraint that all parameters in the model are realistic (Shugart 1984). Both the structure and the emphasis of gap models make it appropriate to have a high level of realism in the model parameters. In the construction of most gap models, the initial verifications involve determining the model's ability to reproduce the general features of forest pattern while constrained to using model parameters that are reasonable.

An example of model verification

It should be clear from the definitions above that model verification and model validations can be the same sorts of test. As an example of a model verification, Fig. 8.8 shows the growth in average diameter of pine trees growing in pine plantations over a 50-year period of time simulated by

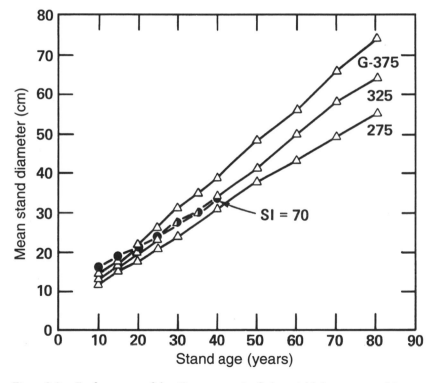

Figure 8.8. Performance of the G parameter in fitting yield data measured by Smalley and Bailey (1974) (●) for Loblolly pine (*Pinus taeda*) plantations compared with that predicted by the FORAR model (△). From Shugart (1984).

the FORAR model developed for forests in southern Arkansas compared with actual data recorded by Smalley and Bailey (1974) for pine plantations. The G parameter in the model that quantifies the growth rate of individual trees can take on values in this particular simulation of between 375 and 275. The question being asked is, 'Can the model simulate what is known about growth in pine plantations using reasonable parameters?'. As can be seen in Fig. 8.8, the model does an adequate job of simulating 40 years of growth in pine plantations of site indices of approximately 70. The site index of 70 indicates that the height of pine trees (in feet in this case) should be 70 feet at the age of approximately 40 years. The model appears to be able to duplicate the pattern observed in real forest plantations with a fair degree of accuracy using reasonable parameters. In this particular example, the G parameter could be varied to match the data, although it could not be varied beyond a certain range. Successful verifications, even though they do not use data that are strictly independent of the model, still represent reasonable tests of the model. Indeed, most models were tested in the verification mode because validation data is so difficult to obtain.

An example of model validation

The degree of independence between data and model output sometimes must be traded off against the need to modify a model to better match a given data set. For example, Fig. 8.9 shows a model validation that is comparable in its character to the verification shown in Fig. 8.8 and just discussed. Figure 8.9 shows change in several important stand structural variables over 60 years: the data measured in the Australian Alps and the model output. The BRIND model (Shugart and Noble 1981) was developed as a collaborative project in 1978. After the model had been verified against several data sets and other accounts of the forest pattern in the *Eucalyptus* forest in what is often called the Australian Alps and a manuscript to the *Australian Journal of Ecology* on the model had been drafted, a new set of data, previously unavailable, was published.

These new data had been collected by E. Lindsay near Bago, New South Wales in the late 1930s and was a summary of thousands of man-years of effort collecting data on growth and dynamics of various sorts of montane *Eucalyptus* forest in New South Wales. The appearance of new and previously unavailable data, even though they had been collected almost 40 years before, created the possibility of validating the model against independent stand data. Unfortunately, Lindsay had collected his

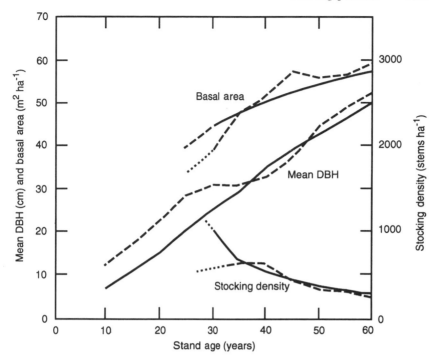

Figure 8.9. A comparison of basal areas, stocking density and mean diameters (DBH) from a yield table (Lindsay's (1939) unpublished data summarised by Borough *et al.* (1978)) with independent output from the BRIND model. Stocking densities are not shown for purposes of clarity at year 10 but are 5000 for Lindsay's data and about 2600 for the example simulation. Solid lines are Lindsay's data; dashed lines are BRIND model output. From Shugart and Noble (1981).

data in forest stands that had what is called adequate to high stocking density. Indeed, he had purposefully collected his data in areas where there were a lot of young trees. Therefore, Lindsay's data started at a different place from any of the model simulations, which were based on starting with forests with moderate stocking densities. Faced with a quandary – should the model be recalibrated to a more appropriate starting point and better match Lindsay's data, or should the two data sets be compared as they were as a model validation (Shugart and Noble chose to do the latter). When one compares forests simulated by the model with what Lindsay actually measured, when the forests are young, Lindsay's stocking densities (the number of trees per unit area) are considerably higher than those simulated by the model. By the time the stands are approximately 35 years of age, the stocking densities on

Lindsay's stands converge to the values simulated by the model (Fig. 8.9), and the stocking densities are similar for the remaining 30 years. In the younger stands, in which the actual data were from overstocked stands, the average diameters of these trees were smaller. This is what one would expect in comparing stands of different densities but similar ages. At the time that the stocking densities converge, the average diameters converge and remain similar to the following 30 years of the simulated model output (Fig. 8.9).

The ability of the model to converge to real data in a validation mode when started at an inappropriate starting point constitutes a very strong test on the model. In fact, many statistical procedures for predicting how things would change over time do not converge once they have been started inappropriately. Therefore, the model–data comparison, which could have been stronger perhaps if the model were started at an appropriate point, is in some respects more appealing because the case that is presented is a validation case (Shugart and Noble 1981).

In general, model validations are difficult to develop. In the case of some of the gap models, model validations have been developed by reserving data sets that are then used after the model development stage as validation tests. This is probably an advisable procedure in the development of any sort of forest model as well as for models of other ecosystems.

Examples of model tests on gap models

Gap models are well suited for examining vegetation response to changing environmental conditions because the expression of plant response to the environment is not limited to reproducing present-day vegetation patterns. The individual-based approach also provides a useful link between more detailed physiological models and the larger-scale responses. The models have also been tested in terms of their ability to reproduce important features of a wide variety of forests and other ecosystems (Table 8.3).

Interest in individual-based models for assessments of the effects of environmental change stems in part from the degree of past model testing and the apparent ability of the models to predict ecosystem patterns and dynamics under novel circumstances (tabulated up to 1983 in Shugart (1984), and for the history of the JABOWA model in Botkin (1993)). Simulation models in general and individual-based gap models in particular have several features that incline scientists to expect a high standard of

Table 8.3. *Examples of model tests on gap models*

Type of model test	Example case	Other analogous cases
With *a priori* parameter estimation for species predict forest-level features (total biomass, leaf area, stem density, average tree diameter, etc.)	Prediction of biomass for floodplain forests (Phipps 1979; Pearlstine *et al.* 1985) Predict effects of timber harvest on nutrient regimes in northern (USA) hardwood forests (Aber *et al.* 1978).	Similar predictions for a diverse array of systems: Estonian conifer forest (Oja 1983), New Zealand southern beech forests (Develice 1988), Pacific Coast Douglas fir forests (Dale and Hemstrom 1984), West African savanna (Menaut *et al.* 1990), etc. Most of the cases shown in Fig. 8.5 have been inspected vis-à-vis this test
Run model for a long period of time, test results against mature forests preserved in a region	Predict structure and composition of relic mature forests in Sweden (Leemans and Prentice 1987)	Predict old growth forest in montane spruce–fir zone in southern Appalachians (Busing and Clebsch 1987), in Mississippi River floodplain (Tharp 1978), upland forests in Arkansas (Mielke *et al.* 1978), and in Chinese mixed forest (Shao 1991). This is a frequently used evaluation for most of the models found in Fig. 8.5.
Calibrate model on stands of a given age, predict stand structure on stands of a different age	Using a model that predicts the structure and composition of old-growth forests in the Great Smoky Mountains National Park (USA), test by independently predicting composition and structure of forest clear-cut 40 and 70 years before (Busing and Clebsch 1987)	Predictions of the composition and structure of Canadian forests (El Bayoumi *et al.* 1984), diameter structure of Puerto Rican rain forest stands of different ages (Doyle 1981), Australian rain forest at different ages after harvest (Shugart *et al.* 1980a)
Use introduction of disease or change in conditions (e.g. fire frequency, etc.) as a 'natural' experiment	Predict the composition of mature forests in East Tennessee (USA) before the introduced chestnut blight eliminated one of the major tree species (Shugart and West 1977)	Predict composition or structure of forest at different fire frequencies in mixed conifer forests (Kercher and Axelrod 1984), for Puerto Rican montane rain forests under different hurricane conditions (Doyle 1981; O'Brien *et al.* 1992)

Table 8.3 (*cont.*)

Type of model test	Example case	Other analogous cases
Predict independent tree diameter increment data	Predict the growth and diameter increment by species and diameters for subtropical rain forest (van Daalen and Shugart 1989)	Predict diameter frequency distributions for Puerto Rican montane rain plots of different ages (Doyle 1981)
Predict forestry yield tables	Test model on its ability to reproduce Swedish yield tables (Leemans and Prentice 1987)	Predict *Eucalyptus* yield tables for montane Australian forests (Shugart and Noble 1981), for Loblolly pine (Mielke *et al.* 1978), for Swiss forests (Kienast and Kuhn 1989), for Chinese mixed forest (Shao 1991)
Predict forest composition change to single environmental gradients	Predict change from deciduous to coniferous forest in the mountains of New England (Botkin *et al.* 1972)	Predict changes in forest composition as a function of flood frequency (Tharp 1978; Pearlstine *et al.* 1985), at a range of locations (Bonan 1989a) in the boreal zone along altitudinal gradients in China (Shao *et al.* 1991). This test and more complex variants are often applied to gap models
Predict forest composition response to multiple environmental gradients	Predict composition and structure of boreal forests on north- and south-facing slopes and for different ages since wildfire in Fairbanks, Alaska region (Bonan 1989a)	Tests against fire and altitude variables in mountains of California (Kercher and Axelrod 1984) and Australia (Shugart and Noble 1981) Tests in complex terrain (slopes, altitudes, soils) in Switzerland (Kienast and Kuhn 1989)
Reconstruct composition of vegetation under past climates (palaeoreconstruction)	Reproduce 16000 year record of forest change based on fossil pollen chronology from East Tennessee (Solomon *et al.* 1980)	Reproduce forest composition under full glacial and other conditions (Solomon *et al.* 1981; Solomon and Shugart 1984; Solomon and Webb 1985; Bonan and Hayden 1990). There have also been several applications of the models in this mode

model testing. The models seem to lack the parsimony that is expected in models; mathematically, the models are not open to formal analysis as are many simpler population models; the high level of detail in model output implies a concomitant degree of detail in testing of models. Since the models are often used in application domains outside the range of conditions used in model development, good scientific practice requires continual model testing and assessment.

The parameters in the gap models are constrained to a degree by the requirement of realistic parameter values (Shugart 1984). As examples, a growth rate parameter must produce a maximum diameter increment of an appropriate size; the maximum heights and diameters of the trees used in allometric functions must be reasonable; and the light extinction coefficient in the models must be within measured ranges. Nonetheless, there is a degree of flexibility in the ranges of the model parameters. The typical protocol for developing these models is to estimate virtually all model parameters prior to any model–data comparisons. The models have been tested in several different forests with a diverse range of model–data intercomparisons (Table 8.3).

Usually a simulator is initially judged through inspection as to whether the model, when parametrised *a priori* with reasonable species parameters for individual plant performance, can reproduce forest-level features such as total biomass, leaf area, stem density or average tree diameter for forests of differing ages. Such tests have been used in a variety of cases (e.g. Estonian conifer forest (Oja 1983), New Zealand southern beech forest (Develice 1988), floodplain forests in the southern USA (Phipps 1979; Pearlstine *et al.* 1985), etc. (Table 8.3). Models are sometimes calibrated by producing appropriate structure for stands of a given age and then tested as to whether they can independently predict the structure for forests of another age (without recalibration).

Shugart and West (1977) predicted composition and biomass for current forest in Tennessee and then projected the expected composition of mature forests. These comparisons were given increased rigour in that the historically documented mature forest had viable American chestnut, *Castanea dentata*, co-dominants that were not present in today's forests owing to the chestnut blight, *Endotheca parasitica* (Fig. 8.10). In this particular model test, the American chestnut was not included in any of the initial model development but was added to a simulation that produced a mature forest. The spirit of the model test was to view the chestnut blight as a major modification of the forest, a 'natural experiment'. The model was developed on the behaviour forest as it now functions (without

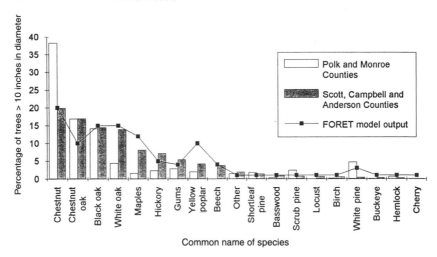

Figure 8.10. Comparison of the prediction of the FORET model of the composition of forests in eastern Tennessee when the American chestnut is included in the model as a viable species. The data are from Foster and Ashe (1908) and consists of an enumeration and a percentage composition of forests in several areas. All trees greater than 10 inches (*c.* 25 cm) in diameter at breast height were counted on 738 acres (*c.* 299 ha) in Polk and Monroe Counties in Tennessee, and on 262 acres (*c.* 106 ha) in Scott, Campbell and Anderson Counties in Tennessee. The original legend in the table read, 'Composition of forests in which chestnut oak forms more than 10% of the mixture – Southern portion of the region – Virgin growth'. Species names are as from the original data. From Shugart and West (1977).

American chestnut) and was tested on its ability to predict the historical condition (with American chestnut).

Busing and Clebsch (1987) tested their spruce–fir forest simulator by inspecting model performance for old-growth stands and then predicting the forest structure for stands of an intermediate age that had been clear-cut 40 to 70 years earlier. Doyle (1981) used structural comparisons with stands of known age as one of several tests on a tropical rain forest simulator, and Shugart *et al.* (1980b) compared model predictions with forest composition in stands of differing ages for Australian rain forest. The comparison of model performance after long periods of simulated time to existing mature forests has been used to test gap models in several cases (Table 8.3). The scientific value of mature forests as a test of the long-term responses of models (and the synthesis of biological understanding that they imply) is considerable.

Independent predictions of tree diameter increment and hence forest

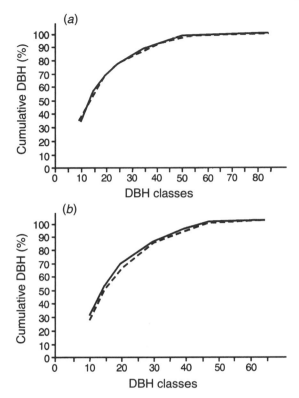

Figure 8.11. Comparison of cumulative percentages of diameter distributions (DBH) between actual observations in 1983 and model simulations initialised with 1972 inventory data and projected to 1983 using the OUTENIQUA model. (*a*) Comparison based on the data from 63 0.04 ha inventory plots located north of the Oliphants River. These data were used to estimate the growth rate for the simulated species by extracting the 95 percentile for growth rates as an estimate of the maximum increment.(*b*). Comparison based on 13 0.04 ha inventory plots located south of the Oliphants River and independent of any model parameter estimation. Measured data (—), simulated data (- - -).

growth was used to test a model of temperate African rain forest (Fig. 8.11, from van Daalen and Shugart (1989)). Some of the most detailed data compilations on forest structural dynamics can be found in forest yield tables. These tables compile the volume, number, mean diameters, heights and other structural quantifiers of forests for stands of different ages growing on sites of varying quality for well over 100 years in some cases. As was discussed in the section above, comparison with yield tables was used as an independent test of the BRIND model of montane *Eucalyptus* communities (Shugart and Noble 1981). Leemans and Prentice

(1987) used Swedish yield tables to inspect the performance of the FORSKA model of Scandinavian forests, and Kienast and Kuhn (1989) made similar use of Swiss yield tables.

One important class of model examination involves the comparison of model predictions for different environmental input conditions followed by an inspection of the responses of forests to natural gradients deemed comparable to the input conditions. The earliest gap model, JABOWA (Botkin et al. 1972), was tested on its ability to predict the approximate altitude of the transition from northern hardwood forest to coniferous forest in New Hampshire. Pearlstine et al. (1985) looked at the responses of simulated floodplain forest composition at different flood frequencies to test the FORFLO model. The BRIND model was tested in its ability to characterise montane zonations of *Eucalyptus* forests in the Brindabella Mountains in Australia (Shugart and Noble 1981) as a function of wildfire frequency and altitude. Kercher and Axelrod (1984) inspected model predictions for fire frequencies, altitude and aspect in the mixed-conifer montane forests of California. Doyle (1981) and O'Brien et al. (1992) used simulations at different hurricane frequencies to assess the performance of the FORICO model of Puerto Rican *Dacroides–Sloanea* forests.

More recent gap models, the simulation of model responses to single controlling variables, have evolved to more complex cases in which the response to multiple gradients and the reconstruction of forest landscapes are assessed. Harrison and Shugart (1990) compared ordinated forest simulation for different slopes and ages with plot data from an Appalachian watershed. Bonan (1989a,b) simulated forest composition, structure, successional response to wildfire and the depth of permafrost (which is under a degree of control from feedbacks related to the forest condition) for boreal forests in the vicinity of Fairbanks, Alaska. The FORECE model was used to simulate the forest patterns on different altitudes, slopes and soils in the Swiss Alps (Kienast and Kuhn 1989). Pastor and Post (1986), building on earlier work by Aber et al. (1978) and Aber and Melillo (1982b), simulated forest pattern in response to soil fertility and texture for forest in the northern hardwood to conifer transition.

Simulations involving forest response to complex gradients and the applications of the models to simulate prehistoric forests (Table 8.3) are germane to the use of the models in assessment of global environmental change. While gap models have been tested to a relatively great degree compared with most ecological models, the models are under a continual revision as new phenomena are incorporated and should be continually tested. Further, any model is an abstraction and as such may not incorporate all the relevant phenomena for a given assessment. Any model extrapolations of a future vegetation should be regarded with caution and compared with other equivalent evaluations from other bases.

Comparisons of different gap models

In general, tests and applications of gap models at a specific site have been relatively successful. The moving of a gap model from sites where it has been tested and developed seems almost always to involve model modifications. These frequently involve adding additional functions to the model to capture unique features of a local environment or biota. Bugmann (1994), Bugmann et al. (1995) and Bugmann and Solomon (1995) tested a model of forest dynamics in central Europe by moving it without adjustment to North America and simulating the expected forest for a number of locations using parameters from the FORENA model (Solomon 1986a). This test was relatively successful for most locations. These and the tests of the models that have been developed for palaeoclimates are evidence for the utility of the models in predicting change under altered conditions. The typical need for recalibration of the models when moved to new areas speaks for caution in the interpretations of these results.

Martin (1992) compared a slightly modified version of the LINK-AGES model (Pastor and Post 1986) with his model EXE. EXE has a more detailed treatment of water and treats energy and momentum exchange with the atmosphere. For Duluth, Minnesota, the two models simulate different forests (Fig. 8.12), with LINKAGES predicting a strong dominance of sugar maple and EXE implying that spruce should outcompete sugar maple (owing to temperature-related effects). Under an altered climate, the compositions (but not the stature) of the forests simulated by the two models are similar, but the causes are different (temperature effects dominate in the LINKAGES simulation; drought effects dominate in the EXE simulation).

Fischlin et al. (1995) developed a similar model experiment in which two variants (based on approach to simulating drought stress) of the FORCLIM model (Bugmann 1994) were compared for locations in the Swiss Alps under current and altered climates. In this model experiment, the two model versions simulated equivalent vegetation for low elevation sites near Berne, Switzerland but produced different responses when subjected to conditions of an altered climate (Houghton et al. 1990).

(*a*)

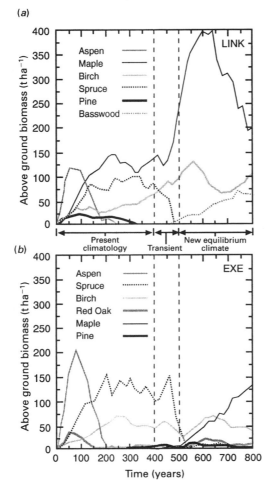

Figure 8.12. Species composition as simulated by the models LINKAGES (*a*) and EXE (*b*) for a site at Duluth, Minnesota. At year 400, climate in the simulation is altered by an increase in the monthly mean temperatures by 3°C and a decrease in monthly rainfall by 10%. This change is put in place over a 100 year interval from simulated year 400 to 500. The figure shows the average biomass from 100 simulations. From Martin (1992).

Bugmann *et al.* (1995) point out that results such as these support the need for a systematic investigation of the inclusion of more detailed physiology and biophysics in the model (see also Bonan 1993; Martin 1996).

Bugmann (1994) compared the behaviour of the FORECE model (Kienast and Kuhn, 1989) with his FORCLIM model in several ways, one of which was prediction of the distribution of Norway spruce (*Picea abies*) as a function of annual mean temperature and precipitation. In a classic paper, Ellenberg (1986) mapped the factors controlling Norway spruce based on his extensive experience with the forests of Central Europe. One can compare directly the range of conditions over which the species occurs (Ellenberg 1986) with the occurrence predicted by the two models (Fig. 8.13). The agreement between the Ellenberg's observations and the FORCLIM model is greater than with the FORECE model. One important feature of this simulation is the ability of the models to simulate what in Chapter 7 was called the realised niche of the species (the distribution shown in Fig. 8.13) from the features of the fundamental niche (the parameters for the species and the interactions with other species and the environment). The relation between the output of individual-based simulators and niche concepts has also been discussed in Shugart *et al.* (1988).

Bugmann *et al.* (1995) summarise the results of several model intercomparisons with gap models by noting that the models need to be inspected for regional analysis of vegetation dynamics by considering whether the level of detail and the appropriate process are included in the model. For global applications, they recommend appropriate simplifications. Such a view implies a nested hierarchy of models with detail appropriate to the immediate question at hand (depending on the spatial scale) and the ability to transfer understanding from one level to the other.

Concluding comments

Individual-based models, in general, attempt to incorporate behaviour, biological and physiological mechanisms in a slightly higher scale in time and space than the scales at which these mechanisms are typically studied. In the case of gap models, the models grow individual trees and also simulate the birth and death of whole populations of trees in a forest. The models see the forests by looking at the (individual) trees. This seemingly impossible task is made possible with the use of modern digital computers. These models would probably not have been practical 30 years ago, before the advent of truly fast computers. The computational demands of these models are no challenge to the new generation of computers. Indeed, they will run on most desk-top 'personal' computers.

Gap models are tools for understanding and predicting landscape dynamics because they have been widely tested. The models include many phenomena of interest and have been relatively successful in tests involving reproducing average compositional and structural patterns.

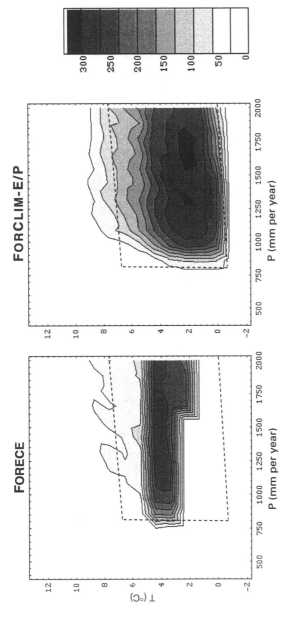

Figure 8.13. Simulated mature forest response of biomass (t ha^{-1}) of Norway spruce (*Picea abies*) in a climatological parameter space with dimensions of annual mean temperature (T) and annual precipitation (P) in central Europe. The FORECE model is from Kienast and Kuhn (1989); the FORCLIM-E/P model is from Bugmann (1994). The dashed line in each graph outlines the area where Norway spruce is a dominant tree according to Ellenberg (1986) except for the vertical line at P = 800 mm year^{-1}, which is not contained in the scheme of Ellenberg. From Bugmann *et al.* (1995).

Bugmann *et al.* (1995) point out that quantitative model comparisons indicate that the behaviour of gap models appears sensitive to the particular formulation used in the models. They further suggest that a systematic investigation of the comparative performance of gap models be undertaken at a variety of locations and under past and present climatic conditions. It is still important to realise that these models are only caricatures of reality and must be tested regularly against information on what real ecosystems do under a different set of conditions.

9 · *Consequences of gap models*

Several applications of individual-based models involve the determination of the larger-scale consequences of individual biology on ecosystem structure and function. Some of these will be discussed in Chapter 11, which deals with applications of such models to the problem of assessing the response of natural landscapes to environmental changes. In this chapter, some of the theoretical consequences of gap models for understanding the dynamics of natural landscapes are considered at the population and the landscape levels. These explorations represent a more general, theoretical exploration of the ecological features of gap models, which complements the more specific applications that follow.

Ecological consequences at the population level

Just as the biology of the species determines regeneration success in gaps of various sizes, it also determines the sizes of gaps created by trees when a large representative of a given species dies. To illustrate this point with an extreme example, consider the leguminous tree *Tachigalia versicolor* as its life history is described by Foster (1977). *Tachigalia* is a very large, highly branched canopy species that is found in evergreen and partially deciduous lowland forest in Panama, parts of Costa Rica and parts of Columbia. It blooms only once in the lifetime of a given tree and the tree typically dies within one year after releasing a wind-dispersed fruit. Several trees in a given area bloom and die at the same time, but not all the trees in an area flower at the same time. The tree, according to Foster (1977), apparently requires a gap opening in the forest canopy for successful maturation and the treefall of the parent greatly increases the opportunity for seedlings in the vicinity of the parent to be able to grow to adulthood. Foster also notes that *Tachigalia versicolor* saplings are often found growing in openings created by the fall of the canopy adult that, in fact, probably produced them. What one sees in *Tachigalia* is an extreme example of coupling the mode-of-birth and mode-of-death life-history attributes of a tree in a way that the tree can successfully prepare regeneration sites for its seedlings. The species creates in its death the conditions needed for its successful regeneration.

The interrelationship between the mode-of-birth and mode-of-death life-history attributes can be extended to produce a simple 'functional types' representation of the manner in which tree species occupy gaps in forest canopies. Considering the development of minimal categories of plant functional types with respect to gap competition in trees, one can divide tree species into:

- species requiring a canopy gap for successful regeneration or otherwise
- species typically generating a gap with the death of a mature individual of typical size or otherwise.

As a product of this pair of dichotomies, there are implied four resultant strategies categorised as species roles (Fig. 9.1). Life-history traits associated with each role are intuitive adaptive syndromes: gap-forming trees are large, while species that do not create gaps are typically smaller; gap-regenerating species are shade intolerant, while species that do not require gaps are shade tolerant (Shugart 1984, 1987).

This classification is intentionally simplistic, and it could be elaborated by subdividing the four roles or by considering other criteria or resources. For example, Grubb's (1977) regeneration niches consider one axis (regeneration) of this dichotomy in detail. As another example, several authors have further divided the 'needs gap' axis to account for subtle differences in the shade tolerance of species (e.g. gap-size partitioning by tropical trees; Denslow 1980; Brokaw 1985a,b). Yet even this two-by-two categorisation can produce a rich array of forest gap dynamics.

Some qualitative features follow from interactions among the four roles (Fig. 9.2). Two of the roles are self-reinforcing: Role 1 species can create the gaps they need to regenerate; Role 4 species can regenerate in the shade and do not drastically open the canopy when they die. Some combinations of roles can replace each other reciprocally: Role 1 species can take over a gap created by Role 2, and vice versa. Other roles tend to give away the space they occupy when they die: Role 3 species cannot create the conditions they need to regenerate but instead favour Roles 2 and 4 in the subsequent cohort of trees. Of course, if trees of the larger roles (Roles 1 and 2) die at a smaller size, they influence the ecosystem dynamics in the same ways as their smaller counterparts (Roles 3 and 4,

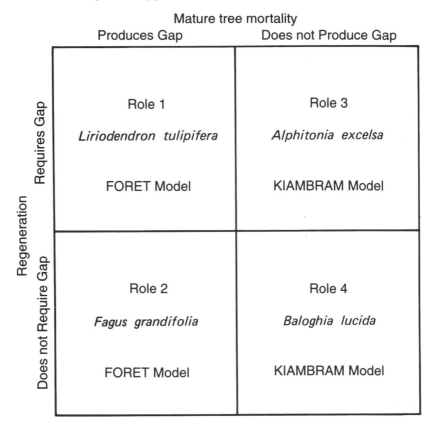

Figure 9.1. The roles of gap-requiring and gap-producing tree life-history traits with example species for each of the four roles. From Shugart (1984).

respectively). A Role 1 tree that dies when small is equivalent to a Role 3 species, hence it does no favour to its progeny in its early death.

These qualitative patterns imply a set of rules governing patterns of gap dynamics (and species replacement). Like any set of rules, these may be followed strictly, loosely or not at all. In the case of forests patterned by gap dynamics (i.e. by competition for and temporary occupancy of space by individual trees), qualitatively different modes of competition may occur. Trees may compete within the rules implied by the foregoing discussion. Under these rules, Role 3 is a loser in local competition involving gap dynamics but can persist by not playing by this set of rules through competing in a different, larger arena. This is the case of fugitive (ruderal) species, which concede gap-scale competitive ability to persist regionally (Marks 1974; Urban *et al.* 1987).

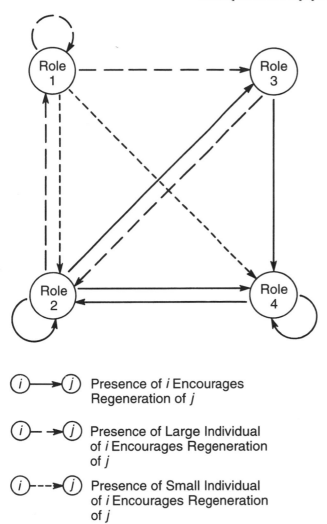

Figure 9.2. The directions of positive or strongly positive interactions among four roles of tree species with respect to population interactions. From Shugart (1984).

Finally, trees have a sufficiently large local effect on the environment that species can alter the rules that govern competitive outcomes. Certain fire-adapted trees (often Role 3 species) may alter mortality by inter-acting with fires in ways that take the advantage away from longer-lived (larger) Roles 1 and 2 trees (Heinselman 1981; Foster and King 1986). In general, frequent disturbances can shift the competitive advantage to Role 3 in many forests. Environmental factors can also change the rules of competition. At high latitudes, low temperatures, a short growing season and low sun angles create a condition under which gap formation by the death of a single tree does not occur. This rules out Roles 1 and 2 as possible strategies. For this reason high-latitude boreal forests should be dominated by trees of Roles 3 and 4.

These simple notions of functional roles of trees has clear implications in interpreting forest dynamics. The basic premise is that life-history traits effectively couple demographic processes of mortality and regeneration. Morphological and physiological considerations impose correlation structure among life-history traits, such that only certain combinations of traits may be realised in any single species: shade-tolerant species tend to be intolerant of moisture stress; long-lived trees tend to grow more slowly than short-lived species, and so on (Huston and Smith 1987). Thus, however functional roles might be defined, there are a limited number of such roles, and these will interact to imply a limited set of rules constraining ecosystem dynamics. These roles, and the implications of their interactions, can be explored conveniently by implementing them in simulation models.

The different roles of trees produce essentially different biomass and numbers dynamics when monospecies stands are simulated with small model plots (c. 0.1 ha). The long-term behaviour of numbers and biomass for typical example species are shown in Fig. 9.3. Each of the roles has a fundamentally different signature of numbers and biomass dynamics in monospecies stands. The dynamics of a Role 1 stand features explosive increases in biomass and numbers following the death of a large canopy tree and an even-aged character to the stand reflecting the tendency for episodic recruitment following the death of a large tree. Role 2-dominated stands are mixed-age and mixed-size structured even at small scales (such as Fig. 9.3b), with episodes of enhanced recruitment associated with drops in biomass from large tree mortality. Role 3 dynamics feature episodes of recruitment that are preceded by the senescence of the previous cohort and an associated drop in biomass. Role 4 dynamics are of mixed-aged and mixed-sized forest with little variance in either numbers or biomass dynamics (Fig. 9.3d), except when simultaneous deaths of several trees produce a transient response.

Forests dominated by trees of the same role

While these modes of stand dynamics are intended as typical cases, there are forests that display similar kinetics. For example, Shugart (1987) found

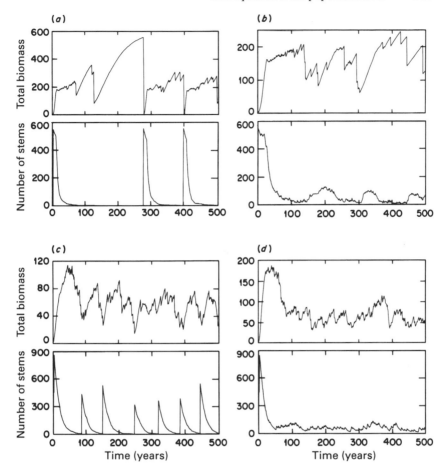

Figure 9.3. Number and biomass dynamics of simulated monospecies forest of trees of four different roles. (*a*) Role 1: *Liriodendron tulipifera* on a 1/12 ha plot as simulated by the FORET model. (*b*) Role 2: *Fagus grandifolia* on a 1/12 ha plot as simulated by the FORET model. (*c*) Role 3: *Alphitonia excelsa* on a 1/30 ha plot as simulated by the KIAMBRAM model. (*d*) Role 4: *Baloghia lucida* on a 1/30 ha plot as simulated by the KIAMBRAM model. From Shugart (1984).

that the Role 3 dynamics of numbers and mass to be in agreement with the dynamics of *Pinus sylvestris* stands at high latitude (Fig. 9.4) reported by Zyabchenko (1982). It is clear from Fig. 9.3 that the numbers and mass dynamics of forest composed of only one role are different one from another. Through the gap creation/gap colonisation process, each of these cases represents a different temporal pattern of environmental variables that co-vary locally with the gap processes.

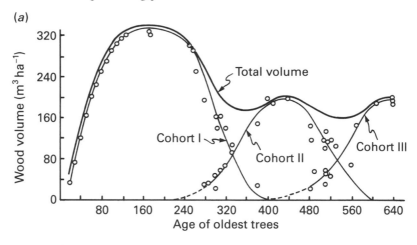

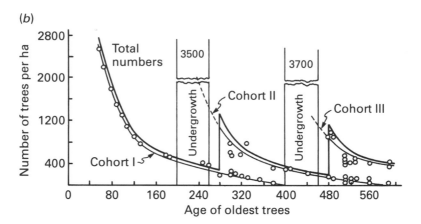

Figure 9.4 Biomass dynamics (*a*) and number of trees (*b*) of *Pinus sylvestris* forests at high latitudes (from 61° to 70° N; longitudes from 28° to 42° E) in the former USSR. Figure redrawn from Zyabchenko (1982).

The phenomenon of wave regeneration in the balsam fir (*Abies balsamea*) in the higher elevation forests in New England described by Sprugel (1976) and the seemingly equivalent phenomena described by Oshima *et al.* (1958) for Mt Shimagee in Japan (see Chapter 7) is essentially a spatial expression in one horizontal dimension of the saw-tooth biomass curve that is expected for the temporal response of a single forested patch of a landscape. The wave regeneration pattern described by

Sprugel (1976) occurs almost exclusively in pure balsam fir stands. Balsam fir functions as a Role 3 species in this ecosystem (i.e. fast growth, short lived, small and regeneration favoured by disturbance). Sprugel (1976) noted that, 'If certain areas are inherently prone to wave-type disturbances due to local topography, prevailing winds, or some other factor, then those forests must almost inevitably become nearly pure fir after a few centuries of waves passing through every 60 to 70 years. In turn, the change toward pure fir probably allows waves to move through the forest more smoothly and evenly, since in a pure stand all the trees reach the same stage of degeneration and susceptibility to environmental stress at the same time'. Thus, the regeneration wave phenomenon allows a Role 3 species that one would expect to be unable to maintain pure stands over landscapes to generate favourable regeneration conditions and to hold dominance. The pattern is striking because it is linear and, thus, is obvious on the landscape.

The orderly wave-regeneration pattern of forested landscapes (Oshima et al. 1958; Sprugel 1976; Sprugel and Bormann 1981) is an unusually regular pattern of organisation of the birth–death pattern across a landscape mosaic. The organising phenomenon that allows this persistence is synchronous mortality along a linear front as a result of environmental conditions. One could also expect synchronous mortality to induce regeneration conditions in two dimensions as well.

A spatial case of synchronous replacement in two dimensions has been studied in detail for the Hawaiian tree 'ohi'a (*Metrosideros polymorpha*), for the island of Hawaii. Between 1954 and 1977, 'ohi'a canopy trees either died or were defoliated over large areas of the montane rain forest on the windward side of Hawaii. This decline or die-back was very rapid in its onset, and it caused considerable concern over the stability of forests of this species. The die-back initially was thought to be an introduced disease that was caused by the fungus *Phytophthora cinnamomi*, which is associated with a similar die-back of the jarrah (*Eucalyptus marginata*) in Western Australia (Podger 1972). Having discovered this pathogen in 'ohi'a die-back stands (Laemmlen and Bega 1972; Kliejunas and Ko 1973; Bega 1974), as well as abnormally high population levels of a boring cerambycid beetle (*Plagithmysus bilineatus*), it appeared that the die-back was caused by an epidemic disease. Studies (Papp et al., 1979) indicated that neither *Phytophthora* nor *Plagithmysus* was universally associated with die-back.

Mueller-Dombois (1980, 1982) and his students at the University of Hawaii noted that a similar die-off of 'ohi'a had occurred on the island of

Maui in the early 1900s. This historical die-back had been initially suspected to be caused by fungal pathogens (Lewton-Brain 1909; Lyon 1909), but no conclusive evidence was found to support this hypothesis (Lyon 1918, 1919). This led to the development of a hypothesis that the die-back was a natural landscape process (Burton 1980). Although the dynamics of 'ohi'a are complex on the landscape, which is partly the result of the ability of this tree to function in a wide variety of soils and topographic positions while displaying a great variation in form, the general pattern of the die-back landscape process is described below.

'Ohi'a is a light-seeded, wind-dispersed tree that occurs as a pioneer on young volcanic substrates (Atkinson 1970). It also is the dominant rain forest tree in most of the rain forests of Hawaii, and it forms monospecies canopies over large areas. 'Ohi'a trees are of small stature and the tree often is in stunted stands on bog soils. It requires high light levels for regeneration (Burton 1980). Therefore, 'ohi'a corresponds reasonably well to what has been abstracted as a Role 3 species. As one would expect for such a species, 'ohi'a does not regenerate well under its own canopy (Gerrish and Mueller-Dombois 1980), which is a feature that led Clarke (1875) and (later) Hosaka (1939) and Egler (1939) to conclude that 'ohi'a forests were decadent.

The apparent mechanisms by which the 'ohi'a maintains itself is outlined in a general form in Fig. 9.5, but the pattern can vary greatly with conditions (Gerrish and Mueller-Dombois 1980; Mueller-Dombois 1980). Starting with a young 'ohi'a stand (Fig. 9.5), tree growth produces a mature 'ohi'a forest. Eventually, competition and nutrient depletion (Kliejunas and Ko 1974) reduce tree vigour and render the stand susceptible to die-back. A triggering event (probably a climatic instability such as excessive rainfall on wet sites, drought on dry sites or, perhaps, lightning (see Mueller-Dombois 1980) causes trees to begin to die. Secondary agents, such as fungus or insects (Kliejunas and Ko 1974), attack the dying trees and help to synchronise the die-back event across the area. 'Ohi'a, which regenerates well as an epiphyte on downed tree trunks (Gerrish and Mueller-Dombois 1980) and requires high light levels (Burton 1980), is able to regenerate in the die-back patch; it produces a young stand to close the regeneration cycle. Because of the synchrony in the mortality event, the landscape biomass response of 'ohi'a is of the same saw-toothed shape that is expected at smaller spatial scales for forests.

Apparently, the die-back mode of regeneration at the landscape scale is not restricted to 'ohi'a but also is found to occur in several tree species – often on islands and typically conforming to the Role 3 abstraction.

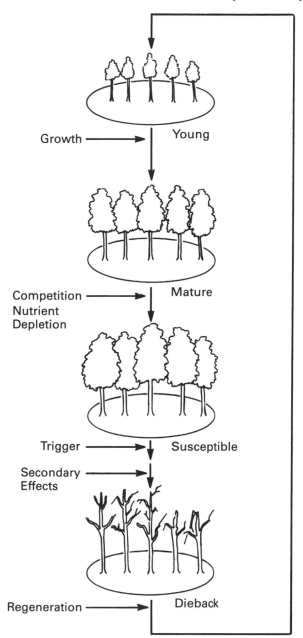

Figure 9.5. The die-back–regeneration cycle for the Hawaiian tree species *Metrosideros polymorpha.* From Shugart (1984).

Paijmans (1976), in referring to forests in the lower montane zone in Papua New Guinea, notes, 'A common feature of mature *Nothofagus* stands are patches of dead or dying trees, for which no obvious cause has been found. It has been suggested (Robbins and Pullen 1965) that groups of even-aged trees die off together on reaching "overmaturity".' Mueller-Dombois (1982) felt that the die-back on New Guinea strongly resembled the die-back phenomenon seen in the Hawaiian Islands. Ash (1981), in discussing the *Nothofagus* die-back in Papua New Guinea, attributed the die-back to similar combinations of weather, substrate, pathogens and successional development that are implicated in the well-studied Hawaiian landscapes.

Mueller-Dombois (1982), in reporting an extensive survey of large Pacific islands, found die-back patterns occurring in several forests. All were dominated by canopy species that are both shade intolerant and have a regeneration pattern that was the two-dimensional patch-analogue to the one-dimensional wave regeneration. These forests included *Eucalyptus deglupta* forests in New Britain, *Acacia* species near Rabaul in New Britain and *Nothofagus* (*N. solaudri*) and *Metrosideros* (*M. umbellata, M. robusta* and *M. excelsa*) forests in New Zealand. Mueller-Dombois (1982) also reported the phenomena in Sri Lanka and New Caledonia. A critical consideration in interpreting these dynamics as a natural feature of a forested landscape is whether or not the species affected by the die-back event is able to regenerate successfully following the synchronised mortality event.

There are any number of factors that can produce synchronous mortality on an even-aged forested landscape. For example, tree-killing bark beetles (Coleoptera, Scolytidae) frequently select weakened trees as hosts and emit pheromonal cues to concentrate dispersing beetles to these hosts (Wood 1982). These beetles are capable of killing living trees over large areas and thus, of re-initialising landscapes with regenerating intolerant tree seedlings. This type of re-initialisation over large areas, even if it is a natural landscape process, makes the landscape subject to radical widespread change. Such vulnerability can occur if the time period for regeneration is short, particularly if some event should alter the ability of the tree to regenerate at this critical time. This appears to be the case in the Maui 'ohi'a die-back (Lewton-Brain 1909; Lyon 1909), in which the areas of die-back were deliberately planted with introduced tree species. As another example, the Australian landscape is undergoing continent-wide episodes of die-back among several species of *Eucalyptus* (Old *et al.* 1981). In some of these cases, *Eucalyptus* regeneration has been altered by human land-use (e.g. grazing – particularly sheep or altered fire regime), which causes regional-scale deforestation. The aetiology of eucalyptus die-backs shows considerable geographic variations and it is a complex consequence of natural processes and human alterations in land-use (Old *et al.* 1981). Nonetheless, *Eucalyptus* die-back illustrates the potential vulnerability of die-back–regeneration landscapes under land-use changes.

Complexity of interactions among roles

Successful species of differing roles can be expected to influence other species on a small patch in different ways. For example, a large mature tree of Role 1 would, upon its death, be expected to create a large gap that would encourage its own regeneration as well as that of other Role 1 species. Further, it would encourage Role 3 species in a community. The death of a small individual of Role 1 would create a smaller gap that would tend to favour the regeneration of Role 2 and 4 species. The other roles would be expected to influence regeneration themselves as a complex interaction of ecological roles and of size and spatial scales. One would expect (Fig. 9.5) that Roles 2 and 4 would tend to be mutually encouraging, whereas Roles 1 and 3 would prosper when the death of canopy trees created large gaps. Roles 2 and 4 would tend to increase in cases where the mortality pattern generated smaller gaps. The interactions among roles are such that the species of any role could be expected to regenerate in a gap left by the mortality of the species of the same or any other role with a fair amount of variation. The pathways shown in Fig. 9.6, therefore, are the most probable paths, but in actual fact, any role could be replaced by another in an actual forest with some small probability.

Forcier (1975) and Woods and Whittaker (1981) studied replacement patterns in hardwood forests in the north-eastern USA. They found a tendency for species replacements to form closed cycles. The closed cycles that they observed in these northern forests are exactly what one would expect if one takes Fig. 9.5 and notes that species in this area are composed only of Roles 2 and 3 species. If one moves to another case where the actual replacement probabilities have been computed for a forest, that is, the study of Horn (1971, 1975a,b, 1976) in the Princeton Forest in New Jersey, one also can obtain a diagram for the three roles present in that forest (Roles 1, 2 and 3) that is also compatible with the interactions shown in Fig. 9.5. The interaction diagram for these three

(*a*)

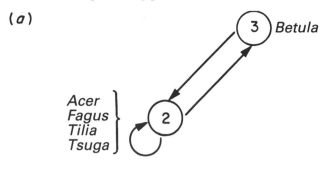

(*b*)

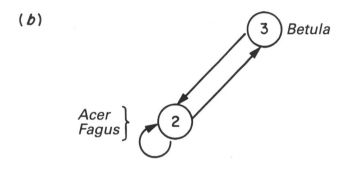

(*c*)

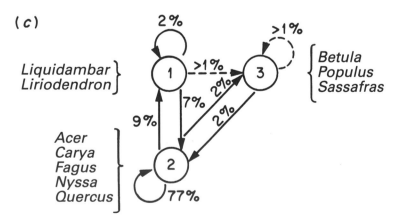

Figure 9.6.   Species role–interaction diagrams for three studies conducted in the north–eastern USA. (*a*) Woods and Whittaker (1981). (*b*) Forcier (1975). (*c*) Horn (1975a). From Shugart (1984).

studies is shown in Fig. 9.6. What is seen in Fig. 9.6 is a subset of the interactions shown in the more general Fig. 9.5.

The use of role-based concepts to formulate large-area models

One of the practical considerations of aggregating species into roles is to
develop a capability to represent the dynamics of natural systems in relatively less complex models. The size of a gap produced by tree mortality is
proportional to the area shaded by the tree. Therefore, the gap size is
related to both tree height and sun angle. In a given region, the average
sun angle can be considered the same, and, therefore, the gap size is controlled principally by tree height. Because a given species can have
different types of role in different forest ecosystems, tree roles have to be
determined relatively, meaning that the size of a gap produced by tree
mortality changes with its relative position in a stand. Two same-sized
trees can produce different sizes of gaps if they are growing in different
forest stands.

Acevedo *et al.* (1995) and Shao *et al.* (1996) have treated the problem of
deriving a computationally fast role-based model based on the four
species roles shown in Fig. 9.1. Acevedo *et al.* (1995) were able to develop
Markovian simulators directly from a gap model (see Chapter 8), and
inspected the behaviour of the resultant model in dynamic simulations of
forest succession.

Shao *et al.* (1996) used a more simulation model-oriented approach to
the same question and emphasised the spatial area simulation of the
models while solving for the equilibrium vegetation composition. Shao *et
al.* assumed that there are n species in a forest stand. Of those n species, m
species can produce gaps, and the size of the gaps produced by the ith such
gap-producing species is the size, PGS_i. Each of the species requires gaps
of some size for regeneration, and the required size is denoted RGS_j for
species j. The difference (D_{ij}) between RGS_i and PGS_j for any given two
species (i and j) is assumed to be related to the possibility that species i
replaces species j, and it is expressed as:

$$D_{ij} = |RGS_i - PGS_j| \qquad (9.1)$$

where the symbol $||$ means absolute value. Thus, $D_{ij} \geq 0$. When $D_{ij} = 0$,
species i has the best chance to replace species j; when the value D_{ij} is
higher, the probability that species j is replaced by species i becomes
smaller. Based on D_{ij}, a gap-phase replacement index (RI) is expressed by
a simple equation:

$$RI_{ij} = e^{-D_{ij}} \tag{9.2}$$

The major advantage of RI_{ij} is that it is bounded between 0 and 1, and its value is proportional to the probability of the success of the species in reproducing itself in gaps produced by the gap-producing species in the forest. For species i, RI_i is calculated by summing RI_{ij}:

$$RI_i = \sum_{j=1}^{n} RI_{ij} \tag{9.3}$$

Presumably, species with higher values of RI would be more abundant. The species biomass in a stand is also related to the size of the trees of the species. Using height and age as indices for size, a species composition (biomass) coefficient (CC_i) for species i is calculated as:

$$CC_i = RI_i H_i^2 A_i \tag{9.4}$$

where H is tree height and A is tree age. The CC value can be normalised by converting to percentages across all species.

Shao et al. (1996) developed procedures for estimating parameters for the role-based model from an existing gap model for China, the KOPIDE model (Shao 1991; Shao et al. 1996). Many of these effects were determined by the influence on tree height of two climate parameters (growing degree days (DEGD) with base temperature of 10 °C and the ratio of annual potential evapotranspiration to precipitation (PET/P)). Changes in tree height subsequently affected size of gap formation (see Eq. 9.1) and, hence, the mixture of species. The model was tested by using data from 81 different climate stations in a study area in north-eastern China bounded by the borders with Russia and North Korea, and covering parts of four provinces: Liaoning, Jilin, Heilongjiang and Neimenggu. The northern part of this area is covered with larch forests, the southern part is basically covered with deciduous–coniferous mixed forest and the western vegetation is an oak–grass complex. The pattern of vegetation simulated by the role-based ROPE model is shown in Fig. 9.7 and corresponds to the actual pattern of vegetation in this part of China (China's Vegetation Editing Committee 1980).

A role-based model reduces the species complexity found in an individual-based model with a functional representation of species roles defined in terms of their influence on the structure of vegetation and the reciprocal effect of the nature of the structure on the performance of the species. The parameters for a role-based model can be obtained from an individual-based simulator (and the responses of the role-based model can

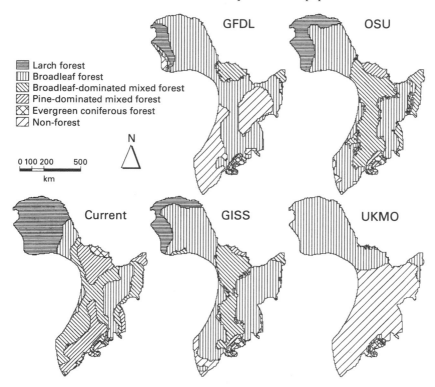

Figure 9.7. Results from the role-based model ROPE, simulating the vegetation pattern for four provinces in China (Liaoning, Jilin, Heilongjiang and Neimenggu) bounded by the borders with Russia and North Korea, using data from 81 different climate stations. Also shown are the expected changes in vegetation under four different climate-change scenarios simulated by general circulation models for doubled CO_2 conditions (Oregon State University (OSU), Schlesinger and Zhao (1988); Geophysical Fluid Dynamics Laboratory (GFDL), Manabe and Wetherald (1987); Goddard Institute for Space Science (GISS), Hansen *et al.* (1988); United Kingdom Meteorological Office (UKMO), Mitchell (1983)). After Shao *et al.* (1996).

be tested for fidelity with the originating individual-based model). Role-based models are computationally very fast and can simulate the vegetation structure and composition over large (subcontinental) areas. The role-based model developed for the northern part of China and discussed as an example here has been used to produce vegetation type maps for climate-change simulations from conditions predicted by four different atmospheric GCMs across China (Shao *et al.* 1996). The results from these model experiments are also shown in Fig. 9.7.

Mosaic size and the rules of the game

One feature that emerges from model experiments is the potential importance of multi-level competition in forests. The species roles can produce patterns of interrole interactions that are asymmetric, competitive or mutualistic, even though the fundamental individual-level interactions are based on competition for space and resources (Shugart 1984). Species can compete within the context of a set of rules. For example, Role 3 species may be extremely competitive for safe sites in forest gaps given a particular forest condition. Species of different roles can in some cases compete to determine the rules that apply in the forest. For example, a successful Role 1 species would tend to reinforce gap dynamics in a given forest, while a successful Role 4 species would reduce the importance of gap regeneration in the same forest. Competition of the first sort (competition within the rules) is the usual case for competitive interactions and one would expect species that were the most similar to be most competitive.

In the second sort of competition (competition to change the rules), one would expect species that are the most dissimilar to be the most competitive. This type of competition is more difficult to study under field conditions. Model experiments involving these sorts of interaction have produced ecosystem dynamics featuring multiple stable states and hysteresis (Shugart *et al.*, 1980b). In these instances, the dynamics associated with one role dominates and installs a self-reinforcing mode of overall system behaviour at one extreme of an environmental gradient; a second role similarly controls the system behaviour at the other extreme of the gradient. In the intermediate range of the gradient, the ecosystem has two quasi-stable modes of behaviour. In this range, the state of the system is a consequence of the prior history of conditions to which the ecosystem has been exposed (Shugart *et al.* 1980b).

Figure 9.8 shows the results of a computer experiment in which *Fagus* (a Role 2 species) competed against *Liriodendron* (a Role 1 species). The competition was conducted under conditions of changing environmental conditions. In the case of these model experiments, the temperature was slowly increased or decreased, and the success of *Liriodendron* expressed in terms of percentage composition of the forest was plotted against the change in this gradient. Under cool climates, the growth rates were slow and the trees all tended to be smaller. Thus, Role 2 species dominated the stands, as is evidenced by the small percentage of *Liriodendron* shown in Fig. 9.8 for cool conditions. Under warm conditions, the trees tended to

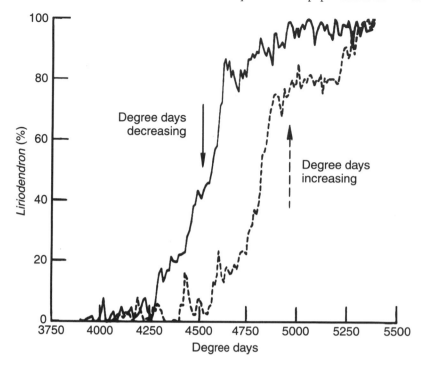

Figure 9.8. Percentage of total stand biomass that is attributable to *Liriodendron* in a two-species mixture with *Fagus grandifolia* under gradually changing degree-day conditions. From Shugart *et al.* (1980b).

grow larger and the *Liriodendron* was able to dominate the stand using gap replacement. However, stand dominance for intermediate degree-day values demonstrated an apparent hysteresis. Either of two possible forests – *Liriodendron* or *Fagus* dominated – could occur depending on the direction of the change of climate. In addition to indicating the possible complexity of constructing dynamics of mixed species forests (even when the roles of the component species are sharply differentiated), the example (Fig. 9.8) also links the simulated gap-model dynamics with applications of non-linear differential equations in forests (Yamamura 1976; Smith 1980).

Trade-offs in individual plant functioning and plant functional types

Factors affecting differential abundance can be coupled to the environment or to one another. In the model experiments that have been conducted by creating forests with more than one species role, the interactions among the species of different roles can produce forests with a variety of successional pathways. Smith and Huston (1989) in a series of model experiments used an individual-based model to inspect the consequences of different species with a spectrum of life-history attributes competing against one another. The fundamental premises in this work were:

1. Plants that photosynthesise at high rates and grow rapidly under conditions of high light are unable to survive at low light levels (shade intolerant). Plants that are able to grow in low light (shade tolerant) have low maximum rates of growth and photosynthesis even under high light levels (see Fig 9.2)
2. Plants that grow rapidly and/or reproduce abundantly under conditions of high available soil moisture are unable to survive under dry, light conditions (Parsons 1968a,b; Orians and Solbrig 1977). Conversely, plants that grow under dry conditions are unable to grow and/or reproduce abundantly even when soil moisture availability is high
3. Tolerance to low light and low moisture are interdependent and inversely correlated.

Smith and Huston used these three premises to generate species (see Fig. 9.9) that they then experimentally manipulated with respect to resource gradients using a modified version of the FORET model (Shugart and West 1977). They found the different combinations of species (and the life-history strategies these species represent) could produce a rich array of successional pathways and gradient responses.

Notable among these patterns of response was a tendency for the successional sequences resulting from competition among different functional types to be simple under low moisture conditions and to become progressively more complex as conditions improve at a site (Fig. 9.10). Patterns of species zonation along a moisture-availability gradient became more sharply differentiated over successional time. When species were simulated in monospecies stands along an environmental gradient, their distributions reflected their physiological tolerances; in complex stands with multispecies competition, the species were distributed along the gradient with unimodal 'bell-shaped' curves (Fig. 9.11). The relative success of species with different roles depended on environmental context. A species with a given set of life-history attributes might be judged successful under one set of environmental conditions and competitors, but a loser under different conditions. The logical trade-offs in life history and physiological features of the species render impossible the emergence of a single set of attributes that is always successful. See Smith

(a)

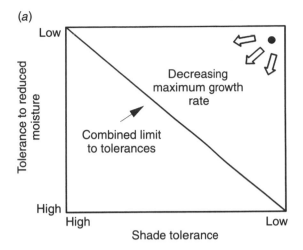

(b)

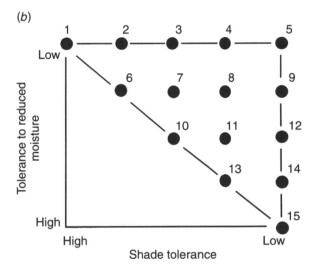

Figure 9.9. (a) Possible woody plant strategies for light and water use with the consequences of trade-offs. The highest growth rates (rates of carbon gain) are in the upper right corner of the figure and growth decreases as the tolerance to either low light or low moisture increases. (b) Conditions of tolerances for 15 plant functional types used in model experiments by Smith and Huston (1989). See also Figs. 9.10 and 9.11 for results of these experiments. From Smith and Huston (1989).

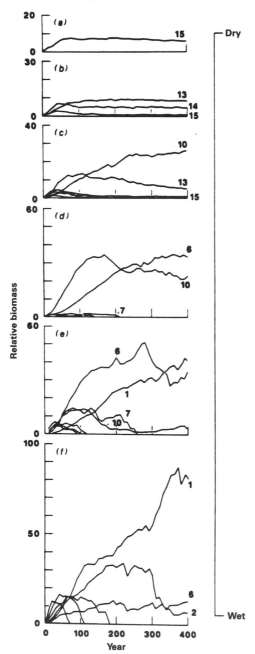

Figure 9.10. Successional sequences resulting from competition among the functional types identified in Fig. 9.9 under different moisture conditions (*a–f*). From Smith and Huston (1989).

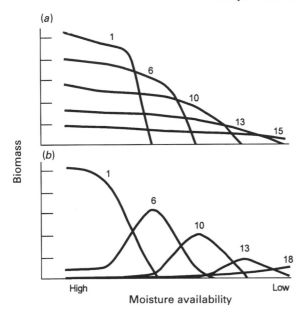

Figure 9.11. Simulated patterns of plant abundance along a moisture gradient. (*a*) The response of five functional types (see Fig. 9.9) simulated under conditions of a monoculture. The plant functional types all have the same physiological optimum in that they all have their best performance under high moisture availability. (*b*) The response of the same five functional types grown in polyculture competition with all other of the 15 functional types shown in Fig. 9.9. The 'ecological optima' of the species are arrayed along the environmental gradient. From Smith and Huston (1989).

and Huston (1989) for a review of the justification of their fundamental premises and for more details on the gradient response expected from plant communities with a spectrum of plant functional types.

Tilman (1982, 1988) has used biological trade-offs as a basis for understanding the diversity of plant species with regard to the division of resource axes. It is refreshing that his results, based on models that seem appropriate to populations that are well mixed and without significant size structure, are in qualitative agreement with those obtained with individual-based models that are not subject to these restrictions.

Ecosystem functional types

That the interaction between two functional types could transform a forest system from one with a single stable state to one with a multiplicity of stable states can be taken as a basis for ecosystem functional types. A related basis is the concept that plants may compete one with another either within a relatively fixed set of rules or by rule alterations through feedbacks onto the environment. Both bases emphasise interactions of ecosystem functional types as altering ecosystem function and ecosystem–environment interactions.

To provide a simple case of ecosystem functional types, consider the example of plant functional types discussed above. In this example case, the four roles of trees (Shugart 1984) also can be used as examples of ecosystem functional types. The dimensions of ecosystem functional types involve alterations on ecosystem dynamics. There are three domains of ecosystem responses:

1. The time domain
2. The frequency domain
3. The spatial domain.

Each of these cases will be discussed using the example plant functional types (and their interactions) for illustration.

Perhaps the most straightforward way in which species can alter the dynamic response of an ecosystem is by changing the rate at which important processes occur. Species with rapid growth rates can shorten the time needed for a forest to recover a closed canopy following a disturbance. An example of a possible change in the eastern deciduous forest of North America might be the demise of American chestnut (*Castanea dentata*) following the introduction of the fungal disease chestnut blight (*Endothia parasitica*). American chestnut once accounted for as much as 50% of the biomass in some widespread forest types and was eliminated as a canopy tree during the early part of the twentieth century. The species among other attributes had leaves with high decomposition rates compared with the oak species (notably white oak, *Quercus alba*, and chestnut oak, *Q. prinus*) that have tended to replace it. One would expect that the demise of American chestnut would result in forests with a slower rate of decomposition and an attendant slowing down of the rate of natural recycling of essential plant nutrients. American chestnut also sustained a large wildlife population that fed on its regularly produced, sweet seeds. The species that have replaced American chestnut have a tendency either to mast (irregular seed production) or to have tannin-laden acorns. Thus, the elimination of this one dominant species even though it was replaced by other species may have resulted in a forest that was different from that before with respect to element cycling rates and seed predation rates.

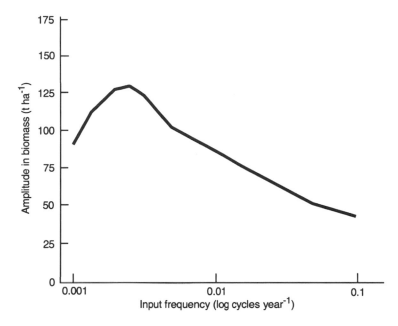

Figure 9.12. The relationship between the amplitude of variation in biomass and the periodicity of a sinusoidal environmental input function as simulated by the ZELIG model. The response is relative to an input of random variation ('Gaussian white noise') of an equivalent amplitude. Note that the forest tends to amplify variation in frequencies of 0.01 and 0.001 cycles per year (typical periods of 250 years). From Shugart and Urban (1989).

There is a direct relationship between the temporal response of a dynamic system and the time responses of the system. Systems with rapidly responding components are affected by the relatively high frequency variation that is attenuated by systems dominated by components with slower responses. Forests have a considerable amount of internal feedbacks among processes and should be expected to attenuate certain frequencies of disturbances or climate cycles, and also to amplify other frequencies. Based on experiments with simulation models, Shugart and Urban (1989) found that forest could be expected to amplify periodic variations in the range of periodicities of 100 to 200 years (Fig. 9.12). Such amplifications would be manifested as cohort dominance in mature forests and periodic synchronous die-backs in forests. These features have been reported for a variety of natural forests.

Emanuel *et al.* (1978a,b) conducted a series of experiments using simulation models in which they inspected the frequency response (the effect on the dynamics of the system to inputs of different frequencies) with the addition and removal of certain species and under conditions of simulated stress (Fig. 9.13). The results of these model experiments were that certain species as a consequence of their life-history strategies in interaction with the environment have a tendency to amplify certain periodicities while other species have the opposite tendency and attenuate periodic inputs.

Amplifiers were often what were categorised as Role 1 species (species that require and generate canopy gaps) and tended to form forests that locally were even-aged. Attenuators were species that were Role 2 and Role 4 species and tended to form forests with 'J'-shaped diameter frequency distributions (even at relatively small spatial scales).

While these results are based on model experiments and involve inspections of actual data that would require hundreds of years of detailed observations, there is considerable evidence for cohort dominance in Role 1-type trees in a variety of forests. A second result of the model investigations of forest responses in the frequency domain – that model forests under stress conditions tended to display a reduction in the richness of the frequency response – has actually been demonstrated experimentally in microcosm-based studies (Van Voris et al. 1980).

The earlier discussions of species roles with respect to gap replacement indicated that two of the roles (1 and 4) were self-reinforcing. Role 1 species require and generate canopy gaps; Role 4 species are the logical converse. Each of the roles competes with the other by changing the spatial grain of the forest light environment. A plenitude of Role 1 individuals favours a forest with a high degree of variation in the spatial pattern of high and low light levels on the forest floor. An abundance of Role 4 individuals would tend to result in forests with a greater degree of homogeneity.

Because the competitiveness of an individual tree is strongly related to its size, the interaction among trees is complex. Size is not typically a state variable used in developing competition theory in animal populations, so that comments regarding tree competition in relation to traditional competition theory are a special case. Competition among trees at the scale of a small patch involves two sorts of interaction. First, trees can compete with one another for a right to determine the 'rules of the game'. For example, Role 1 and Role 3 species prosper when the typical tree in the forest is large. However, if Role 3 species are successful in their competition for sites with Role 1 species, they tend to install a small canopy forest that in turn disadvantages the Role 1 species. Thus, one has

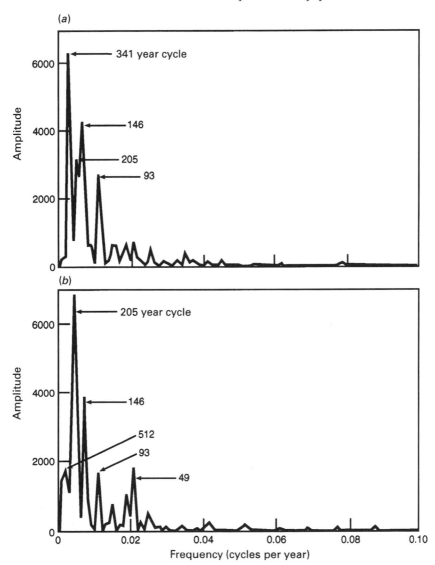

Figure 9.13. Power spectral density for the biomass dynamics of the FORET model at a state of quasi-equilibrium (average of 100 simulated 1/12 ha plots). (*a*) Base case with 32 species of tree potentially in the simulated forest: thought to approximate the condition of mid-slope forests in eastern Tennessee: (*b*) Case with American chestnut (*Castanea dentata*) added as a viable species. Data from Emanuel *et al.* (1978a,b); figure from Shugart *et al.* (1981).

the situation in which the presence of a Role 1 species is advantageous to a Role 3 species, but the presence of a Role 3 species is disadvantageous to Role 1 species. These relationships, if diagrammed by a zoologist, would resemble predation in some cases, parasitism in others and mutualism in still others (Fig. 9.2). Rather than competing for resources such as food and energy, the trees compete for space.

The interactions among the roles of trees do include several interactions that are competitive between roles. Some interactions are mutualistic in that roles tend to help one another. The important point is that interactions among plants are as complex as the interactions among entire food webs of animals competing with and feeding upon one another (see Connell (1990) and Goldberg (1990)). Presumably, within a role, competition resembles that described in animal communities. That is, Role 1 species would compete strongly against other Role 1 species for the opportunity to be the dominant Role 1 species in the forest. It is not clear at this point whether competition among trees within roles is as strong as the variety of interactions across roles. It is clear that species that can change the spatial grain of the ecosystem to either more or less heterogeneity could potentially manifest an extremely large effect on ecosystem function.

Consequences of gap models at the landscape level

As a basis for developing theories concerning landscape functioning, a landscape can be thought of as a mosaic of different patches in which the 'within-patch' changes strongly control the change in the patch over time. This is the logical initial case to be considered in understanding landscape dynamics. This case is relatively simple to conceptualise, and there is a surprising richness in the dynamics and patterns that can develop on such mosaic landscapes. There has been a considerable development of model-based theory for this type of landscape for forest ecosystems. For this reason, this chapter will focus initially on the biomass dynamics of forested landscapes. Then, the general implications of these results for other landscape systems will be discussed, particularly with respect to the dynamics of animal populations inhabiting mosaic landscapes.

When one considers the dynamics of biomass in a small area of the order of 1/10th of a hectare in a forest landscape matrix, the expected pattern is a 'saw-toothed' curve with increases of biomass associated with growth, rapid drops in biomass when large trees die and re-initiating of growth after the regeneration phase. The expected pattern of biomass dynamics for a changing mosaic landscape was discussed in Chapter 7 and is revisited here because the application of individual-based models has contributed to ideas about the shape of the landscape biomass-change curve for different forests.

What are the biomass dynamics of the entire landscape? If the landscape is a collection of a very large number of independent elements, then the shape of the biomass curve of an entire landscape is the sum of the shapes of each of the biomass curves for the small pieces of landscape. The idealised or fundamental response is for a landscape with similarly behaving elements that have all been simultaneously altered by a single event, for example a forest fire or timber harvest. Gap models have been used to develop such of these idealised landscape dynamics (Fig. 9.14) for a range of forest systems. These curves can then be compared to determine the effects of species diversity, external disturbances and other features in altering the idealised landscape dynamics curve.

Biomass dynamics on forested mosaics

The dynamics of relatively simple forests (forests that are dominated by one or two species) represent a fundamental or intrinsic biomass response for the landscape as described initially in Chapter 7 (see Fig. 7.11). Graphs of such biomass curves are shown in the upper left-hand portion of Fig. 9.14. The two cases are Arkansas forest with wildfire and Australian eucalyptus forests without wildfire. Under the appropriate fire regime, these forests (and the model simulations of these forests) are strongly dominated by one species of tree (*Pinus taeda* in the case of the Arkansas forest and *Eucalyptus delegatensis* in the Australian forest). In both simulation cases, starting with a landscape composed of 50 patches of completely cleared ground, the initial response of the summed biomass curve is to increase to a maximum and then to drop to an oscillating equilibrium level of biomass after a period of time. This result is caused by the synchrony (and subsequent desynchronisation over time) of the 50 saw-tooth curves describing the biomass of each individual point and summed to produce the landscape response. Initially, all of the saw-teeth in the component curves are synchronised. Their sum, the landscape biomass curve, increases rapidly, because each of the first 'teeth' in each of the 50 saw-toothed curves are added together.

When the component curves are all approaching their maximum value, their summation reaches its maximum value. The large trees on each of the plots mature and eventually die. This mortality is a chance event, and, as the large trees on each of the plots die, the synchrony of the saw-teeth in each of the small plot biomass dynamics curves is lost. The

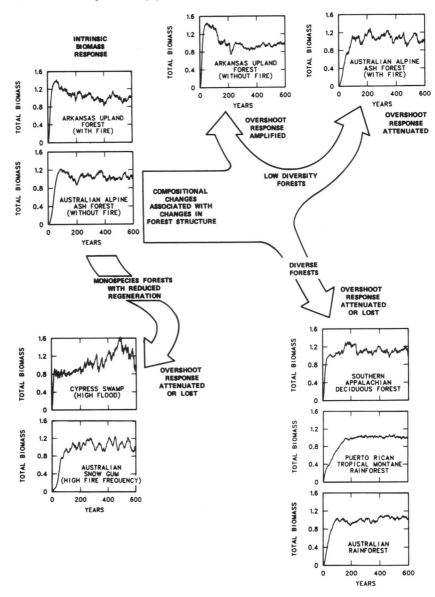

Figure 9.14. Biomass dynamics of several forests as simulated by several different gap models. Each simulation is the sum of 50 plots starting with open plots in year 0 and simulated for 600 years. For aid in comparison, each graph is scaled so that the biomass in year 600 has a value of 1.0. From Shugart (1984).

sum of these desynchronised curves for biomass settles to a lower value because the summation is over a mixture of landscape elements, some with mature canopy trees and, therefore, high biomass and others with smaller trees in the recovery phases of the mosaic dynamics.

In the right side and upper corner of Figure 9.14 are the biomass dynamics for forests that have relatively few species but a sequence in the replacement of the species through succession. In the case of the Arkansas forests without fire, there is a very large increase in biomass and a very large drop in biomass. The initial response of the dynamics of the component curves is for a landscape element dominated by pine trees (*Pinus taeda*); the regeneration and recovery on the landscape elements following the death of these dominant trees is mostly to species of oak (*Quercus* spp.) of relatively smaller stature. The compositional change in this simulated forest succession amplifies the overshoot behaviour of the biomass dynamics curve with respect to the eventual equilibrium.

In Australian eucalyptus forests with fire (the furthest right corner of Fig. 9.14), one finds the opposite case. The species composition changes in this successional sequence tend to replace smaller trees with large trees later in the sequence. Thus, the biomass overshoot response is relatively attenuated.

For low-diversity forests, the landscape biomass curve can be altered in its shape by the shift in its composition of species during succession. Some mixtures of species in a landscape produce very large initial flushes of standing biomass over time, with the more mature forests being of lower stature. In other cases, the opposite can occur. The response is intrinsically case specific.

In the lower right-hand portion of Fig. 9.14, there are biomass dynamics curves for simulated landscapes for diverse forest ecosystems. There appears to be a loss in the magnitude of the pronounced overshoot seen in less complex forests. Diverse forests have species with different growth rates, sizes and longevities. Consequently, the synchrony of the saw-tooth curves that make up the biomass dynamics of the landscape is lost early in the successional sequence and the biomass curve for the forest does not feature the overshoot behaviour.

The quasi-equilibrium landscape

The idea of a forested landscape as a dynamic mosaic of changing patches was well expressed by Bormann and Likens (1979a) in what they call the 'shifting mosaic steady-state concept of ecosystem dynamics'.

...the Shifting-Mosaic Steady State may be visualised as an array of irregular patches composed of vegetation of different ages. In some patches, particularly those where there has been a recent fall of a large tree, total respiration would exceed GPP (gross primary production), while in other patches the reverse would be true. For the ecosystem as a whole, the forces of aggradation and of decomposition would be approximately balanced, and gross primary production would about equal total ecosystem respiration. Over the long term, nutrients temporarily concentrated in small area in fallen trunks would be made available to large area by root adsorption and redistribution by litterfall.

The structure of the ecosystem would range from openings to all degrees of stratification, with dead trees concentrated on the forest floor in areas of recent disturbance. The forest stand would be considered all-aged and would contain a representation of most species, including some early-successional species, on a continuing basis.

In an essential sense, this is an old concept in ecology (Watt 1925; Aubréville 1933, 1938; Whittaker 1953; Whittaker and Levin 1977). The idea that a landscape could be viewed as a statistical average of the mosaic elements that make it up united with the idea that the mosaic elements themselves are non-equilibrium systems was implied in discussions on the mosaic nature of ecological landscape (Chapter 7). Figure 7.12 (p. 200) illustrated the summation of individual-patch, saw-tooth curves for biomass at small scales producing a landscape biomass curve. If the number of patches making up a landscape is large, the landscape dynamics (a statistical average) will become predictable.

In landscapes composed of many patches, the proportion of patches in a given successional state making up the landscape will also be relatively constant. As Bormann and Likens (1979a) point out in the above quotation, the resulting theoretical landscape will contain a mixture of patches of different successional ages. In small landscapes or landscapes composed of relatively few patches, the stabilising aspect of averaging large numbers is lost and the dynamics of the landscape and the proportion of patches in differing states making up the landscape become less predictable. Some of the properties of large and small landscapes are summarised in Table 9.1.

If a landscape is small, it takes on many of the attributes of the non-equilibrium mosaic patches that make it up. One can classify such a landscape as being an effectively non-equilibrium landscape. If a landscape is large, then the averaging of the non-equilibrium behaviour of the patches that make it up tends to make it act as an equilibrium system. Such a landscape is referred to as a quasi-equilibrium landscape.

An important attribute of landscapes involves the size of the patches

Table 9.1. *Properties of effectively non-equilibrium and quasi-equilibrium landscapes*

Property	Effectively non-equilibrium landscape	Quasi-equilibrium landscape
Disturbance size	Large	Small
Landscape size	Small	Large
Time since last disturbance for component landscape patches	Irregular distribution of ages of patches since last disturbance	All ages of patches caused by disturbance
Total landscape biomass	Unpredictable	Regular
Age distributions of component populations	Unstable for long-lived organisims	Stable

Source: Modified from Shugart (1984).

that make up a given landscape. The disturbances that operate on a landscape determine, as least to some degree, the patch size. One can combine the operation of disturbance scale with the condition that a landscape must contain a certain number of patches to become a predictable entity in a statistical sense. This allows landscapes to be categorised with respect to their size and the size of the disturbances that pattern them (Fig. 9.14).

For example, in Australia, the amount of land burned each year by fires approaches the size of the actual species ranges of a large number of the commercial forest species. Therefore, for many of the species that are harvested in Australia, the entire species population does not have a stable age distribution. Some ages of trees can be represented more than others because they are of individuals regenerated in a particular fire and not subsequently destroyed by later fires. For example, *Eucalyptus delegatensis* tree populations were virtually destroyed in a tremendous forest fire in Australia in 1939 that burned over the range of most of the entire species. For this reason, there are (in 1998) few trees of this species older than 55 years. A large number of trees regenerated following the 1939 fire. There were other fires that also created big mortality events followed by big birth events since 1939 (notably in 1984). Thus, for *Eucalyptus delegatensis* throughout south-eastern Australia, most of the trees are only of a few age classes. This situation has important consequences. One of these is that several species of animal that require old *Eucalyptus delegatensis* trees as habitat are now considered endangered species.

If one plots the scale of the disturbances that drive a forest and the size

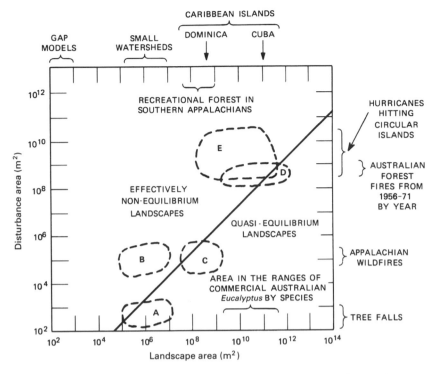

Figure 9.15. Scale of disturbance and scale of landscape for several example ecosystems. From Shugart and West (1981).

of a forest (Fig. 9.15), one can separate landscapes into effectively non-equilibrium landscapes and quasi-equilibrium landscapes. In Figure 9.15, landscape area is plotted along the x-axis; disturbance area is plotted along the y-axis. The line on this graph is a 1-to-50 ratio of disturbance area to landscape area. The 1-to-50 ratio was derived (see Shugart and West 1981) from the number of samples of simulated plots that needed to be averaged to obtain a statistically reliable estimate of landscape biomass. About 50 plots, taken on average, tend to produce a fairly predictable landscape level response. On this graph, one sees that the situation described above for forest fires and the range of tree species in Australia is as dashed area D. Therefore, very many of the Australian forests dominated by *Eucalyptus* species are effectively non-equilibrium landscapes with respect to their biomass dynamics.

If the fall of the tree is the disturbance of interest (gap-scale disturbances), then small watershed-sized areas correspond to area A of Fig. 9.15. These are quasi-equilibrium landscapes. A watershed biomass level for watersheds of first-order streams in the Appalachian Mountains is dis-equilibrated by treefalls. However, if Appalachian wildfires are the focal disturbance (area B in Fig. 9.15), these same watersheds are relatively small and the dynamics of their biomass would be unpredictable without a known fire history. Such systems are categorised as effectively non-equilibrium landscapes. Indeed, only in the largest parks in the south (indicated by C) are the landscapes large enough to average away the effects of the disturbance from forest fires in the total systems biomass dynamics.

In some cases, entire biotas may inhabit effectively non-equilibrium landscape. One continental-scale example has already been discussed for *Eucalyptus* forest biomass dynamics under the Australian fire disturbance regime (D in Fig. 9.15). As another example, the size of the hurricanes that disturb West Indian forests are large when compared with the size of the islands in the Caribbean (E in Fig. 9.15). The Caribbean islands are small with respect to the spatial scale of a major climatological feature that disturbs them and for this reason might function as effectively non-equilibrium landscapes.

From a practical point-of-view, it is extremely difficult to manage effectively non-equilibrium landscapes. Landscapes that are small with respect to the forces that change them in nature can be expected to have an erratic intrinsic behaviour. Such systems are difficult to manage toward a goal of constancy because they are regularly disequilibrated by disturbance events. Busing (1991) points out that to manage a landscape for a particular habitat type (or for a particular species that uses one of the several habitat types that occur on a dynamic mosaic) requires a landscape area much greater than the biomass-based 50-to-1 ratio of landscape size to disturbance size used to develop the two classes of landscape in Fig. 9.15. As will be discussed in the following section, habitat dynamics on small landscapes increase the extirpation rate of resident species. These considerations point to the need for very large land areas for nature reserves or parks that are intended to preserve habitat and biotic diversity. Further, as Kessell (1979a) points out, the natural landscape manager needs the capability to project the future response of the landscape to the particular regime of disturbances and habitat types as a prerequisite to rational management.

Mosaic dynamics and habitat dynamics

As the availability of habitat suitable for a given species waxes and wanes with the dynamic responses of the landscape under a particular disturbance regime, the resident population would be exposed to habitat 'bottlenecks' in which the population is forced to low levels in times of habitat shortages. Therefore, one might expect small landscapes driven by relatively large disturbances (called effectively non-equilibrium landscapes in the section above) periodically to present the populations of species that inhabit them with high likelihoods of extinction. This suggests a general investigation of patterns of species diversity on mosaic landscapes under disturbance regimes of different relative sizes.

The response of a plant or animal population whose habitat is one of the elements making up a dynamically changing mosaic is related to the rate of increase of the population with respect to both the rate of change and the frequency of disturbance of the landscape mosaic. In large landscapes with landscape dynamics that are slow with respect to the dynamics of the population, then the population response is strongly controlled by the birth and death rates of the population. In this case, factors such as the availability of resources or space in suitable habitat are relatively constant factors, and the traditional modelling approaches of population dynamics (viz. autonomous non-linear differential equations, Chapter 4) are more likely to be appropriate to project the expected population response.

When the landscape processes are similar to (or fast relative to) the population's dynamics, one would expect the population dynamics to reflect an influence from the landscape processes. From theoretical arguments, the potential for population oscillations, instabilities and extinctions are greatly increased when population systems are driven with periodic inputs of a frequency near their natural frequency (see May 1973).

Chapter 5 provided several examples of species being distributed in space according to their Grinnellian niche. From a Grinnellian perspective, one would expect the locations of the individuals constituting a species population on a mosaic landscape to be associated with suitable microsites. Further, on a dynamic mosaic landscape, habitat microsites could be expected to vary in both space and time. If such a system were represented by a logistic equation (Eqs. 4.8 and 4.9), the K term (representing the carrying capacity for a given landscape) would be expected to fluctuate as the proportion of the landscape comprising suitable habitats changes with time. One could predict the occurrences of a species (in space) with increasing confidence when the landscape processes (manifested as suitable habitat availability over time and/or space) dominate the population processes. This would be the case for highly mobile species (with the ability to disperse to suitable microhabitats) or over a sufficiently long period of time.

Diversity of species as a response to size of habitats

In a landscape that behaves as a shifting mosaic of habitats, diversity patterns observed by community ecologists can be described by seemingly simple models relating the species–carrying capacity to habitat availability on the mosaic landscape. One of these models is the species–area curve: an important relationship in the development of the theory of island biogeography (Fig. 9.16).

Seagle and Shugart (1985) used a model of the vegetation dynamics of the island of Tasmania (Noble and Slatyer 1980) to simulate the patch dynamics of islands of different sizes. Ten habitat types were recognised, based on Noble and Slatyer's analysis of the successional pattern in Tasmanian wet-sclerophyll and temperate rain forest landscape. The transition probabilities between habitat types were calculated as the reciprocal of the successional time required to change from one habitat type to another. The resultant 10×10 matrix of transition probabilities was implemented as a first-order Markov process (see Chapter 10). The age of each patch in the simulated landscape was also computed as the length of time that the patch had been classified as a given type. The landscape area was taken to be the number of patches in a particular simulation. For each simulation, the starting conditions for each patch in the landscape were allocated in proportion to the equilibrium distribution of patches for the underlying Markov matrix. The successional pattern in Tasmania was characterised by Noble and Slatyer (1980) as being influenced by the effects of wildfires of different intensities and by landslides. These disturbances caused the later successional habitat types to be reset to earlier types. At equilibrium, the landscape simulated by the model has all habitat types represented to some degree on the landscape because of the inclusion of these disturbances.

To simulate the colonisation of these model island landscapes by animals, a pool of 150 hypothetical species was generated from uniform statistical distributions of different species-level attributes. These characteristics included:

1. The range of habitats utilised by each species
2. The range of habitat ages utilised by each species
3. The intrinsic rate of increase of each species

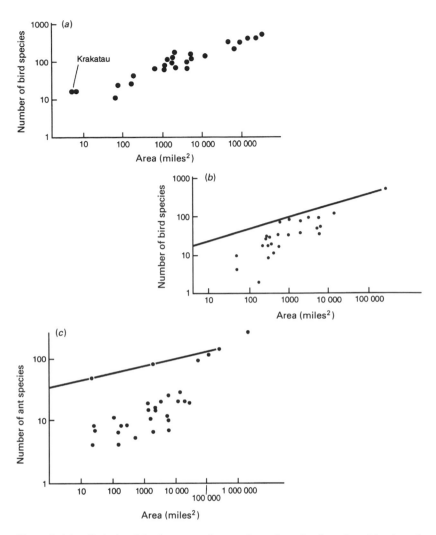

Figure 9.16. Relationships between the number of species found on islands and the sizes of islands. The relationship (sometimes referred to as Darlington's law) is for a power relation (where *S* is the number of species, *A* is the area of an island and *C* and *z* are model parameters). (*a*) Numbers of species of land and freshwater birds on islands in the Sunda group in southeast Asia, the Philippines and New Guinea (from MacArthur and Wilson 1967). (*b*) Numbers of species of land and freshwater birds on islands in the South Pacific including the Moluccas, Melanesia, Micronesia, Polynesia and Hawaii. The line is drawn through the two islands (Kei and New Guinea) nearest to the source regions for the species and indicates lower species richness for islands of comparable size but with a great degree of isolation (from MacArthur and Wilson 1967). (*c*) Number of species of ponerine ants in the faunas of various Mollucan and Melanesian islands. The line in this case represents the number of species in quadrats of increasing size on New Guinea. The islands support fewer species than a land area of equivalent size on New Guinea (from Wilson 1961).

4. The number of habitat patches needed to support an individual of a given species

5. The species-specific efficiency (used to determine the competition coefficient), for cases in which one species competed with another, of utilising patches that overlapped in their potential occupancy (owing to 1 and 2, above) with other species.

Points 1 and 2 are consistent with allowing the carrying-capacity term in the logistic equation to vary with the habitat availability. Points 1, 2 and 5 are directly related to features of the niche-hyperspace concept (see discussion in Chapter 5) relating competition with niche overlap.

During each year of a model run, all populations on the simulated island were allowed to grow or decline according to the logistic equation, where the carrying capacity was the sum of the patches suitable for occupation by a given species divided by the number of patches required to support an individual of the species. When the species were allowed to compete (see Lotka–Volterra equation, Eqs. 4.10 and 4.11), the competition coefficients were the proportion of patches a species shared with the competitor multiplied by the ratio of their utilisation efficiencies. The probability of a population (of size N) becoming extinct was the inverse of 2 raised to the N^{-1} power (the probability of all individuals in a given generation being the same sex). Population simulations from the model (Fig. 9.17) were initialised with no species occupying the simulated landscapes. Colonising species were selected randomly from the pool of available species to allow extinct populations to re-establish themselves on the simulated landscape (mean = 12 invasions; standard deviation = 2.0).

The effect of increasing landscape area on the number of established species (Fig. 9.18) is a rapid increase in the number of species. This rate of increase in diversity with landscape size slows as the landscape area becomes large. The shapes of the species–area curves are not different. However, there are fewer species on a landscape of a given size when the effects of competition are included in the simulation. In general, smaller landscapes support smaller populations that are sensitive to extinction because of fluctuations in suitable habitat. These habitat fluctuations are a product of statistical variation in disturbances and the rates of successional processes on small landscapes. Small landscapes with the high extinction rates, and consequently a low richness of animal species, are what were called 'effectively non-equilibrium landscapes' in the section above.

The model,

$$S = CA^2 \tag{9.5}$$

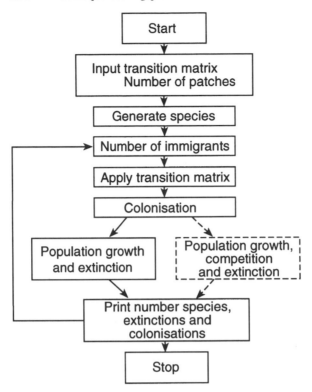

Figure 9.17. Flow chart of a model of animal abundance on landscapes of different size. From Seagle and Shugart (1985).

where S is the number of species; A is the area of an island; and C and z are model parameters, has been used to describe the relation between the size and the species richness of oceanic islands (Preston 1962; MacArthur and Wilson 1967). There is considerable debate on the interpretability of the parameter z (Connor and McCoy 1979; Sugihara 1980), but its value has consistently been reported from empirical studies to be in the range of 0.20 to 0.35. Fitting the Eq. 9.5 to estimate values of z for the curves shown in Fig. 9.18, Seagle and Shugart (1985) obtained values of 0.29 for z in the no-competition case and a value of 0.31 in the case with competition. While these results are correlative rather than conclusive, they indicate the possibility that patterns of animal richness reported on oceanic islands could be, at least in part, a direct consequence of the size-related statistical variations in landscape dynamics. One important feature of this result is that the shape of the species–diversity-to-area relation for species on a mosaic landscape is not particularly different in cases in

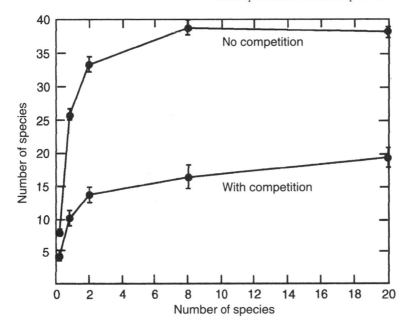

Figure 9.18. Number of species after 300 simulated years on landscapes of different size and with or without interspecific competition. From Seagle and Shugart (1985).

which the species compete, and in those in which they do not.

Diversity of species as a response to successional processes

In a similar model experiment using an individual-based forest simulator (ZELIG), Urban and Smith (1989) simulated a 9 ha forest stand for 750 years and sampled the simulated landscape at 50-year intervals to characterised 225 0.04 ha microhabitat patches. These samples are equivalent in their spatial scale to microhabitat samples frequently used in bird-habitat studies (James and Shugart 1971). On each of the simulated microhabitat patches the diameters of the trees in five size classes (0–10, 10–20, 20–30, 30–60 and >60 cm DBH) were collected to quantify the microhabitat conditions.

Urban and Smith (1989) were interested in determining what patterns observed in animal populations could be associated with the changes in patterns of microhabitats during succession. The variability in micro-habitat was measured by the variability in the tree diameter distributions in successional time (Fig. 9.19). Starting with recently abandoned land (and

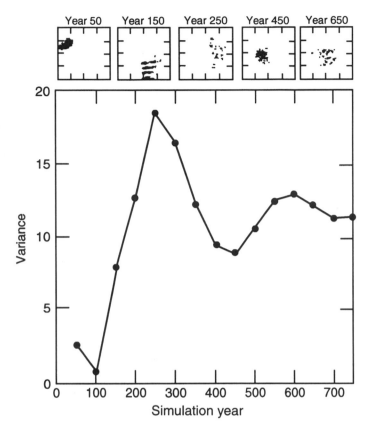

Figure 9.19. Pattern of variation of microhabitat structure through time based on the area of the 95% confidence ellipses of two principal components for each of the 15 sampling periods. From Urban and Smith (1989).

assuming that the plants all have an adequate nearby seed source), the variance in microhabitats initially decreases until about year 100, followed by an increase in microhabitat heterogeneity for 200 years and eventually settling to an intermediate level of variability over a long period of time. These changes are directly related to processes that produce similar responses in the biomass curves discussed in the sections above and in Chapter 7.

Urban and Smith (1989) used what is called a 'null model' approach to inspect the patterns of animal population associated with ecological succession. In a null model approach, one designs a simple model to detect whether given observed patterns in nature require a more complex explanatory model. They generated 10 sets of 50 species niches with

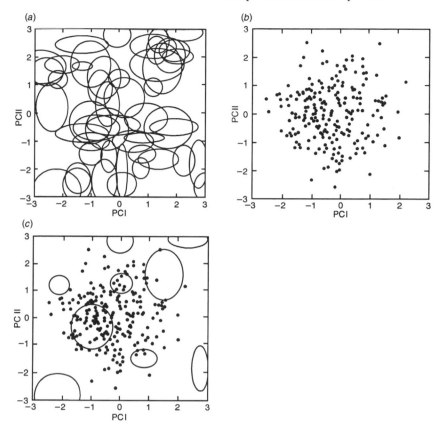

Figure 9.20. Schematics of procedures used to generate species abundances through succession. (*a*) Set of 50 niche ellipsoids in microhabitat space. The axes are the first two principal components (PCI and PCII) computed, and account for the variation in habitat structure simulated by an individual tree-based succession model (see Chapter 5 for a discussion of the application of multivariate statistical procedures as an analogue of the Grinnellian niche). (*b*) Statistical distribution of microhabitat samples in principal component space in year 750. (*c*) Several of the niche ellipses from (*a*) overlaid with the sample distribution of the habitats from (*b*). Samples falling within each ellipse are tallied as abundance for each species. From Urban and Shugart (1989).

respect to statistically obtained variables (principal components) that represented the patterns of variation in the simulated microhabitat on a 9 ha forest plot over successional time. The sizes of these niches were typical of those measured in the field for birds with respect to the simulated habitat variables. An example of one of these sets of niches is shown in Fig. 9.20*a*. To estimate the related success of a species, the quantified

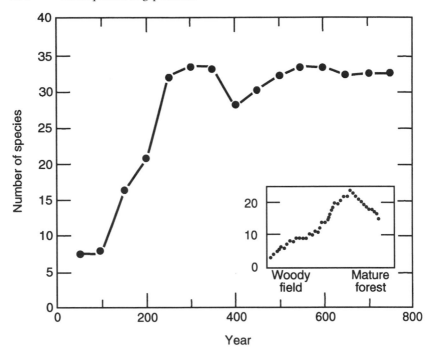

Figure 9.21. Number of hypothetical bird species supported by the simulated forest through time (mean of 10 species assemblages). Inset: number of bird species censused in second growth forest in southern Illinois and graphed on an ordinated successional gradient (unpublished data from Urban (1981)). The axis in the inset corresponds to the first 400 years of the simulation. From Urban and Smith (1989).

value of the simulated microhabitat at a given time (Fig. 9.20*b*) was matched with the niches of the species (Fig. 9.20*c*). Microhabitats falling in a given species niche were classified as contributing to the abundance of that species.

The number of bird species measured by actual field studies in forests of different ages (shown as an inset in Fig. 9.21) shows a strong pattern of similarity to those obtained by applying a random niche model to the simulated habitat data. Diversity of bird species initially increases over forest succession as the forest builds, then decreases when the forest crowns are closed and thinning processes actually reduce the microhabitat diversity (as measured by tree structure). As the canopy breaks up as a result of tree mortality and gap formation, the microhabitat diversity increases and the expected diversity of birds goes up.

Urban and Smith (1989) distinguish two levels in the study of bird habitat relationships. At the stand level, variation in microhabitat is seen as accounting for the diversity and abundance of species in the stand. At a larger, landscape level, the variety of microhabitat present for birds is related to the area of the landscape. Consequently, at the landscape level the diversity of bird species is strongly related to area. Birds are highly mobile animals, potentially able to locate suitable habitat across large landscape mosaics. One might expect these effects to be moderated in plants and animals less able to disperse themselves. Hunter (1990), in a major text on the topic of managing forests for wildlife diversity, notes a similar distinction to that of Urban and Smith (1989). Hunter elaborates what he calls a 'macro approach' to diversity management that is focused on maintaining landscape habitat heterogeneity and a 'micro approach' to diversity management that involves the creation of microhabitat patterns in forest stands that are necessary for the needs of particular species.

Concluding comments

The trade-offs that appear to arise from biochemical, physiological and morphological functions in individual plants can be used as a basis to identify plant functional types. These plant functional types are useful for understanding the overall patterns of dynamics of ecosystems. In the example of shade-tolerant and shade-intolerant plants, a simple categorisation of plants into two functional types is a traditional explanation of dynamics in (particularly) closed-canopy systems. Individual-based models can then be used as theoretical tools to explore the consequences of different patterns of physiological trade-offs in plants and to explore the complexities of plants interacting with one another indirectly by means of environmental modifications at local scales. The basis for plant functional types is in the manner in which the environment interacts with the plants to produce consistent patterns of physiological, morphological and life-history response in the distribution patterns of the species. For a functional types cosmology to produce satisfactory results in predicting the change of vegetation in a world with altered climate (temperature, moisture and water concentration in the atmosphere), the goals and the time and space scales of the application must be stated very clearly. One important determination is the degree to which demographic aspects of plants (reproductive features, establishment, mortality rates) are tied to the functional aspects (metabolism, physiology, morphology).

A recent comparison of formulations of functional types for a range of different biomes (Smith *et al.* 1997) found that the classification of plants into functional types based on demographic dimensions (regeneration habit, mortality rates) and functional attributes (growth rates, shade tolerance) appears to be decoupled. Given the seed of a plant, the process of inferring the attributes of the plant from which it came (without a prior knowledge of the flora) is relatively uncertain. Barring the trivial cases (coconuts would not be expected on a herbaceous plant), a light-seeded plant could be a herb, tree or shrub.

If environmental changes excite processes that involve demography (migration) as well as function, then the necessary functional typology will likely need to be two-dimensional (demography × function). It appears to be equally enigmatic to guess the response of plants to increased levels of water from their other attributes: implying an important third dimension to the typology (Smith *et al.* 1995).

The basis for ecosystem functional types discussed in this chapter also arises from the manner in which plants modify their environments and, hence, the ecological systems in which they are found by changing the temporal, spectral or spatial fabric of the system response. In the earlier sections, a simple plant functional type categorisation (the four roles of trees) was used to identify different ecosystem functional types (attenuators, amplifiers, etc.). The relationship between plant and ecosystem functional types may not be a one-to-one mapping. Species that are quite different at the plant functional type level might alter ecosystem response in similar ways and, therefore, be considered similar ecosystem functional types. Strong reinforcing feedback between alterations in the environment caused by the success of a group of species of the same ecosystem functional type that also favours the plant functional types found in the group represents an ordering principle for ecosystems. The implication is that not only do species get sorted by the environment (e.g. Box 1981) but that ecosystems, in a sense, also sort the components of which they are composed. Such a principle(s) could explain why not only the structure and life history of plants with different phylogenetic origins are often convergent at the species level in different environments, but also why the ecosystems themselves appear similar.

The importance of mosaic processes discussed initially in Chapter 7 as a fundamental consideration for understanding larger-scale ecology is reiterated in the more theoretical discussions in this chapter. Large-scale aspects of the expected dynamics of mosaic landscapes include the 'overshoot' response of biomass dynamics of forests, and considerations that arise from the equilibrium or non-equilibrium nature of landscapes as a consequence of disturbance size. Smaller-scale aspects of the dynamics of mosaic landscapes relate to issues of diversity of species on mosaic landscapes as a function of size and of time.

10 · *Landscape models*

In the previous chapter, theoretical dynamics of landscapes assumed to function as dynamic mosaics were presented with relatively little direct consideration of the consequences of environmental change on such landscapes. However, landscapes should change in response to either local or widespread environmental variation. The prediction of these responses for landscapes (or larger regions) is a challenge both in terms of model development and model testing. Landscape models require large data sets for initialisation and parameter development. In many cases, the logistics of conducting experiments on entire landscapes (and the time needed for the responses of landscapes to experimental manipulation) make the direct testing an impossibility.

One source of information for testing and formulating landscape models is the spatially extensive data sets collected by satellites and other monitoring systems over the past decades. While these images may not be at the resolution required to test a particular model, the increased ability to characterise large areas has heightened the interest in the function of landscape systems. Despite problems, the need to predict better the consequences of change on natural landscapes has created an imperative to develop and test landscape-level models.

A basic consideration in modelling the dynamics of landscape systems is whether or not the landscape can be divided reasonably into a number of discrete, smaller units (Weinstein and Shugart 1983). For different types of model, landscapes are assumed to be:

1. Mosaic landscapes: the models are based on an assumption that the landscape can be divided into discrete, non-interactive elements that change dynamically with time
2. Interactive landscapes: the models are constructed with a recognition that the landscape elements constituting the modelled landscape are interactive and can interact in ways that alter another's dynamics

3. Homogeneous landscapes: the models are based on an assumption that the landscape functions as a uniform dynamic unit.

Each of these views of landscapes is based on the nature of interactions and dominant processes and has an associated modelling paradigm. Some of the features of mosaic landscapes were discussed in the previous chapter. Models of mosaic landscapes as well as the other two types of landscape conceptualisation will be discussed in this chapter. The applicability of models of each of three types of landscape is dependent on the particular phenomena as well as the landscape under consideration.

Mosaic landscape models

For mosaic landscapes, Weinstein and Shugart (1983) divided cases of simulation models into four cases based on the number of elements in a landscape and the time each was considered to be discrete or continuous (Fig. 10.1). Markov models, differential equation models and difference equation models are obtainable from the same sorts of data set (re-measurement data indicating change of the state of some number of landscape elements). The example cases indicated on Fig. 10.1 will be discussed initially. Cases in which the elements of the landscape system are interactive or in which the landscape is assumed to be homogeneous will be discussed in later sections of this chapter.

Markov and semi-Markov models of vegetation

Markov models of succession are mathematically and conceptually the most straightforward of the succession models presently in use. These models are constructed by determining the probability that the vegetation on a prescribed (usually relatively small) area will be in some other vegetation type after a given interval (van Hulst 1979). These probabilities are referred to as transition probabilities. Markov models have two important properties.

1. The transition probabilities at time t_n depend only on the immediate past value at time t_{n-1} and are independent of the state of the system at any time earlier than t_{n-1}. This condition describes what is sometimes referred to as a first-order Markov process.
2. The system is stable if the transition processes do not change (are homogeneous) in time.

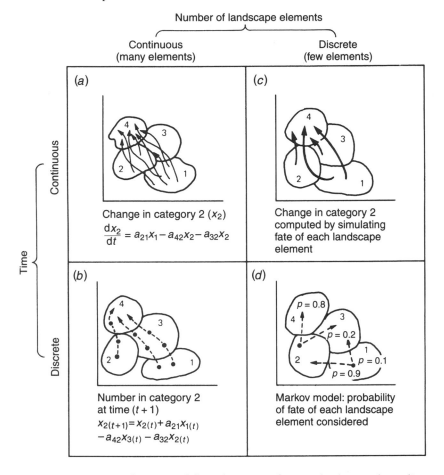

Figure 10.1. Approaches to modelling dynamics of mosaic landscapes depending on the number of landscape elements and time. From Weinstein and Shugart (1983).

It is an essential requirement of these models to have a scheme for classifying the vegetation into identifiable categories.

The manner in which the vegetation states are classified has varied across applications of Markov models. Horn (1975a,b, 1976) used the species of a canopy tree as the state of a Markov model developed for a forest near the Institute of Advanced Studies in Princeton, New Jersey (Fig. 10.2). The time interval of this particular model was the generation time of canopy trees. Waggoner and Stephens (1971) categorised the forest types according to the most abundant species (of individual trees over 12 cm DBH) on 0.01 ha plots located on the Connecticut

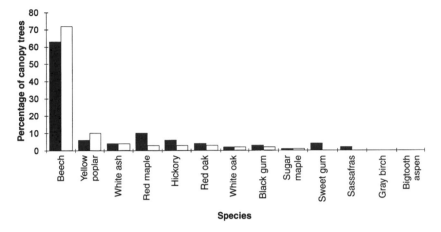

Figure 10.2. Predicted (■) and actual (□) composition of canopy trees in a forest near Princeton University's Institute of Advanced Studies. Simulated results are from a Markov model developed by Horn (1975b).

Agricultural Experiment Station and applied a Markov model over uniform time intervals. Other Markov models of ecological succession in which the dominant species on a small sample quadrate are used to classify the landscape into discrete elements are those of Hobbs and Legg (1983), Lippe *et al.* (1985) and Gimingham *et al.* (1981) for heathlands; Isagi and Nakagoshi (1990) for fire-dominated landscapes; and Acevedo (1980, 1981) for tropical rain forests.

One can use multivariable state classification schemes in developing Markov models. For example, one could categorise a small plot of land both by the species of the largest individual and by the number of individuals (e.g. highly-stocked white oak-dominated type, understocked Loblolly pine stands, etc.). Hool (1966) used this approach in developing a Markov model of stand change over a large area. Austin (1980) and Austin and Belbin (1981) used statistical classification procedures to designate Markov states of lawn vegetation, as did Usher (1981, 1992) for grasslands.

In a Markov model, because one must determine the probability of transition from any state of the vegetation to any other state, the number of model parameters is a function of the square of the number of states (or categories in the model). Therefore, in the development of a Markov model, one is forced to trade-off between the increased resolution in being able to enumerate many different system states and the parameter estimation problems that necessarily attend this greater resolution. Relatively uncommon transitions from one state to another need to be estimated with equivalent precision to those of the other more common transitions. This feature creates a need to observe the frequency of occurrence of the rare transitions between some of the states and implies large re–measurement data sets.

An alternative to direct measurement to determine the parameters of a Markov model is to develop a theory that allows the estimation of the model parameters on some other basis. For example, Horn (1975a,b, 1976) assumed that the proportion of trees of a given species found growing below a canopy tree indicated the transition probabilities. Noble and Slatyer (1978, 1980) have developed a theoretical concept called the 'vital attributes' concept, which uses regeneration, response to disturbance and longevity of plants to determine the parameters of a Markov model (Moore and Noble 1990, 1993).

The vital attributes approach of Noble and Slatyer (1978, 1980) has historical antecedents in Humboldt (1807), Grisebach (1838) and, particularly, Warming (1909) and Raunkiaer (1934). It is philosophically allied with Gleason's (1939) emphasis on the individuals' properties as important determinants in succession. The application of the method in predicting the temporal sequence of community change is analogous to the geographer's applications that involve prediction of spatial pattern of community change as a climate/physiognomy response (e.g. Raunkiaer 1934; Box 1981). The development of theoretical methods of estimating the transition probabilities creates the possibility of developing larger Markov models (in which parameter estimation from data normally would be proscribed owing to logistic difficulties (e.g. Cattelino *et al.* 1979; Kessell 1976, 1979a,b; Potter *et al.* 1979; Kessell and Potter 1980)).

The FATE model (Moore and Noble 1990) is a general model of vegetation dynamics oriented toward an individual-plant-based simulation of vegetation. The modelling approach is to categorise species into functional groups according to their life history, response to the environment and response to disturbances. The attributes used to categorise plants in the model are summarised in Table 10.1, which also provides an indication of the level of detail in the modelling approach. As a demonstration of the application of this class of model, Table 10.2 provides a listing of parameters used in the FATE model to simulate the response of Tasmanian *Eucalyptus* forests to wildfires and Table 10.3 provides an example of output from this model. The useful features of this modelling approach are in the relatively modest information demands with respect to species attributes and the ability of the resultant models to produce the expected pattern of response from vegetation exposed to novel disturbance regimes.

A Markov model can be used as a 'summary model' for more detailed simulators such as an individual-based model by applying the output from the detailed model to estimate the parameters for a Markov model. Alternatively, one can interpret the individual-tree-based model parameters in the context of a 'vital-attributes' modelling approach to obtain a Markovian representation of the species mixture (Moore and Noble 1990). Since both these approaches produce a Markov model, good practice would probably involve applying both techniques and then checking the resultant models for consistency. A Markov model, so derived, can then be used to simulate the expected regional dynamics.

If the interest is in the average behaviour of the vegetation expected for a large area, a first-order Markov model can be modified to create a system of linear ordinary differential equations (Shugart *et al.* 1973). Such Markovian models (or the linear differential equation analogues) are simplifications of any individual-based model that is their basis and, as such, may not have the ability to reproduce the total range of dynamics in the base model. In particular, non-linear dynamics such as hysteretic responses to environmental change or multiple stable equilibria would not be found in this class of model (see Chapter 9).

Differential equation models

Differential equations have been the dominant formulations used to develop models of a variety of ecological systems. Some of the earliest models used in systems ecology (e.g. Garfinkel 1962; Olson 1963) and population dynamics were developed using differential equations as a mathematical paradigm (see Chapter 4). In the case of succession models, one application of differential equations is specifically intended for cases involving the simulation of changes of large areas of land (Shugart *et al.* 1973). These models are related to the Markov formulations that have just been discussed in that the sorts of information typically needed to apply either approach are similar.

The underlying assumptions for developing large-area or regional succession models using differential equations are as follows.

1. The vegetative cover can be divided into a finite number of 'cover-states' (Shugart *et al.* 1973) or vegetation types.
2. The effects of spatial heterogeneity on the dynamics of aerial extent of these cover-states are reasonably constant over the time segments of interest.

Table 10.1. *Parameters used in the FATE model*

Plant Function	Process	Model parameters	Parameter domain
Life history	Dispersal	Is functional group widely dispersed?	Yes/no
	Dormancy/propagule storage	Does functional group have innate dormancy? If so then:	Yes/no
		(i) What is longevity of dormant pool?	Years
		(ii) What is proportion of propagules moving from the dormant to the active propagule pool as a result of a disturbance?	Proportion by levels: none = 0% low = 10% half = 50% most = 90% all = 100%
		(iii) What is longevity of the active propagule pool?	Years
	Germination/enforced dormancy	Germination rates at each resource level	Levels: none, low, moderate or high
	Establishment	See environmental response (below)	
	Growth	Size of immature plants as a proportion of mature plants	Proportion
	Maturation	Maturation time	Years
	Senescence	Life span	Years
Environmental response	Tolerance of environmental conditions	Number of life stages and strata occupied by these life stages	Small integer for germinants, immatures and matures
		Survival or death of individuals in each life stage	Yes/no
	Niche relationships	Maximum abundance at a site	Low, medium or high
Disturbance response	Fate: escape, death, resprouting	Proportion of each life stage meeting each of the three fates	Proportion (also for resprouting plants the age of plants that can resprout)

Table 10.2. *Functional attributes of plants in a forest community of south-western Tasmania used to simulate the response of the systems to wildfire. Table 10.1 provides a description of the parameters and Table 10.3 provides example output*

	Eucalyptus	Nothofagus	Atherosperma	Acacia	'Sedges'
Maximum abundance	High	High	High	Med.	High
Maximum time to maturation	10	30	30	10	2
Life span	400	400	250	70	20
Size of immatures to matures	Low	Low	Low	Low	Low
Stratum of immatures	2	2	2	1	1
Stratum of matures	3	3	3	2	1
Widely dispersed propagules	No	No	Yes	No	Yes
Innate dormancy	No	No	No	Yes	No
Active propagule life span	0	0	0	5	0
Dormant propagule life span				200	
Proportion of dormant propagules activated by fire				High	

	Eucalyptus			Nothofagus			Atherosperma			Acacia			'Sedges'		
Resource level	Low	Mod	High	Low	Mod	High	Low	Mod	High	Low	Mod	High	Low	Mod	High
Germination rate	Nil	Nil	High	Mod	Mod	High	Mod	Mod	High	No	No	High	Nil	Nil	High
Germinant survival	No	No	Yes	Yes	Yes	Yes	Yes	Yes	Yes	No	Yes	Yes	No	No	Yes
Immature survival	No	Yes	Yes	Yes	Yes	Yes	Yes	Yes	Yes	Yes	Yes	Yes	No	No	Yes
Mature survival	No	Yes	Yes	Yes	Yes	Yes	Yes	Yes	Yes	Yes	Yes	Yes	Yes	Yes	Yes
Propagules unaffected (by fire)	All			All			All			All			All		
Immatures unaffected (by fire)	None			None			None			None			None		
Adults unaffected (by fire)	None			None			None			None			None		
Immatures resprouting	None			None			None			None			None		
Adults resprouting	None			None			None			None			None		

Source: From Moore and Noble (1990).

Table 10.3. *Stages of succession following a wildfire for forests in south-western Tasmania simulated by the FATE model. On the left of the table is a summary of the community composition at different times following a fire. On the right is a schematic[b] providing more information about each functional group*

Time	Community composition[a]					
	Eucalyptus	Nothofagus	Atherosperma	Acacia	Sedges	
0	High	Low	Medium	Prop.(H)	Prop.(H)	
Fire	Prop.(H)	Prop.(H)	Prop.(H)	Prop.(H)	Prop.(H)	
1	Imm.	Imm.	Imm.	Imm.	Imm.	
3	Imm.	Imm.	Imm.	Imm.	Medium	
5	Imm.	Imm.	Imm.	Imm.	High	
11	Imm.	Imm.	Imm.	Medium	Prop.(H)	
31	High	Low	Low	Medium	Prop.(H)	
71	High	Low	Low	Low	Prop.(H)	
73	High	Low	Low	Prop.(H)	Prop.(H)	
80	High	Low	Medium	Prop.(H)	Prop.(H)	
138	High	Medium	Medium	Prop.(L)	Prop.(H)	
205	High	Medium	High	Prop.(L)	Prop.(H)	
271	High	Medium	High	Prop.(L)	Prop.(H)	
330	High	High	High	—	Prop.(H)	
401	—	High	High	—	Prop.(H)	

Notes:

[a] Low, Medium and High indicate the relative abundances of mature plants when they are present at a given time; Imm. indicates that only immature plants are present at the time indicated; Prop.(H) indicates that propagules in high abundance are the only life stage present; Prop.(L) indicates propagules in low abundance; — indicates the functional group is locally extinct.

[b] Symbols indicate the abundance of mature plants (large cartoons) and immature plants (small cartoons): one symbol represents low abundance; two represent medium abundance; and three, high abundance. Open circles represent dormant propagules; solid circles, active propagules (one circle indicates low abundance; two circles, high abundance). The range of ages of each sequence of consecutive cohorts is shown along with the size of the cohort at the time of recruitment (l = low, m = moderate, h = high). As an example, m 0–149 indicates a sequence of moderate-sized cohorts aged from 0 to 149 years.

Source: From Moore and Noble (1990).

3. The dynamics of the amount of area in each cover-state can be thought of as the consequence of the input of land area into the cover-state category and the output of land from the category. This input–output behaviour is considered to be determinate (the amounts of land in each cover-state in the regional successional system provide sufficient information to allow the computation of the future behaviour of the system; this is similar to the assumption in the case of first-order Markov models that the state of the system is sufficient to determine the probabilities of the next system state). The assumption that a system is determinate does not preclude the consideration of statistical fluctuations or sampling error.

4. If the differential equations used to represent the successional system are linear, then the input–output relations are superposable (see Chapter 5) implying that the response of a cover-state to two inputs summed together is the same as the sum of the responses taken separately.

The essential concept in these models is that for a large landscape the change in the vegetative cover can be thought of as a flow of area from one category to another (Fig. 10.3). The development of these models (e.g. Shugart *et al.* 1973; Johnson and Sharpe 1976) involves devising a vegetation categorisation (cover-states) and determining the rates of change and the transfers of area among these cover-states based either on re-measurement data or upon assumptions about the rates of ecological succession (Fig. 10.4). This is analogous to the procedures discussed in the section above for Markov models.

Given the initial conditions (the amount of land area in each of the cover-states at the beginning of the simulation), one can use these models to project the dynamics of a region over time. Figure 10.5 shows an example of such a case. In this instance, the aerial extent of forests in Michigan is projected forward for 250 years starting with the amounts of forest cover in 1970. The simulation case is in the absence of wildfire and harvest and the simulation shows a pronounced decrease in disturbance-generated vegetation types and a regular increase in the extent of fir–spruce forests.

GIS applications: an example using a gap model

GIS use computers to overlay and interact spatial data sets to develop maps that can be changed over time to produce a dynamic representation of a landscape. These applications are most easily developed for mosaic

(a)

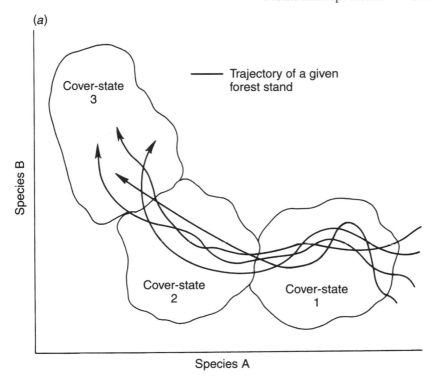

(b)

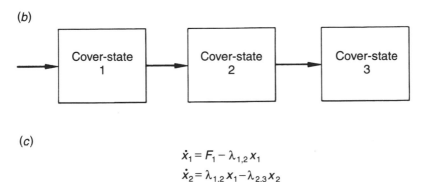

(c)

$$\dot{x}_1 = F_1 - \lambda_{1,2} x_1$$
$$\dot{x}_2 = \lambda_{1,2} x_1 - \lambda_{2,3} x_2$$
$$\dot{x}_3 = \lambda_{2,3} x_2$$

Figure 10.3. Three abstractions representing forest succession over large regions. (*a*) Succession represented by a number of stand trajectories through a species–attribute hyperspace (see also Fig. 10.1). (*b*) Compartment diagram with arrows representing transfers between cover–states. (*c*) System of differential equations indicating transfers between cover–states. From Shugart *et al.* (1973).

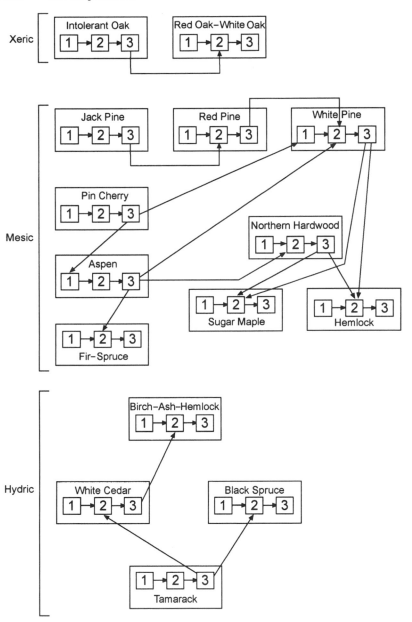

Figure 10.4. Model topology of western Great Lakes regional succession model. The labelled blocks (modules) indicate forest types identified by the dominant tree species. The three blocks within each module indicate the dominant size category of trees (1, seedlings and saplings; 2, pole timber; 3, saw timber) within each forest type. Arrows represent transfers of areas of land from one block to another. From Shugart *et al.* (1973).

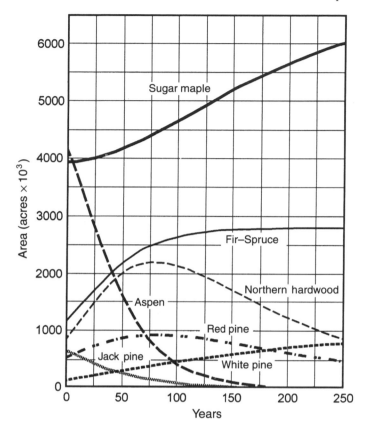

Figure 10.5. Simulation over 250 years of the mesic portion of the western Great Lakes successional model. From Shugart *et al.* (1973).

landscapes. One example of such an application using a GIS to project the successional changes of an area forward in time and based on information derived from a gap model is for the mixed broad-leaved/*Pinus koraiensis* forest of China.

The mixed broad-leaved/*Pinus koraiensis* forest is one of the most complex and valuable forest ecosystem types in north-east China. Once it covered the whole eastern mountainous area between 500 and 1200 m in altitude. It is still one of the largest timber reserves in China. The forest is rich in wildlife and provides the habitat of several endangered species, such as the tiger. Because of changes in land-use (i.e. woodland clearing for cultivation, and excessive timber harvest), the area covered by mixed broad-leaved/*Pinus koraiensis* forest has been greatly reduced and is being replaced by plantations and secondary forests, where forest remains at all

(Cheng *et al.* 1986; Wang 1986; Yang and Wu 1986). At present, only several forest islands (reserves) remain, and the exact nature of succession and climax composition are still a topic of controversy (Chen 1982; Miles *et al.* 1983).

The natural development of mixed broad-leaved/*Pinus koraiensis* forests in the Changbaishan Biosphere Program, China is simulated by means of the gap model KOPIDE (Shao *et al.* 1991). The model derives from the FORET model (Shugart and West 1977; Shugart 1984). As is the case for many gap models, light and other resource availability are assumed to be homogeneous within patches. KOPIDE simulates growth and succession of Korean pine forest under three initial conditions: (a) open ground following clearcutting, (b) secondary forest and (c) mature forest. In general, Korean pine is simulated as the mature phase forest species in the highlands of north-east China. This finding supports one of the two principal succession theories about the dynamics of these forests.

KOPIDE has been linked with a GIS containing site and stand data sets (Shao *et al.* 1991). In this case the KOPIDE gap model was used to project the expected changes of different types of vegetation, and the GIS was used to organise fundamental data on soils, site history and other information used in the KOPIDE model. Through this linkage, one can predict landscape dynamics as a dynamically changing set of maps with a relatively high level of detail. An example is shown for a test area where several succession serial stages are located (Fig. 10.6). In this case, a relatively small, forested landscape patch 800 m above sea level on the north slope of Changbai Mountain was thoroughly investigated and mapped into a GIS using a combination of infrared aerial photographs and field measurements. The area was selected for the study because all of the main forest types in the succession series of the mixed broad-leaved/*Pinus koraiensis* forest can be found in this area. Figure 10.6 shows the simulation results of landscape development for the next 150 years with a time interval of 50 years. Over the 150 year time period all of the stands are eventually dominated by *Pinus koraiensis* forest.

GIS capabilities to display and manipulate data coupled with the capability of a variety of mosaic models to project the consequences of ecological succession at points forward in time provide land managers with a tool for management decisions. Early versions of this approach were developed and applied to management problems in several parks in the USA and Australia (Cattelino *et al.* 1979; Kessell 1979a) and numerous subsequent applications have been developed for natural landscape management in parks and elsewhere.

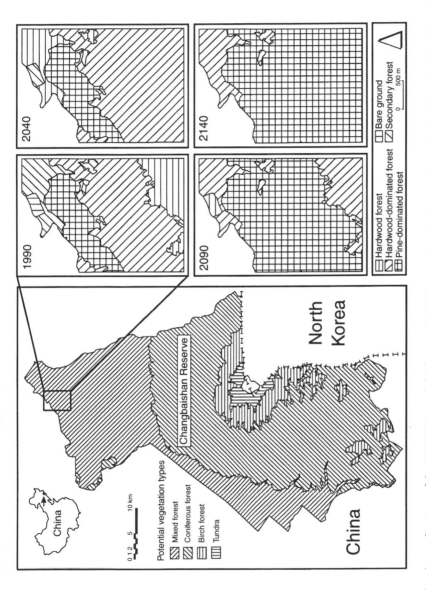

Figure 10.6. An application of the KOPIDE model (Shao *et al.* 1991) to project landscape change in the Changbaishan Biosphere Reserve in China. In this application, the simulation model was used to develop an expected change in different types of vegetation and a GIS program was used to map these changes over the area at different intervals of time.

In very large-scale estimations of the effects of climate change or other global-scale environmental changes, relatively simple geographer's algorithms (see Chapter 6) have been used in conjunction with GIS data sets for climate, soils and expected changes to produce maps of potential regional and global change. These cases will be discussed in more detail in Chapter 14. The principal concern with such evaluations is that they do not provide insight into the dynamics of the transient aspects of the changes in the terrestrial surface. Applications such as the one provided above provide insight into the information (and computer power) needed to develop more dynamic maps of expected environmental change through time.

Interactive mosaic models and spatial models

The assumption that one can treat a landscape as a mosaic of locally interactive elements is a simplification that has proved useful in a variety of terrestrial systems. It is likely that the success of such models is in part a result of the patchy or mosaic nature of many terrestrial systems and of the local nature of interactions with light and nutrient elements. There are other influences on terrestrial ecosystems that are not so easily localised in landscape systems. These include such effects as wildfires, insect outbreaks, migration of species in conjunction with an invading species, etc. Developed to deal with these spatial landscape phenomena, there is a spectrum of models that 'relax' the assumption that landscapes are mosaic. At one end of this spectrum are models that recognise a landscape mosaic and allow the parts of this mosaic to interact with one another. At the other end of the spectrum are explicitly spatial models that characterise a landscape as an interactive whole. In the section that follows, examples of this spectrum of models (from interactive mosaic models to fully spatial models) will be presented.

Markovian approaches to spatial dynamics

The most straightforward approach to introducing spatial effects in a mosaic model is in the case of spatially interactive Markov models. In the classic Markovian approach to simulating the dynamics of landscapes (see the earlier section in this chapter), the state of an element of the mosaic landscape determines the probability that the element will be in a given state in the next time interval. In a spatial Markov model, these transition probabilities are not only a function of the state of the landscape element but are also a function of the states of the surrounding elements. The adjustment of transition probabilities according to the spatial context of a cell is a logical model addition for many phenomena. For example, the likelihood of a patch of a forest landscape being burned by wildfire is greatly increased if the adjacent cells of the landscape are on fire. The chances of a particular species of plant growing on a given landscape element is higher if there are seed trees on adjacent cells.

Spatial Markov models have been used to simulate the dynamics of heathland vegetation (van Tongeren and Prentice 1986) and a transect Markov model has been implemented to simulate the vegetation dynamics of coastal dune ecosystems (Shugart *et al.* 1988; Rastetter 1991). This transect model, DUNE, maintains the computational efficiency of a Markov model but also incorporates both a mechanistic formulation of the important population processes and the realism of spatial heterogeneity. The model is based upon a Markov chain representation of the life-stage development of each plant species at intervals along a transect. If a species has 'n' life stages that are ecologically important, then that species is represented at each interval by an n-bit word, each bit signifying the presence or absence of a respective life stage and there are, therefore, 2^n possible simulation states for the species. In its current implementation on the computer, the number of life stages represented for each species and the spatial resolution (interval width along a spatial transect) can be defined by the model user. This allows the adjustment of the resolution of spatial patterns and the life history detail to optimise computational efficiency.

The state transition probabilities in the DUNE model are calculated based on seed availability and environmental factors affecting sprouting, growth and mortality. For n as the number of plant life stages for a particular species, there are 2^n possible state transitions for each species at each location during any particular time step. For $n > 3$, the dimensionality of the problem can be reduced by considering each life stage individually. This also facilitates a more mechanistic formulation of the transition probabilities that incorporate the growth characteristics of the species.

There are four possible transitions for the individuals of a particular life stage at a particular location: (1) they can all die, (2) they can all remain unchanged, (3) some can mature to the next life stage and some remain the same, or (4) they can all mature to the next life stage. Three other possibilities involving some plants dying, and some either remaining the same and/or maturing are indistinguishable from possibilities 2, 3 and 4 because only the presence or absence of individuals in each life stage is followed in the model. Since these four transitions represent all possibilities, the probabilities associated with them must sum to one. It is, therefore, only necessary to calculate three of the probabilities, the fourth can be calculated by difference. Possibilities 3 and 4, however, do not exist for the oldest life stage, consequently only two probabilities must be calculated for this last life stage and one of these can be calculated by difference.

In addition, a recruitment probability must be calculated. Thus, a total of $3(n-1)+2 = 3n-1$ probabilities must be calculated at each time step, for each species, at each location. The 2^n state transition probabilities can be calculated from the life-stage transition probabilities by (a) cross-multiplying the probabilities associated with each life stage and the probabilities at each of the other life stages, and (b) summing the probabilities of all redundant outcomes.

The DUNE model has been implemented to simulate the vegetation dynamics of coastal dune ecosystems in the south-eastern USA (Shugart et al. 1988; Rastetter 1991). Because of the transect formulation of the model, several important physical variables can be simulated dynamically: notably the height of the water table, the height of the sand mass at any point and the salinity of the water table at each point. The model is driven by the position of the beach front (which has a height of 0 for sea level). The successional dynamics of an example simulation (Fig. 10.7) feature the development of a horizontal gradient of vegetative pattern, a reduction of the height of the sand dune through aeolian erosion, a landward displacement of the dune system (a consequence of vegetation-mediated aeolian transport of sand) and the eventual development of a back–dune marsh as the water table moves to the surface behind the dune. This particular model is presently in a prototype form, but the example (Fig. 10.7) is indicative of the richness of behaviour that can be developed even from relatively simple transect models.

Individual-based spatial models

Like gap models, spatial forest models simulate forest dynamics by modelling the establishment, growth and mortality of individual trees within a defined area. An inherent difference between gap models and spatial forest models is in the form of the competition functions. The spatial forest models differ from gap models in their explicit consideration of tree position in the horizontal plane. Because of the explicit consideration of horizontal position, spatial forest models generally use a measure of competition that is a direct function of the proximity and size of

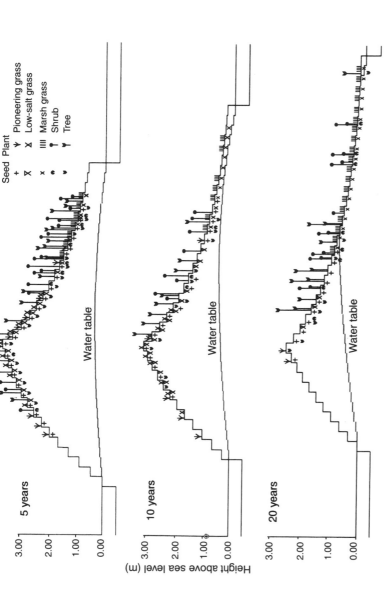

Figure 10.7. Example output from a computer model (Shugart *et al.* 1988; Rastetter 1991) simulating the pattern and dynamics of vegetation along a transect through a barrier island such as might be found along the Atlantic coast of North America. The open ocean is to the right of the figure. Over the time of the simulation the island moves landward and the water table on the right side of the island forms a brackish marsh with scattered shrubs. Trees and shrubs occupy a position behind the sheltering dune and the dune flattens. From Shugart *et al.* (1988).

neighbouring individuals. The competition indices used in spatial models vary greatly in their design, they can be classified into three major categories: (a) distance-based ratios, (b) influence-zone overlap indices, and (c) growing-space polygons.

Distance-weighted size ratios (Hegyi 1974; Daniels 1976) define the degree of competition between a given tree and a neighbouring individual as a function of the ratio of the sizes of the two trees (competitor/subject tree) multiplied by the inverse of the distance between the two individuals.

The influence-zone indices (Gerrard 1969; Bella 1971) are based on the assumption of a circular zone of influence around every tree, wherein direct competition occurs (Staebler 1951). The extent to which this area overlaps the influence zone of neighbouring trees represents a measure of encroachment and crowding of a tree's optimal functional environment. These indices vary with regard to the type of overlap expressions used (i.e. linear, angular, aerial).

Growing-space polygons (Brown 1965; Moore *et al.* 1973; Alard 1974; Pelz 1978; Doyle 1983) represent geometrical designs to calculate non-overlapping crown area of a tree as limited by the proximity and size of neighbouring individuals.

Each of these indices is based on the relative horizontal position of individuals on the plot. They vary in the methods that are used to determine which neighbouring individuals are to be considered as potential competitors, in defining the size of the zone of influence for a given individual, in consideration of size of competing individuals relative to the target tree and in their consideration of potential differences in competitive ability among species.

With the exception of those of Doyle (1983), the above-mentioned competition indices are based on statistical models and the calculated values of the competition indices are regressed against the observed growth rate of individuals to determine the functional relationship between competition and growth. As a result, the functional form of the relationship between the competition index and the growth rate is site specific (i.e. related to site factors such as nutrient and moisture availability) and data intensive. For these reasons, most spatial models have been developed for managed forests (e.g. Mitchell 1975) and, with a few exceptions (e.g. Ek and Monserud (1974a,b) FOREST model), simulate monospecific stands.

Traditionally, gap models have been used to characterise the dynamics of a location by averaging the simulations for a number of independent plots to provide a composite view of temporal dynamics for a given set of site conditions and climatic parameters. The approach is analogous to a random sampling design to characterise the structure and composition over a larger area. Recent research, particularly in tropical forest systems, has focused on the importance of gap size in the subsequent dynamics of vegetation replacement following disturbance (Denslow 1980; Brokaw 1985a,b) and provided an ecological interest in more spatially explicit models. Also, the availability of spatial data obtained by remote sensing from aeroplanes and satellites and the development of large tree stem maps have generated an increased interest in the simulation of spatial pattern in forests and other systems.

The ZELIG model

A spatial version of a gap model, ZELIG, has been developed to extend the gap modelling paradigm to address questions involving spatial phenomena larger than a single plot. ZELIG (Smith and Urban 1988) is an individual tree simulator of the JABOWA (Botkin et al. 1972) and FORET (Shugart and West 1977) genre. It is designed to simulate a contiguous forest area of user-defined size. It differs operationally from the JABOWA/FORET approach in that the simulated forest changes through time as a spatially interactive unit rather than as a series of independent plots. The model is based on a grid system, typically with the grid defined as $10\,m \times 10\,m$ cells. Grid dimensions are defined by the investigator and may range from a one-dimensional transect to a rectangle of any length or width. Vegetation pattern is spatially explicit to a resolution of 0.01 ha, in that each tree is assigned to a grid cell, but positions within the cell are not accounted.

Just as with the conventional forest gap models, establishment and growth of individuals on each grid cell are based on the assumption of horizontal homogeneity (see gap model description above), in this case at the scale of 0.04 ha. In ZELIG, this zone is defined by aggregating the adjacent cells to produce a 0.04 ha quadrate centred on the grid cell of interest (Fig. 10.8). Following the gap modelling approach, woody biomass and leaf area at defined height intervals are calculated for the 0.04 ha quadrate. These sums represent neighbourhood (gap scale) constraints used to modify the establishment, growth and death of trees on the centre grid cell.

The ZELIG model can be applied in three modes. The first is using the spatially interactive grid system. Interactive grids can be used to simulate

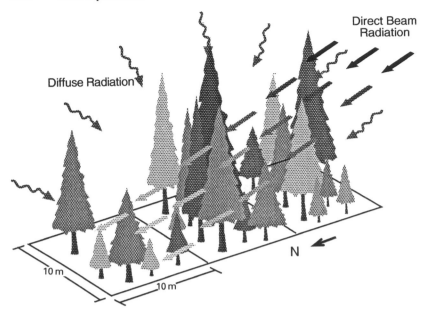

Figure 10.8. Schematic diagram of the transect version of the ZELIG model depicting the interactive shading among plots. Unlike the figure, the model is based on an assumption of horizontal homogeneity within each plot. From Weishampel *et al.* (1992).

contiguous forest stands up to a few tens of hectares in size. For larger-scale concerns, the grid can be collapsed to a transect, which is especially appropriate for applications over long gradients, for topographic pattern or for the incorporation of seed dispersal effects on species migration under changing environmental conditions (Urban *et al.* 1991). The grid can be made non-interactive (i.e. each grid cell representing a single plot), in which case the ZELIG modelling shell becomes a traditional gap model.

As an initial test, ZELIG was used to simulate vegetation succession as a function of gap size. Five functional types of tree were defined to represent the range of light response functions and life-history characteristics (maximum size, longevity, growth rate, etc.) from high-light-demanding gap species to shade-tolerant non-gap species. A contiguous grid of 5 ha was simulated for 500 years to define a set of initial conditions prior to disturbance. These initial conditions were then used to simulate a series of disturbances (i.e. episodic mortality of canopy individuals) resulting in gaps of various sizes. This process was replicated to provide a sample of 10

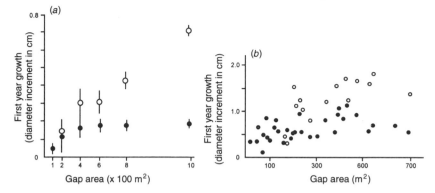

Figure 10.9. Comparisons of diameter increment of shade-tolerant and shade-intolerant trees growing in gaps of different sizes. O, pioneer/shade intolerant; ●, primary/shade tolerant. (*a*) Experimental data from Brokaw (1985a) based on experimentally created gaps in tropical rain forests in Panama. The vertical lines are 95% confidence limits. (*b*) Response of ZELIG model simulating the equivalent condition with different functional types of tree for tropical forests.

gaps of each size. The model was able to simulate the observed patterns of growth for primary and pioneer types as a function of gap size (Fig. 10.9).

Remote sensing provides a potential means to obtain the spatial data needed to develop and test spatial models (Running *et al.* 1989) such as the ZELIG models. Cohen *et al.* (1990) analysed the spatial patterns of digital imagery from aerial videography of a chronosequence (80, 140 and 450 years) of *Pseudotsuga menziesii* stands in the Willamette National Forest in the Cascade Range of western Oregon. Image grey-scale values represented sunlit crowns and shadow patterns formed by discontinuities of canopy features. These patterns resulting from sun-angle geometry and biological factors (e.g. tree height, canopy dimensions and gap dynamics) were highly correlated to stand age and tree size variability (Cohen and Spies 1992).

Using the transect version of ZELIG (Weishampel *et al.* 1992), a 500-plot, south–north transect initialised from bare ground was simulated for 80, 140 and 450 years for comparison with the independent videography data (Cohen and Spies 1992). Biomass, leaf area index (LAI) and maximum tree height, thought to correspond to percentage cover, canopy layering and shadow patterns represented by the digital imagery, respectively, were generated by the model for each plot for each time interval. These modelled attributes were analysed at 10 m resolution and also at 30 m resolution.

Semivariances, half the average of the squared differences between all possible pairs of points separated by a chosen distance (Robertson 1987), were computed for these structural features over a range of distances to generate semivariograms. Semivariances for each variable for both the 10 and 30 m resolutions were standardised to z-scores for comparative purposes using the mean and standard deviation of the semivariances over all lags at the 10 m resolution. This geostatistical technique has been used to describe graphically the spatial variation of digital imagery (Curran 1980, 1988; Cohen *et al.* 1990) and vegetation structure (Palmer 1988) along a transect.

Young Douglas fir stands in the western hemlock zone of the Cascade Range frequently possess a relatively continuous, single-layered canopy composed primarily of *Pseudotsuga* sp. With age, tree size variation and between-tree spacing (i.e. gap size) increase as mortality becomes more conspicuous. Shade-tolerant *Tsuga heterophylla* (western hemlock), *Thuja plicata* (western red cedar), and *Abies amablis* (Pacific silver fir) colonise the understory producing multiple canopy layers (Spies and Franklin 1989). Though the model consistently underestimated canopy height as measured by Spies *et al.* (1990), the simulations captured the successional increase of structural diversity in both the vertical and horizontal dimensions exhibited by these forests (Fig. 10.10). This temporal increase of interplot variation was suppressed at the coarser 30 m resolution as information was lost through averaging from increasing the grain size (Atkinson and Danson 1988; Turner *et al.* 1989a,b). These general spatio-temporal patterns were mirrored by the semivariograms from the digital imagery and the simulated forest transects.

The FOREST model

Probably the most elaborate forest dynamic model using individual tree growth as an underlying paradigm is the FOREST model (Ek and Monserud 1974a,b). The model simulates growth and reproduction of mixed-species and even- or uneven-aged stands and includes natural regeneration and growth as well as a variety of management applications. The model treats tree growth and competition in a manner that is analogous in mechanism and detail to the spatial models of forest plantations. For example, competition among nearby trees is computed by determining the crown overlap with all adjacent trees. One very detailed part of the model is the computation of seed rain (both production and dispersal) from the trees simulated on the plot.

Table 10.4 outlines the principal subroutines in the model and provides

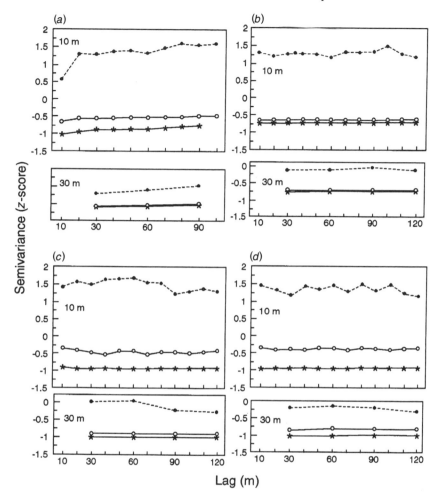

Figure 10.10.   Semivariograms from (*a*) remotely sensed digital matrices (grey-scale pixel values) compressed to a transect (adapted from Cohen *et al.* 1990) and from ZELIG model predictions for (*b*) biomass, (*c*) leaf area index and (*d*) maximum tree height. These attributes were generated for each plot at 10 m and 30 m resolution for time intervals of 80 (✱), 140 (○) and 450 (●) years. From Weishampel *et al.* (1992).

an indication of the level of detail in this model. The model is currently restricted in its application to the northern hardwood forest in or near the state of Wisconsin. The model requires considerable amounts of information about the growth and habits of the trees simulated in a given forest, and it provides output at a level of detail much beyond that collected in

Table 10.4. *Principal subroutines in the FOREST model in order of implementation*

Subroutine	Role
MAIN	Determines height, diameter and crown development of overstory trees
INPUT	Accepts parameter values for each species, primarily for overstory development
STANGN	Accepts real tree input data or generates spatial patterns (clustered, random, uniform, etc.) and tree characteristics for each species
HOWFAR	Determines distance between points on main plot and buffer zone needed for evaluation of competition and seed and sprout distribution (eliminates plot edge effects)
COMPE	Evaluates tree competition
YIELD	Calculates timber product yields based on individual tree dimensions, specific gravity and bark characteristics
STAT	Computes parameters of distributions of tree and stand characteristics for summary output
CUT	Orders trees by size or increment for pruning or harvest treatments and implements these treatments on individual trees by species. Harvest options include row thinning, selection according to specified criteria, spacing rules, cuts to basal area levels and combinations of the above. The timing and degree of cutting may be set by the user or allowed to vary as dictated by stand development
OUTPUT	Prepares table, stem map and graphical output describing stand development
REPRO	Accepts input of initial reproduction status, reproduction parameters for each species and specifications for degree and timing of any changes in reproduction parameters to be implemented during the run
PSEED	Determines seed and sprout production for each overstory tree as a function of species, size and threshold age
SEEDYR	Generates seed year multiplier for each species, i.e. frequency of good, moderate and poor seed years
DSTRIB	Distributes seeds and sprouts (root suckers and basal sprouts) from each overstory tree to subplots within main plot
GRMIN8	Calculates seed germination as a function of microsite and overstory cover conditions
GROW	Controls growth and mortality of reproduction until surviving individual reproduction stems reach overstory status
MAIN	Assumes control of stem development

Source: From Ek and Monserud (1974b).

most ecological studies (e.g. maps of exact positions and sizes of trees by species). Application of the FOREST model to the problem of assessing the effects of fragmentation of forested landscapes will be discussed in Chapter 12.

Homogenous landscape models

Another approach to simulating the response of landscapes is to assume that the internal workings of the landscape are sufficiently well mixed to allow the landscape to be simulated in its entirety without resorting to consideration of the dynamics of the various component parts. Models based on an assumption that the important processes in an ecosystem can be approximated in aggregate without explicit consideration of spatial heterogeneity originate from initial attempts to formulate ecosystems models. Thus, these models have a rich history of development and application.

Many of this class of model scale-up the response of processes as they are understood at the smaller scale to the landscape or regional system response. For example, one might assume that the fluxes of heat, water and CO_2 associated with the functioning of a single leaf are duplicated by the sum of the responses of the billions of individual leaves comprising a vegetated landscape. In some cases, the underlying assumption of homogeneous landscape models is that the mathematical structure of particularly biophysical and chemical reactions at the landscape level resemble those observed at a detailed level (perhaps with some differences in the model parameters). These models are usually the type of ecological model linked to other models of ocean or atmospheric dynamics to assess the feedbacks among these major Earth systems (Ojima 1992).

There is a rich array of models of this class. For example, most models of element cycling in watersheds or other ecosystems tend to view the processes as being homogeneous within the system of definition. Indeed, traditional ecosystem models were often referred to as point models because they simulated the dynamics of ecosystems with no explicit references to spatial heterogeneity. Two types of this model that have been widely applied in the context of changing environmental conditions and the feedbacks between vegetation change and other global changes are those that simulate a regional plant canopy and those that are focused on the storage and transfer of material. Examples of each of these will be discussed in the remainder of this section.

Canopy process models

Plant physiologists have made considerable progress in understanding the dynamic response of individual leaves to their environment. Much of this work was initially developed looking at the response of leaves (over minutes or seconds) in small chambers in which the temperature, vapour pressure deficit and other important variables for understanding leaf function were measured and/or controlled. The biophysical response of a leaf (in terms of flux of water, CO_2 and energy) is strongly tied to the changes in the leaf stomata as influenced by the state of the leaf and the environmental condition. The communication between the inside of the leaf and the environment is controlled by the aperture of the stomata, microscopic holes in the leaf surface. If the stomatal openings are relatively large, resistance to molecular diffusion is low. Water diffuses out from the moist spaces inside the leaf and CO_2 diffuses into the same internal spaces to compensate for the CO_2 taken up by the plant. As the stomatal aperture closes, the resistance to these diffusion processes is increased. The sizes of these stomatal openings change under different conditions and are controlled by the plant.

The balances of CO_2, water and energy in the plant are strongly interwoven and interactive at the stomatal level. For example, the rate at which CO_2 diffuses into the leaf to supply an essential component of photosynthesis is controlled by the stomatal resistance, as is the cooling evaporative flux of water out of the leaf. In canopy process models, equations relating the CO_2, water and energy fluxes for leaves are used to simulate the CO_2, water and energy fluxes of areas that are considerably larger than single leaves.

Canopy process models (e.g. Running and Coughlan 1988) simulate the flux of CO_2 and water from plant canopies over time scales of seconds to a day. In general, these models are extensions of leaf-level models of photosynthesis (e.g. Farquhar *et al.* 1980) and transpiration (e.g. Penman 1948; Monteith 1972) applied to whole canopies. The models do not consider individual plants but view the canopy as a single, multi-layer unit with a fixed structure (i.e. leaf area). Photosynthesis and transpiration are simulated by estimating microclimatic variation and stomatal conductance for the canopy (or canopy layers).

Woodward (1987b) produced a straightforward model designed to scale the ecological physiology of plants to some of the global-scale consequences of vegetation/environment interactions. Some of the elements of Woodward's model were discussed earlier in Chapter 6, and one aspect

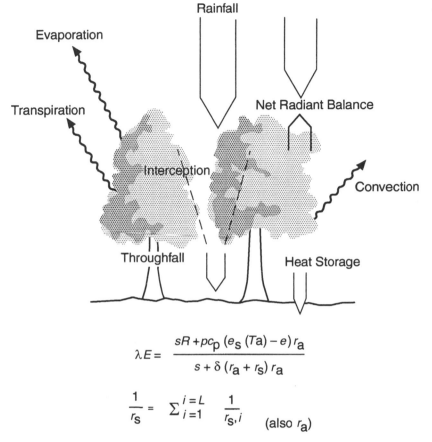

$$\lambda E = \frac{sR + pc_p \, (e_s \, (Ta) - e) \, r_a}{s + \delta \, (r_a + r_s) \, r_a}$$

$$\frac{1}{r_s} = \sum_{i=1}^{i=L} \frac{1}{r_{s,i}} \quad \text{(also } r_a\text{)}$$

Figure 10.11. Schematic diagram of the components of the hydrological and energy balances of a forest used by Woodward to simulate the vegetation water use as a function of the climate, site conditions and leaf area. Woodward systematically varied the leaf areas at locations to determine the maximum leaf area that used the annual water input to the system. In the equations: λE is evapotranspiration; s is the slope of the saturation vapour pressure density versus temperature curve at air temperature (Ta); R is net radiant flux density; p is water vapour pressure; c_p is specific heat of the air; $e_s(Ta)$ is saturation vapour pressure density at air temperature; e is the vapour pressure density; $\delta \, (r_a + r_s)r_a$ is the apparent psychometric constant. See Woodward (1987b) for details.

of the model (the changes in life forms of plants associated with different temperature conditions) is shown in Table 6.5 (p. 171). Woodward developed a simple model of the energy and hydrological balance of a plant canopy (Fig. 10.11) using the Penman–Monteith equation to determine canopy transpiration (Penman 1948; Monteith 1981b; see also Woodward

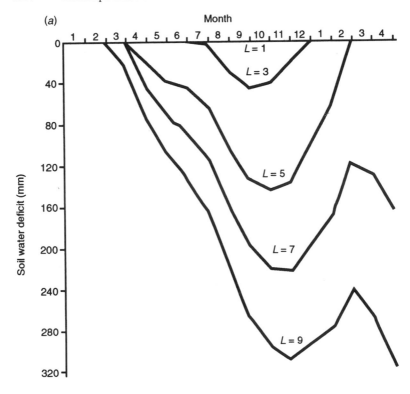

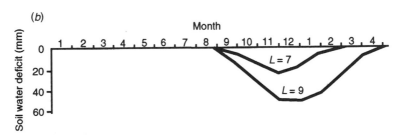

Figure 10.12. Predicted monthly soil water deficit for soils at two locations in Australia using the water and energy balance model shown in Fig. 10.11. In the case of Brisbane (*a*), a leaf area (*L*) of 3 is expected because the water in the soil is not drawn below the 80 mm rainfall equivalent soil deficit that is needed to supply the transpiration and the soil is able to recharge over the course of the year. Leaf areas of 5 dry the site below the 80 mm rainfall equivalent soil deficit (although the soil can still recharge) and leaf areas of 7 or greater appear to dry the soil beyond its ability to recharge over the annual cycle. In Sydney (*b*), the site appears to have sufficient precipitation to sustain leaf areas of as high as 9. From Woodward (1987b).

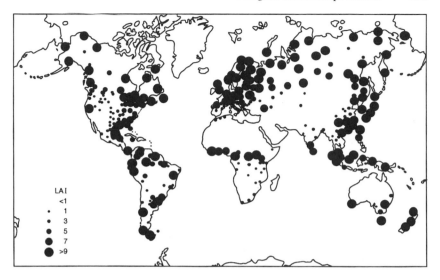

Figure 10.13. Global map of leaf area-based climate and soil interactions with a dynamic plant canopy model. From Woodward (1987b).

1987b). The model functions by solving for the water transpired by the canopy of a given leaf area and subtracting this evapotranspiration from the water held in the soil. If the deficit of water in the soil was drawn below the equivalent of 80 mm of rainfall or if the soil is unable to recharge with water over the course of a year, then the leaf area was assumed to be too high and a lower value for leaf area was used until the maximum leaf area was determined (Fig. 10.12). This approach can then be applied to a number of points with weather data to obtain a global map of leaf area (Fig. 10.13).

Several models have been developed to include canopy effects and other surface process effects in GCMs of the atmosphere. One of the first such models was the BATS model (Table 10.5). BATS is an acronym for Biosphere–Atmosphere Transfer Scheme. The model is designed to join other models simulating the dynamics of the atmosphere in response to conditions on the land surface. In part because it is designed to connect with atmospheric GCMs, the BATS model represents a very large land area ($c.\ 20\,000\ km^2$) that is assumed to be homogeneous with respect to its dynamic response. Calculations in the model are made at subhourly intervals. Some of the surface interactions resemble those for leaves with respect to evapotranspiration, heat and CO_2 dynamics. Stomatal resistances (associated with the size of the stomatal opening that allow exchanges between the interior and exterior or leaves) are related to

Table 10.5. *Principal features and assumptions of models simulating landscape dynamics based on leaf processes*

Type of model	Applications	Major assumptions
Land surface interaction models Examples: BATS (Dickinson and Henderson-Sellers 1986); SiB (Sellers *et al.* 1986)	Used to couple the land surface to the atmosphere in the context of large dynamic models of the atmosphere (GCMs). Models attempt to represent the energy, water and momentum transfers (principal variable in GCMs) between atmosphere and the terrestrial surface	Very large land area (*c.* 20 000 km^2) assumed homogeneous. Calculations are made at subhourly intervals. Surface interactions resemble those for leaves with respect to evapotranspiration, heat and CO_2 dynamic interactions. Stomatal resistances (associated with the size of the stomatal opening that allow exchanges between the interior and exterior of leaves) related to temperature, humidity (vapour pressure deficit) and light. Features of the land surface (vegetation, soil moisture–holding capacity, etc.) vary geographically but in most cases do not change dynamically
Regional ecosystem simulation models Example: FOREST-BCG (Running and Coughlan 1988)	Developed in the context of reproducing and enriching information obtained over large areas by remote sensing. Simulating effects of topography and climate on regional vegetation. Models would convert remote sensing information to information useful in obtaining parameters for land–surface interaction models (above)	'Point models' applied to points along a landscape. Daily time steps for water and carbon fluxes; biomass calculated annually. Contain submodels for hydrologic processes (snow melt, evaporation, runoff, transpiration). Carbon fixation modelled as light, water and CO_2 altered the leaf response in a layered plant canopy. Allocation of photosynthate to plant parts indicates plant growth (as does uptake and allocation of nitrogen)

Global leaf area simulation models
Example: Woodward (1987b)

Developed to reproduce global pattern of leaf area as a function of environmental variables

Model of a layered plant canopy (heat, water, CO_2 fluxes) that produces the estimated leaf areas for points (distributed over the Earth) under the assumption that vegetation processes will produce an optimal leaf area. The response is for a vegetation at equilibrium with respect to leaf area

Table 10.6. *Vegetation/land cover parameters used in the BATS model. Parameters were developed for a computer experiment on the effect of forest clearing in South America on the global climate*

Parameter	Cropland	Evergreen broad-leaved forest	Deciduous shrubland
		Example land cover types	
Maximum fractional vegetation cover (%)	0.85	0.90	0.80
Difference between maximum fractional vegetation cover and cover at temperature of 269 K (%)	0.6	0.5	0.3
Roughness length of vegetation (m)	0.06	2.0	0.1
Depth of the total soil layer (m)	1.0	1.5	1.0
Depth of the upper soil layer (m)	0.1	0.1	0.1
Rooting ratio (upper divided by total soil layers)	3	12	5
Vegetation albedo for wavelengths $<0.7 \mu m$	0.10	0.04	0.08
Vegetation albedo for wavelengths $>0.7 \mu m$	0.30	0.20	0.28
Minimal stomatal resistance (sm^{-1})	150	250	250
Maximum LAI (leaf area index)	6	6	6
Minimum LAI	0.5	5.0	1.0
Stem (and dead matter) area index	0.5	2.0	2.0
Inverse square root of leaf dimension $(m^{-\frac{1}{2}})$	10	5	5
Light sensitivity factor (Wm^{-2})	0.01	0.03	0.01

Source: From Dickinson and Henderson-Sellers (1988)

temperature, humidity (vapour pressure deficit) and light using empirical or theoretical functions. The larger scale pattern of the Earth's surface is represented by including a range of land-cover types (crops, desert, tundra, evergreen needleleaf forest, short grass prairie, etc.). Each of these cover types has attributes that are involved with equations in the BATS model (Table 10.6). Features of the land surface (vegetation, soil mois-ture-holding capacity, etc.) vary geographically but in most cases do not change dynamically (Table 10.6) in response to disturbance, climate change or other non-atmospheric factors. The responses to these longer-term surface changes are typically investigated in model experiments or in model sensitivity studies using GCMs (e.g. Dickinson and Henderson-Sellers 1988; Mintz 1984; Sellers et al. 1986; Shukla and Mintz 1982).

An alternative application of a general model representation of the leaf's biophysical processes to large areas is that of Running and his colleagues (Running and Coughlan 1988; Running and Nemani 1988; Running et al. 1989). This model, called the FOREST-BCG model, was developed in the context of enriching information on canopy processes using remote sensing. Unlike the land-surface models mentioned in the section above, FOREST-BCG incorporates topographic heterogeneity (Running et al. 1989) and uses this information to adjust climatic conditions and to simulate effects of topography and climate on regional vegetation. This and similar models are not viewed as competitors to the land-surface models. Rather, one of their applications is the conversion of remote sensing information (that can be obtained over large areas) to a form useful in obtaining parameters for land-surface interaction models (Fig. 10.12).

FOREST-BCG uses daily time steps for water and carbon fluxes; biomass is calculated annually. The model contains submodels for hydrologic processes such as snow melt, evaporation, runoff, transpiration. Carbon fixation is modelled as the leaf-level response for a layered plant canopy with dynamically changing gradients of light, water and CO_2. Allocation of photosynthate to plant parts gives an indication of plant growth (as does uptake and allocation of nitrogen).

Running et al. (1989) have used the FOREST-BCG model to simulate seasonal variation in primary productivity across a range of sites in North America (Fig. 10.14). The simulated patterns of productivity closely matched the remotely sensed values of NDVI (Tucker 1979) for the sites, an index believed to be related to photosynthetic activity (Tucker et al. 1985a).

Material transfer models

One 'standard' ecosystem model is the compartment model of the transfer of material or energy into an ecosystem, through the components of an ecosystem and, eventually, out of the ecosystem. The compartment models discussed in Chapter 4 were linear models, but a relatively large set of similarly structured non-linear models of material transfer have been developed as well. Since most formulations of material transfer models by necessity conserve matter, the non-linearities in these models largely arise from using complex functions of more than one state variable to control the transfer of material from one compartment to another. The material

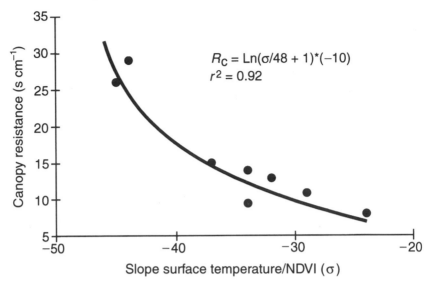

Figure 10.14. The relationship between the simulated canopy resistance from the FOREST-BCG model and the slope of the surface temperature/NDVI for 8 days during the summer of 1985 for a 25 × 25 km² forest area in Montana. The NDVI is a corrected ratio of red (0.58–0.69 μm wavelengths) and infrared (0.725–1.1 μm wavelengths) on board the NOAA series of polar-orbiting satellites (see Sellers 1985, 1987). This same satellite system was discussed in Chapter 5 in an application involving remote sensing of bird habitat over Africa. The relation shown between the canopy resistance and relatively easily measured climate and satellite-derived information could be used to obtain wide-area representations of canopy biophysics. R_c is canopy resistance. (From Nemani and Running 1989).

flow models are properly considered landscape models because many of the functions in the models represent average or 'typical' transfers of material for an ecosystem. The models sometimes are referred to by their developers as 'point models', emphasising that the models represent an average ecosystem behaviour without a specific spatial reference or any effects of spatial heterogeneity.

Historically, two large research programmes greatly extended the application of material flow models in ecology. The first such programme was the largely (but certainly not exclusively) United States Atomic Energy Commission (USAEC) sponsored programme on the movement of radioactive isotopes in natural environments in the 1950s and 1960s. The second was the International Biological Programme (IBP), an international programme emphasising the understanding of productivity of ecosystems that extended from the mid-1960s into the 1970s. Each of these programmes produced rich data sets that are still invaluable in understanding element cycles in a wide variety of ecosystems.

The USAEC radioecology programme (Jordan 1986) produced whole ecosystem experiments and direct measurements of rates of transfer of isotopes through natural environments. The larger goal of the programme was to understand how radioactive isotopes from nuclear weapons testing and from nuclear reactors would be transferred to humans through the environment. The models developed to predict the ecological transfer of these materials were by-and-large linear compartment models (Chapter 4). The models were relatively successful for several reasons. First, linear models (even complex models) can be solved at equilibrium by algebraic manipulations, which was an advantage given the computational power available when they were developed. Second, in linear models in which the transfer of material is controlled by the amount of material in a source compartment (called donor-controlled flow), the equilibrium levels of materials are the maximum amount for all compartments in the case of a constant supply. Importantly, radioactive decay of a given amount of an isotope is a linear, donor-controlled process. Third, for radioactive isotopes the amount of material transferred is a very small proportion of the overall material transfer. Linear models have proved useful in such cases in a range of biological and industrial applications.

The predictions that the linear models were able to make with a great deal of success were largely of two types. One, for isotopes with relatively short half-lives (the time necessary for half of an initial amount of a radioisotope to be lost through radioactive decay), was in the identification of the principal rapid pathways that would carry radioactive dose to humans in the event of a nuclear release. Second, for long-lived isotopes, was the determination of where and to what levels particular isotopes would be most concentrated as the material moved through food chains.

The IBP models were a much more heterogeneous collection of models and modelling approaches that emphasised understanding natural productivity. The roles of ecosystem modelling in the IBP were several-fold and varied significantly in different parts of the international programme (Table 10.7). In retrospect, the IBP models were developed in an atmosphere of large goals: ecosystem prediction, project co-ordination and quantification of important processes. The programme provided a cross-biome, quantitative and comparative directions for large-scale ecosystem work that persists to the present. In terms of understanding the response of ecosystems to environmental changes, the systematic investigation of processes controlling major transfers of major elements

Table 10.7. *Variety of modelling approaches used to represent material flows in several different ecosystems in the Biome programs of the USA IBP*

Biome program: model name	Model formulation	Ecological considerations	Author
Grassland: PWNEE	Non-linear differential equations	A group of 40 state variables with emphasis on functional relationships among components of a grassland ecosystem	Bledsoe *et al.* (1971)
Grassland: LINEAR	Linear differential equations with time-varying coefficients	A group of 40 state variables with elaborated decomposition section (28 of the state variables)	Patten (1971)
Grassland: ELM	Non-linear difference equations	Model with five sections: abiotic, producer, consumer, decomposer and nutrients. Model intended to predict primary productivity for grasslands under perturbed conditions; determine consequences of responses to perturbations	Anway *et al.* (1972), Innis (1975)
Eastern deciduous forest	A collection of 70+ modules intended to be connected to form larger models to answer research questions. Mostly non-linear differential equations	Emphasis on general formulations to capture the biology, physiology, biophysics or other appropriate mechanism for processes such as leaf photosynthesis, decomposition, etc.	O'Neill (1975), Shugart *et al.* (1974)
Desert	Differential equations with time steps varying according to the rates and activity of different processes	Large number of state variables (100 to 200+). Some emphasis on statistics for coupling of the state variables	Goodall (1975)
Tundra	An evolving set of difference and differential equation models	An initial orientation on population processes (lemming population crashes) evolving through a series of models to consider plant and nutrient-related processes.	Miller *et al.* (1975)
Coniferous forest	Discrete time-step simulations developed for several cases (~20) of ecological interactions	A computer simulation language (FLEX) for simulating hierarchically structured systems (based on general systems theory of Klir (1972)) was used to solve behaviour of models and submodels	Overton (1975)

(e.g. carbon, nitrogen, phosphorus) in the form of experiments and dynamic models may be the IBP's most important contribution.

As an example of the approaches founded in the IBP and of the application of material flux models to problems in assessing environmental change, consider the CENTURY model of carbon, nitrogen and phosphorus transfer in natural systems. Isotope studies in the early 1970s were the initial genesis of the CENTURY model (Fig. 10.15), which was developed as a soil process-related section of the IBP Grassland Biome ELM model (Parton 1978). The CENTURY model (Parton et al. 1987) was developed to simulate soil organic matter dynamics and plant production in grazed grasslands and agroecosystems. The data used to develop the model came from long-term incubation studies of ^{14}C-labelled plant material in different soil types (e.g. Ladd et al. 1981; Sorenson 1981), soil carbon dating (Martel and Paul 1974), soil particle size fractionation data (Tiessen et al. 1982; Tiessen and Stewart 1983) and modelling studies at different levels of resolution (Cole et al. 1977; Van Veen et al. 1984; Parton et al. 1993).

The CENTURY model simulates the dynamics of carbon, nitrogen and phosphorus in a soil–plant system using monthly time steps. The input data required for the model include soil texture, monthly precipitation, maximum and minimum air temperatures and plant lignin content. By the late 1980s, the CENTURY model had been used to simulate regional patterns of soil carbon, nitrogen and phosphorus and plant production for the USA central grasslands region (Parton 1978; Fig. 10.16) and the impact of management practices on agroecosystems (Parton et al. 1987, 1988).

In the CENTURY model, the soil organic matter is divided into: (a) an active soil fraction consisting of live microbes and microbial products (1–2 year turnover time); (b) a protected fraction that is resistant to decomposition (20–40 year turnover time); and (c) a fraction that is physically or chemically isolated (800–1200 year turnover time). The plant residue is divided into structural (2–5 year turnover time) and metabolic (0.1–1.0 year turnover time) pools as a function of the lignin to nitrogen ratio of the residue. Decomposition is calculated by multiplying the decay rate specified for each state variable by the combined effect of soil moisture and soil temperature on decomposition. The decay rate of the structural material is also a function of its lignin. The active soil organic matter decay rate changes as a function of the soil silt plus clay content (low values for high silt and clay soils). The respiration loss for carbon is fixed except for active soil organic matter, the respiration of which decreases

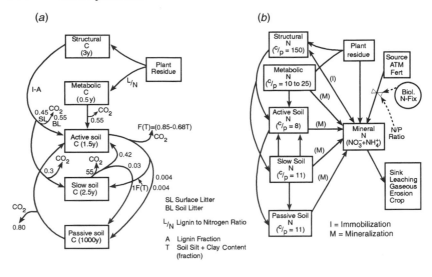

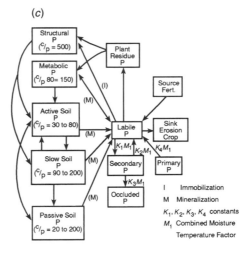

Figure 10.15. Flow diagram for carbon, nitrogen and phosphorus dynamics in the CENTURY model. (*a*) Carbon pools and fluxes; (*b*) nitrogen pools and fluxes; (*c*) phosphorus pools and fluxes.

with the soil silt plus clay content. Submodels for nitrogen and phosphorus parallel the carbon model structure with modifications as appropriate.

The model also includes a plant production submodel, which simulates the monthly dynamics of carbon, nitrogen, phosphorus and sulphur in the live and dead aboveground plant material, live roots and resistant (structural) and labile (metabolic) surface and root detritus pools. Maximum

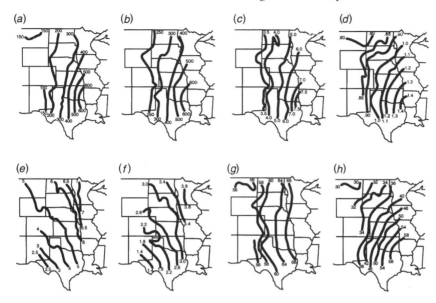

Figure 10.16. Regional patterns simulated by the CENTURY model for: (*a*) above-ground production (gm^{-2}year^{-1}); (*b*) below-ground production (gm^{-2}year^{-1}); (*c*) nitrogen mineralisation (gm^{-2}year^{-1}); (*d*) phosphorus mineralisation (gm^{-2}year^{-1}); (*e*) soil carbon on fine-textured soils (gm^{-2}); (*f*) soil carbon on sandy soils (gm^{-2}); (*g*) organic phosphorus on fine-textured soils (gm^{-2}); (*h*) organic phosphorus on sandy soils (gm^{-2}). Maps are based on running the CENTURY model at 56 sites in the region and using a contouring computer program (The S Package, Bell Labs, Murray Hill, New Jersey, USA) to generate the maps. The CENTURY model was run for 10 000 years at each of the 56 sites and the resultant output represents the state of the systems after 10 000 years of soil formation. At each site, the model response was computed for a fine-textured soil (25% sand, 30% clay and 45% silt) and a sandy soil (75% sand, 10% clay and 15% silt). Only the results from the sandy soil are shown in (*c*) and (*d*).

potential plant growth is estimated as a function of the annual precipitation and is reduced if sufficient nitrogen, phosphorus or sulphur is not available.

Combined models of change in structure and material flow

The FOREST-BCG model (Running and Coughlan 1988; Running and Nemani 1988; Running *et al.* 1989) for hydrology and plant canopy function has been welded to an individual-tree-based forest model to provide a capability to simulate the change in forest structure over time (Friend *et al.* 1993). The resultant model, HYBRID, also uses a more detailed photosynthesis model, PGEN (Friend 1991), to improve the representation of mechanisms for photosynthesis production and allocation. The main simplifying physiological assumptions used in HYBRID (at the leaf level) are that the ratio of the CO_2 partial pressure inside the leaf to the outside partial pressure is constant, that the activity of the enzyme Rubisco and its concentration in the leaf is unimportant and that leaf and air temperatures are the same. In allocation of photosynthate it is assumed that leaf growth has a priority over stem growth and that the parameters of a given species do not change during the simulation. The model parameters are intentionally restricted in number. Table 10.8 lists the model parameters for two species. Results from the HYBRID model for simultaneous effects of climate change and increased CO_2 levels in the atmosphere will be discussed in Chapters 12 and 14.

The DOLY model (Woodward *et al.* 1995) is a global scale terrestrial model that combines predicting primary productivity and the phytogeographical problem of predicting leaf area across continents. The model has features associated with canopy process models, such as a model representation of the bulk biochemical features of photosynthesis; the dependence of CO_2 exchange, temperature and moisture changes on the stomatal conductance; and the role of canopy conductance on evapotranspiration and soil water losses. The involvement of nitrogen with the photosynthesis process and the dynamics of the uptake and allocation of nitrogen are also simulated. The long-term balance of these processes also determines the ecosystem's leaf area using an approach like that illustrated in Fig. 10.12.

The model can be used to simulate the global distributions of leaf areas and net primary productivity. An example of a model test involving the prediction of net annual primary productivity at 19 sites is shown in Fig. 10.17. In this illustration the solid lines are regressions of predicted values at different locations against observations. The closeness of fit of the regression is $r^2 = 0.95$. The dashed line in the figure indicates a perfect fit. The ability of the DOLY model to predict processes involved with the major fluxes of heat, water and CO_2 and simultaneously to predict leaf area, makes it a useful model for inspecting large-scale responses of the terrestrial vegetation. Applications of the DOLY model, as well as other canopy process models and material transfer models, will be discussed in Chapter 13.

Concluding comments

The richness of the modelling approaches and applications that have been developed to predict the response of landscapes to environmental change is great. Many of these models have been under development for over a

Table 10.8. *Species parameters used in the HYBRID model. The allometric constants are for functions of the form aP^b, where P is either tree diameter (DBH) or stem dry weight*

Parameter	Lodgepole pine (*Pinus contorta*)	White oak	Units (*Quercus alba*)
Bark thickness conversion factor	0.0075	0.033	cm cm^{-1}
Stem wood dry weight = $a(DBH)^b$	$a = 0.07194$ $b = 2.449$	$a = 0.0914$ $b = 2.2537$	kg
Branch dry weight = $a(DBH)^b$	$a = 0.00912$ $b = 2.244$	$a = 0.0274$ $b = 2.3371$	kg
DBH = a(stem dry weight)^b	$a = 0.1605$ $b = 0.4132$	$a = 0.1044$ $b = 0.4579$	cm
Percentage live sapwood	5	27.9	%
Specific leaf area	21.9	26.7	m^2 kg(C)$^{-1}$
Total/projected leaf area	2.5	1.0	m^2 m^{-2}
Canopy light extinction coefficient	−0.5	−0.4	Dimensionless
Optimal temperature for electron transport	21	31	°C
Effective root hydraulic conductivity	0.000215	0.00023	mol(H$_2$O)g^{-1}MPa^{-1}s^{-1}
Leaf dark respiration at 25°C	1.05	1.75	μmol m^{-2}s^{-1}
Nitrogen in chlorophyll	0.6321	2.6743	mmol m^{-2}
Assimilation CO$_2$ versus light response curvature	0.45	0.45	Dimensionless
Assimilation CO$_2$ versus light response initial slope	0.28	0.28	Dimensionless
Years leaves remain on tree	3	1	year
Root to leaf dry weight ratio	4.33	13.64	kg(C) kg(C)$^{-1}$
Total leaf area to sapwood leaf area coefficient	0.35	0.4	m^{2}cm^{-2}
Height = $137 + b_2(DBH)$ $- b_3(DBH)^2$	$b_2 = 48.63$ $b_3 = 0.1216$	$b_2 = 43.63$ $b_3 = 0.1091$	Dimensionless Dimensionless
'Stomata'[a]	2	1	Dimensionless
Percentage fine roots of total roots	23	7.3	%

Note:
[a] 1, hypostomatous leaves (stomata found on lower leaf surface); 2, amphistomatous leaves (stomata on all leaf surfaces).
Source: From Friend *et al.* 1993

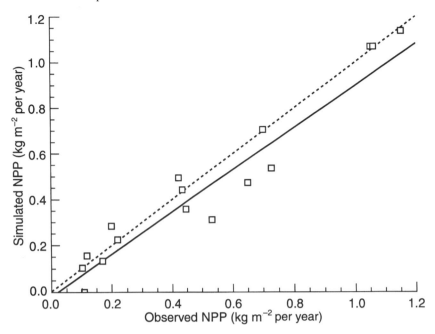

Figure 10.17. Annual net primary productivity (NPP) as simulated by the DOLY model. Simulated values are compared at 19 sites (Raich *et al.* 1991; McGuire *et al.* 1992). The solid lines are regressions of predicted values versus observations. The goodness of fit is $r^2 = 0.95$. The dashed line indicates exact agreement. From Woodward *et al.* (1995).

decade and represent an elaborate synthesis of laboratory and field data. However, there are no shortages of challenges in the area of landscape modelling. Cross model comparisons and the development of data sets with sufficient content to test the models are obvious immediate objectives.

In their application to larger space-scale problems, the different landscape models have strengths and weaknesses that are, to a degree, complementary. Many of the models that simulate the landscape elements for mosaic landscapes have demonstrated a capability to predict changes in the structure of vegetation associated with environmental change but are limited in continental- and global-scale applications by a lack of the species-specific information that they require. Interactive landscape models have provided cautionary results that point to the potential importance of spatial interactions as factors that can alter the rates of processes to a significant degree, but these models tend to be data and information demanding to a degree that currently limits their applications to case studies. Models of ecosystem material flows scale up to larger scales relatively easily (given the appropriate base data sets) but have little in their internal mechanisms to change the ecosystem structure and the feedbacks of structure onto the underlying processes.

In part because of the increase in cheap computational power, chimeras that combine different models (sometimes with different underlying assumptions) in ways that backstop one another's weaknesses have been produced as solutions to these problems. The HYBRID model, which uses a gap model to change the physical structure of the plant canopy (based on the allometry and the performance of individual plants) and a canopy model to capture the productivity of the canopy to determine the water-use and carbon fixation, is an example of such a development. Other mixed models, such as the DOLY model, are intended to tackle the difficult problem of mixing a material flow model with phenomena associated with canopy processes. The data to test such models, as well as the understanding of the numerical and analytical features of these models, are still a work in progress.

Nevertheless, we are beginning to apply a variety of landscape models to the problem of understanding the ways that natural ecosystems might respond to large-scale environmental change. Some examples of these findings will be the topic of the next three chapters and are part of an ongoing research agenda.

In a practical sense, the two most important issues to be addressed in understanding the response of the terrestrial surface of our planet to environmental change are the understanding of the present and future role of humans in altering landscape pattern and function, and the effect of the altered concentration of CO_2 in the Earth's atmosphere on terrestrial processes — either indirectly through modification of the Earth's climate or directly through alterations in fundamental plant processes related to photosynthesis.

Part 4
Evaluation of global change

11 · *Mosaic landscape models*

Just as different phenomena may vary in their importance at different time scales (Chapter 3), they also vary in the manner they propagate over space. The representation of spatial responses and dynamics of ecosystems over large areas may, in particular cases, be better represented by one mode of model formulation than others. With respect to representing spatial pattern, ecosystem responses are typically modelled as one of three cases. These are cases in which:

1. The responses to environmental change are local and relatively independent of the landscape configuration and spatial effects
2. The responses to environmental changes have a pronounced spatial component that either amplifies or attenuates the local responses
3. The responses are such that the landscapes respond as homogeneous units (and the local effects are less important or similar for all landscape elements).

Spatial ecosystems models (or landscape models) within this range of structural and conceptual differences can, in some cases, produce the same results in terms of the expected responses to change. Because one has more confidence in their predictions, robust results (same predicted responses from models with differing underlying assumptions) are consequential in evaluating possible responses of ecosystems to large–scale environmental change.

In this and the two chapters that follow, the predicted effects of environmental changes on natural landscapes will be discussed. The discussion will be organised according to the types of model that have been used. The initial discussions in the present chapter will treat models of mosaic landscapes (Case 1, above). Chapter 12 will then consider the results obtained from models of interactive landscapes (Case 2, above). Finally, Chapter 13 will deal with the predictions of models that have been applied to homogeneous landscapes often over very large spatial domains (Case 3, above).

The application of phytogeographical models to assess climate change effects

Perhaps the simplest of models for relating vegetation pattern to climate on a mosaic landscape are based on using classic phytogeographic climate/vegetation classifications to interpret the potential magnitude of climate changes. Bioclimatic models (see Chapter 6) relate the distribution of vegetation or plant life forms with biologically important features of the climate: often measures that integrate moisture and temperature conditions (see Smith *et al.* 1993). These classification models have a history of applications in predicting the distribution of vegetation under changed climate conditions, both for past climatic conditions associated with the last glacial maximum (Manabe and Stouffer 1980; Hansen *et al.* 1984; Prentice and Fung 1990) and predictions of future climate patterns under conditions of doubled CO_2 (Emanuel *et al.* 1985a,b; Prentice and Fung 1990; Smith *et al.* 1992a,b).

Prediction of alterations in the patterns of biomes for changed climates

Smith *et al.* (1995) evaluated the shifts in Holdridge life zones (Chapter 6) globally in response to four different climate change scenarios produced by four different GCMs of the atmosphere for a doubled-CO_2 condition (Table 11.1). The global distributions of Holdridge life zones for the current climatic conditions were mapped using an interpolated climate data base of mean monthly precipitation and temperature at a 0.5° latitude × 0.5° longitude resolution (Leemans and Cramer 1991). Changes in mean monthly precipitation and temperature were calculated for each GCM scenario (Table 11.1) for each computational grid element by taking the difference between current climate simulated by each simulation model and for the equivalent doubled-CO_2 climatic condition. These relative changes were then interpolated to 0.5° latitude and longitude resolution. The implied changes in monthly precipitation and temperature were applied to the global climate data base to provide a climate change scenario. The altered data bases corresponding to each of the four GCM scenarios (Table 11.1) were then used to reclassify the expected vegetation in each of the grid cells (0.5° latitude × 0.5° longi-

Table 11.1. *Resolution and mean global changes in temperature and precipitation for four GCMs used to construct alternative climate change scenarios*

GCM (acronym)	Resolution (latitude × longitude) (°)	Change in Mean Global	
		Temperature (°C)	Precipitation (% change)
Oregon State University (OSU)[a]	4 × 5	2.84	7.8
Geophysical Fluid Dynamics Laboratory (GFDL)[b]	4.5 × 7.5	4.00	8.7
Goddard Institute for Space Science (GISS)[c]	7.8 × 10	4.20	11.0
UK Meteorological Office (UKMO)[d]	5 × 5	5.20	15.0

Notes:
[a] Schlesinger and Zhao (1988).
[b] Manabe and Wetherald (1987).
[c] Hansen *et al.* (1988).
[d] Mitchell (1983).
Source: From Smith *et al.* (1995).

tude) using the Holdridge system and based on the interpolated climate data.

All of the GCMs imply a global climate with higher average temperatures and increased precipitation, but the spatial pattern of these changes vary among the GCMs. However, the climate models all produce a greater degree of warming in the higher northern latitudes and tend to feature a drier interior for the interiors of North America and Eurasia.

The GCMs differ from one another with respect to the manner in which physically important phenomena are represented by the models (e.g. formation of clouds, formation of sea ice, etc.) and in their spatial resolution (Table 11.1). The differences in the climates simulated by the models reflect these differences in model formulation, as well as differences arising from spatial resolution and numerical considerations.

Maps of the distribution of potential biome types (grouped Holdridge life zone categories based on the general structure of the implied vegetation) under current climate conditions and climate change scenarios based on the four GCMs are presented in Fig. 11.1 (colour plate). All four climate change scenarios investigated show a significant shift in the

Table 11.2. *Changes in the aerial coverage of major biome types under current and changed climate conditions*

Biome type[a]	Current climate (area in $10^3\,\mathrm{km}^2$)	OSU (area change in $10^3\,\mathrm{km}^2$)	GFDL (area change in $10^3\,\mathrm{km}^2$)	GISS (area change in $10^3\,\mathrm{km}^2$)	UKMO (area change in $10^3\,\mathrm{km}^2$)
Tundra	939	−302	−515	−314	−573
Desert	3699	−619	−630	−962	−980
Grassland	1923	30	969	694	810
Dry forest	1816	4	608	487	1296
Mesic forest	5172	561	−402	120	−519

Notes:
[a] Holdridge life zone categories included in each biome type. **Tundra**: polar dry tundra, polar moist tundra, polar wet tundra, polar rain tundra; **desert**: polar desert, boreal desert, cool temperate desert, warm temperate desert, subtropical desert, subtropical desert bush, tropical desert, tropical desert bush; **grassland**: cool temperate steppe, warm temperate thorn steppe, subtropical thorn steppe, tropical thorn steppe, tropical very dry forest; **dry forest**: warm temperate dry forest, subtropical dry forest, tropical dry forest; **mesic forest**: forests not listed above: moist boreal forest, wet boreal forest, boreal rain forest, cool temperate forest, warm temperate forest, subtropical and tropical forest zones.
Source: From Smith *et al.* (1995).

distribution of these biome types. Changes in the global aerial coverage of major biome types (aggregated Holdridge life zones: see the key to the maps in Fig. 11.1) are shown in Table 11.2.

There is a general qualitative agreement among the scenarios in some of the directions of vegetation change in response to a climatic warming. For example, the extents of tundra and desert are decreased, and the potential areas of grasslands and forests are increased under all four scenarios. Despite the agreement in increased forest cover, the scenarios differ in the degree to which the increase is attributable to mesic and xeric forest components. Mesic forest cover increased under the GISS and OSU scenarios but decreased in the GFDL and UKMO scenarios. The predicted decreases in mesic forest by GFDL and UKMO are offset by larger increases in dry forest, therefore, forest cover increased overall in all the scenarios.

The changes in coverage of the biome types presented in Table 11.2 are the outcome of spatial changes in the climate pattern predicted by each GCM, which produces spatial changes in the distribution of life zones. The decline in tundra observed under all scenarios is primarily caused by a shift from tundra to mesic forest. This transition is a result of the warming at higher latitudes and the subsequent northward movement of potential boreal forest zones into the areas now occupied by wet tundra. A second major component of change in tundra is a conversion to polar desert in areas where warming and/or decreases in precipitation occurs. A decrease in the global extent of desert seen in all four scenarios is the result of the conversion of polar desert to tundra in the higher latitudes, and from desert to grassland in the temperate and tropical regions. Furthermore, there is a significant conversion from desert to mesic forest under the GFDL and UKMO scenarios. These shifts occur in the northern latitudes where cold desert/dry tundra zones increase in both temperature and precipitation.

The increased cover of grassland under all scenarios is a function of shifts from desert to grassland with increased precipitation in areas of the temperate and tropical regions, and the transition of dry and mesic forests to grassland as a result of drying in forested regions. The extent of dry forest increases with increasing precipitation in grassland regions, and with increased temperatures and/or decreased precipitation in mesic forests. The later transition occurs primarily in the subtropical and tropical regions and is most pronounced in the UKMO scenario, resulting in a doubling of the global extent of dry forest.

The changes in potential mesic forest vary with climate change scenarios. While all the scenarios have increased precipitation (see Table 11.1), the increase in the evapotranspiration associated with the higher temperatures causes the GFDL and the UKMO scenarios to be for an effectively drier earth (more precipitation but even more evapotranspiration). There is a loss of mesic forests in both the UKMO and the GFDL scenarios (Table 11.2). The OSU and the GISS scenarios also have increased temperature and precipitation, but in these scenarios, even with the increased evapotranspirative demand, there is an increased supply of available moisture. The major difference in the potential aerial coverage of mesic forest between the scenarios is in the degree of mesic forest change to dry forest in subtropical and tropical regions.

Comparisons of static model output between current climatic conditions and climatic scenarios derived from GCM experiments must be interpreted with two important caveats in mind. First, changes in the spatial distribution of potential vegetation zones under altered climatic conditions do not represent expected changes in vegetation structure. Rather, these changes are a mapping of the expected distribution of climates associated with existing vegetation. Shifts in the patterns of climatic variables (and the additional factor of altered CO_2 concentrations in the atmosphere) may cause ecological transitions to novel vegetation patterns. It is known from palaeoecological studies that vegetation associated with past climates can be quite different from extant vegetation (see Chapter 2). Areas of change between two potential vegetation maps derived from static models can be reasonably interpreted as areas of expected ecological stress, where only a directionality of potential vegetation change may be inferred from the changes in equilibrium zones mapped. For example, a shift from a climate associated with moist boreal forest to a drier climate now associated with steppe grasslands could be expected to engender changes toward lower levels of woody biomass over long time periods at a given site. The magnitude and temporal dynamics of this change must be modelled through more detailed, dynamic approaches (see examples below). Used in this provisional role, static models serve a useful purpose in identifying sites of expected directional change where more mechanistically based approaches can be employed to examine the expected temporal dynamics and magnitude of forest change.

The second important caveat required for the reasonable employment of static ecoclimatic models concerns the fact that static models are established to predict natural vegetation complexes associated with current climate variables. This expected relationship assumes that natural vegetation groups are currently in equilibrium with present environmental conditions and that areas to be analysed are generally free of human land-use interventions. Both of these features affect the direct applications of static model outputs to modern landscapes. There are several potential shortcomings in interpretations of results based on these mapping techniques. This led to an inspection of ecological responses at a variety of levels (total vegetation, species level, individual plant level, plant tissue level) to determine if there is a consistency and robustness to these results. These will be discussed in the sections below.

Application of plant energy balance models to map leaf area

Woodward (1987b) developed an approach for predicting leaf area (area of leaf surface per area of ground surface in $m^2 m^{-2}$) and associated physiognomy using a model of plant energy balance. This model is described in brief in Chapter 10 (see Figs.10.11, 10.12 and 10.13). The model uses the Penman–Monteith evapotranspiration model (Penman 1948; Monteith 1972) to simulate the biophysics and physiology of water-use at the plant canopy level. Using a set of parameters describing the environment of the canopy, combined with a functional relationship between leaf environment and stomatal conductance, the model predicts evapotranspiration for a given leaf area and climatic conditions. By solving the model iteratively for varying values of leaf area, Woodward (1987b) used the model to solve for the maximum leaf area which could be sustained under the climatic conditions at any given location (Fig. 10.13).

This approach provides a process-based alternative to the vegetation–climate classification models described above. By using this approach to predict leaf area under current and changed climate conditions, it is possible to predict changes in the leaf area that can be sustained under the changed climate conditions (Woodward 1987b; Smith et al. 1992b). The comparison of current and changed leaf area index can then be used to infer potential changes in the composition and structure of vegetation that may relate to the predicted shifts in leaf area index.

Like the Holdridge system, the energy balance-based approach of Woodward (1987b) represents an equilibrium solution to a dynamic process. The predicted patterns of leaf area index under current and changed climate are equilibrium solutions to the corresponding spatial changes in climate patterns. The model does not explore the temporal dynamics of vegetation response to changed climate, rather it assumes that vegetation can respond by increasing/decreasing leaf area to equilibrate with new climate conditions, even though this may require major shifts in species or even life-form composition.

For example, Fig. 11.2 (colour plate) shows the leaf area index predicted for Africa under the current climate using the same interpolated climate data base (mean monthly precipitation and temperature at a 0.5° latitude × 0.5° longitude resolution (from Leemans and Cramer (1991)) used to develop the current climate map shown in Fig. 11.1 (colour plate). Under two climate change scenarios (UKMO and OSU from Table 11.1), the model is in broad general agreement with the Holdridge projections developed as in Fig. 11.1. The change in climate under the UKMO scenario shows a reduction in leaf area attributable to the general drying in the scenario. Much of the area currently in mesic forest and rain forest in central-west Africa has a pronounced decline in leaf area, corresponding to the shift to dry woodland and savanna implied by the Holdridge analysis (Fig. 11.1). The wetter OSU scenario moves much of this same region into the tropical forest zone (Fig. 11.1) and one sees a substantial potential leaf area increase predicted in this area by the biophysical model (Fig. 11.2). The patterns of change seen in the phytogeographical Holdridge analysis (Fig. 11.1) is in reasonable agreement with the response of the more biophysical Woodward (1987b) model for Africa under present conditions and for the two climate change cases.

Effects of biome shifts on the global carbon budget

Vegetation changes in response to climatic change of the magnitudes shown in Table 11.1 imply potential alterations of the amounts of carbon stored on the terrestrial surface. One question is, 'What are the fundamental properties of the feedback loop from changes in: atmospheric $CO_2 \rightarrow$ change in climate $\rightarrow$ change in vegetation pattern $\rightarrow$ change in terrestrial carbon stored $\rightarrow$ change in atmospheric CO_2?' This is a straightforward question of considerable importance to which we do not now have a clear answer. If the feedback loop is a positive feedback loop then one is faced with the possibility of a 'runaway greenhouse effect' in which the climate change, induced by higher CO_2 levels in the atmosphere, changes the vegetation in ways that produce more atmospheric CO_2 and, consequently, still more warming. If the feedback loop is negative feedback, then a climatic warming from higher CO_2 levels in the atmosphere would increase the carbon shortage on the terrestrial surface and, thus, act to reduce the climate change effect. One consideration that makes it difficult to answer the 'homeostatic feedback' problem from earth and atmosphere is the complexity of the system dynamics involved (one needs to understand the dynamics of the climate system, the terrestrial surface and the global carbon cycle).

Prentice and Fung (1990) took up one aspect of this problem by considering whether or not the changes in the terrestrial vegetation at equilibrium with a changed climate might be a source or a sink for carbon. By inspecting the long-term response to change one can apply phytogeographical algorithms to predict the new vegetation. Prentice and Fung (1990) used the GISS GCM (also used to develop the scenario in Table 11.1) (Hansen et al. 1988) to develop alternative global climates. One was for 18,000 BP for conditions of a glaciated Earth with sea levels about 120 m lower than today (and consequently a greater land area). The other was for a doubled-CO_2 atmosphere with a sea level rise of 2 m. In their application, Prentice and Fung used a modified version derived from Holdridge (1967) to cover the surface with vegetation appropriate to the simulated climates.

Several different authors (see footnotes in Table 11.3) have attempted

Table 11.3. *Global terrestrial carbon reservoir changes in percentages relative to today's estimate for past (18 000 BP) and future, double-CO_2 conditions*[a]

Climate	Carbon reservoir[b]	Estimated size of present carbon reservoir (Pg C)	Percentage change in carbon reservoir at sea level changes (m) of			
			−130	−100	0	+2
18 000 BP	Biomass					
	Estimate 1	748	6	3	−7	
	Estimate 2	834	5	2	−8	
	Soil					
	Estimate 3	1143	−4	−6	−13	
	Estimate 4	1313	−1	−3	−10	
Future, double-CO_2	Biomass					
	Estimate 1	748			31	31
	Estimate 2	834			35	35
	Soil					
	Estimate 3	1143			1	1
	Estimate 4	1313			3	3

Note:
[a] Climatic changes are from the GISS GCM (Hansen *et al.* 1988). Sea level change in the past was evaluated for sea levels reduced 130 m and 100 m to bracket the 121 m sea level reduction (relative to current sea level at that time).
[b] 1, Olsen *et al.* (1983); 2, Whittaker and Likens (1975); 3, Post *et al.* (1982); 4, Atjay *et al.* (1979) and Schlesinger (1984).
Source: From Prentice and Fung (1990).

to estimate carbon concentrations by biome for soils and vegetation for purposes of computing global terrestrial carbon budgets. Prentice and Fung (1990) used these information sources in combination to obtain an estimate of the range of variation in their estimates. The carbon in the soils and vegetation for each vegetation type were computed by using different estimates of carbon pools in soils and vegetation within each type. This amounts to assuming that the carbon stored in a vegetation type such as 'boreal forest' is the same in the past, present and future. The results of their study (Table 11.3) are for the two climate warmings (from 18,000 BP to present times; from present times to a warmer Earth). In the warming since the last glaciation, they estimated the terrestrial surface to have stored a net of 200 Pg (1 Pg=10^{15} g) carbon in vegetation and soils (for the case in which sea level rise was set at 130 m). Much of this was the result of increased carbon storage on the new land surface that emerged from the seas. Changes in the locations of biomes on the surface had a relatively negligible effect on the carbon storage on the terrestrial surface (18,000 BP to present).

By the Prentice and Fung (1990) calculations, shifts in the terrestrial biomes under a putative future warming store between 245 and 338 Pg carbon with little significant effect owing to sea level rise. Therefore, from these calculations, the terrestrial surface appears to have a potential negative feedback effect on a warming climate. A significant amount of this increased storage is in the expansion of the Prentice and Fung category called 'tropical forests'.

Smith *et al.* (1992b) computed changes in global terrestrial carbon pools for the four different GCMs (see Table 11.1) (as well as an additional case with a modified version of the GISS model). They used a fully expanded version of the Holdridge life zone classification (37 life zones) calibrated against the Olson *et al.* (1983) vegetation carbon-storage data set and the Post *et al.* (1982) soil carbon-storage data set. Maps in Fig. 11.1 display the pattern of life zones for the current climate and for the four scenarios listed in Table 11.1 that are used as a basis for the vegetation changes in these calculations.

Smith *et al.* (1992b) found that even though some of the climate change scenarios were for a warmer and wetter Earth and others for a relatively warmer and effectively drier Earth (owing to evapotranspiration increases with temperature overshadowing the effects of increased precipitation), the terrestrial surface appeared to be a sink for CO_2 under the conditions of a climatic warming. However, the strength of this feedback was considerably less (adding between 8.5 and 180.5 Pg to the estimated current terrestrial carbon storage) than that calculated by Prentice and Fung (additional storage of 245 to 338 Pg more terrestrial carbon in response to a doubled-CO_2 warming). These differences appear to be a result of primarily higher differentiation of tropical forest types in the Smith *et al.* calculations, which separate the 'tropical forests' of Prentice and Fung into a larger number of categories (e.g. tropical dry forests, subtropical dry forest, etc.), some of which have relatively lower carbon storage in vegetation and soils. The pattern of carbon source or sinks on the terrestrial surface are quite heterogeneous (Fig. 11.3, colour plate).

It is significant that the terrestrial surface appears potentially to be a weak sink for CO_2 during a warming. These results (Prentice and Fung 1990; Smith *et al.* 1992b) and other similar studies since are all based on some relatively limiting assumptions about the response of the vegetation to climatic change. These include the reliability of vegetation–climate correlations in the future (particularly with the potential effects of elevated CO_2 operating on plant processes as a factor that might alter these relationships) and the equilibrium evaluations that must be used in such studies.* Developers of these global carbon budget calculations are well aware of these shortcomings and describe these results as being motivations to develop better data sets and models for the estimation of the global terrestrial carbon.

The application of Grinnellian niche concepts to assess climate change effects

From palaeoecological studies of the changes in vegetation over the past 10 000 to 20 000 years, it appears that species respond independently in the changes in their continental ranges with major climatic change (Chapter 2, see particularly Davis 1976, 1981a,b, 1983, 1986; Huntley and Birks 1983; Huntley 1988; Webb 1988). This leads to an interest in understanding the shift in species distributions in response to environmental change as an alternative to the evaluation of large-scale potential vegetation zone changes listed above. One straightforward way to predict the change in the range of a species in a different climate is to assume that the conditions under which it occurs and then to use this information to determine where the species might be expected to occur in the future. This procedure is a direct application of Grinnellian niche concepts (Chapter 5). The approach has been used to reconstruct past climates from various sorts of fossil evidence (such as pollen found in lake sediments) and to evaluate future climate effects. In some applications, spatial change in an assemblage of species or of a vegetation type can also be inferred from analogous procedures.

These predictions are for shifts in ranges after some relatively long interval of time during which the species distribution has come to an equilibrium with the climatic conditions. Presumably, the temporal scales are sufficiently long that there is no need to be concerned with whatever delays arise in the species redistribution owing to migration lag-times, the time required for the species already occupying an area to be displaced, and for the successional changes involved with the range shifts. However,

---

* Chapter 12 will address the potential transient effects, including migration and successional delays on changes in the global carbon budget for two of the scenarios (GISS and GFDL) shown in Table 11.1. Owing to the dynamics of the terrestrial vegetation and soils under a climate change, the initial evaluations indicate the Earth's surface is a relatively substantial source for carbon through the first 200 years of its response to a climatic change.

these approaches also involve assuming that the species involved have not evolved to be different: even though major shifts in the ranges of a species could also be a source of considerable selection.

Zabinski and Davis (1989) mapped the changes in the distribution of four tree species (eastern hemlock, *Tsuga canadensis*; American beech, *Fagus grandifolia*; yellow birch, *Betula lutea*; sugar maple, *Acer saccharum*) in response to the GISS and GFDL scenarios for a doubled-CO_2 condition (see Table 11.1). The distribution of these species tends to conform to isotherms of mean monthly temperatures and/or heat sums (such as growing degree days). The basis for the mapping of changes in species ranges by Zabinski and Davis was to determine the mean January temperature associated with the northern range boundary of the species and the mean July temperature corresponding to the southern boundary. Such mapped changes reflect correlations between climate and species distributions. The expected species ranges were obtained by mapping the change in these climatic variables (Fig. 11.4). Under the two climate-change scenarios (GISS and GFDL), species were displaced from much of their North American range (with the GFDL case being somewhat more extreme). The magnitude of the displacements predicted for these four tree species for North America are of the order of the displacements of Holdridge life zones mentioned in the section above.

In a similar application, Huntley (1994) mapped the distribution of two important European tree species (beech, *Fagus silvatica*; Scots pine, *Pinus sylvestris*) in response to climate change under the present climatic conditions and under two climate change scenarios (OSU and UKMO, see Table 11.1). Climate variables used were the temperature sum (degree-days) above a 5°C threshold, the mean temperature of the coldest month (computed as the mean temperature of that month which, averaged across the period of record (usually 30 years), had the lowest mean temperature) and an index of moisture availability estimated by the ratio of actual to potential evapotranspiration (Prentice *et al.* 1993). These variables were interpolated to a $50 \times 50 \text{ km}^2$ grid for Europe and calibrated against the current distribution of the two tree species from *Atlas Florae Europaeae* (Jalas and Suominen 1973, 1976). As has been reported for the North American case (above), the displacements in the ranges of the species are significant and imply that the potential future range of the species, under a climate change, would not overlap with the current range of species over areas the size of large countries. Huntley (1994) also identified several species of European birds that one might expect to be displaced in their ranges in response to shifts in the distribution of species with which they

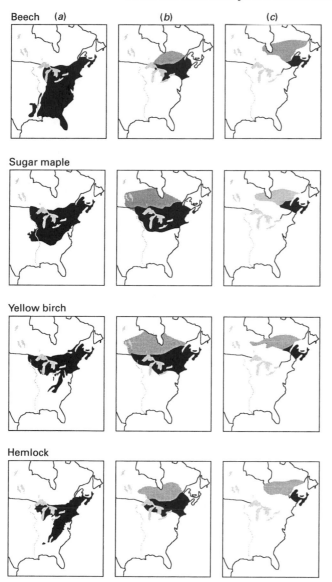

Figure 11.4. Present and potential future geographic ranges for four tree species (eastern hemlock, *Tsuga canadensis*; American beech, *Fagus grandifolia*; yellow birch, *Betula lutea*; sugar maple, *Acer saccharum*) under current climate conditions and under conditions predicted by the GISS and GFDL GCMs for a CO_2 doubling (see Table 11.1). Present range of each species is shown in dark shading in (*a*). In (*b*) and (*c*), changes in species ranges implied by the climate changes for the GISS (*b*) and GFDL (*c*) scenarios are shown. In the (*b*) and (*c*) maps, the dark shading is the part of the former species range that still has appropriate climatic conditions for each species, and the light shading indicates potential new area in the species range. From Zabinski and Davis (1989).

are normally associated. For example, the parrot crossbill (*Loxia pytyopsit-tacus*) is highly dependent on *Pinus sylvestris* as a food plant (Newton 1972) and one would expect the bird to be shifted from its present range. Using the BIOCLIM model (see Chapter 5) to determine the range of climate conditions associated with each of 57 of Australia's threatened vertebrate species to examine the potential effects of a 1°C warming, Dexter and Chapman (1995) found 84% of these species experienced a reduction of the range of their climatically suitable habitat.

Webb *et al.* (1987) related the distribution of the taxa normally found in palynological (fossil pollen) reconstructions of past climates by com-puting a response surface relating pollen abundance with climatic factors. The response surface (Fig. 11.5) is based on a large data set of recent pollen deposition at a large number of sites across North America (Bernabo and Webb 1977; Webb 1987). The response surfaces allow the determination of the relative abundance of pollen from a given taxon with interrelated environmental variables. Bartlein *et al.* (1986) initially computed a response surface (using polynomial regression) for eight species with respect to mean July temperature and precipitation. Webb *et al.* (1987) included these two variables plus mean January temperature to compute the surfaces for six species shown in Fig. 11.5. Note that the two variables used in the Zabinski and Davis (1989) study are included in the three climate variables used by Webb *et al.* (1987). These methods are directly analogous to the techniques used in statistical determination of the Grinnellian niches of species (see Chapter 5 and Austin *et al.* 1990).

One of the more striking applications of Webb *et al.* (1987) was the development of an expected palaeontological reconstruction of the dis-tribution of spruce pollen abundance (as well as several other categories of pollen sources) expected in sedimentary deposits in lakes for North America that could be compared with actual observations at 3000 year intervals (Fig. 11.6).

This prediction was developed using climate simulated by a GCM (from the US National Center for Atmospheric Research) incorporating data for conditions (e.g. sea surface temperatures as inferred from ocean sediments, location of the Laurentide ice sheet) at 3000 year intervals from 18,000 years BP to the present condition (COHMAP, 1988). Comparison between the observed and simulated pollen percentages for spruce shows that, while there are specific cases and locations where the predicted and observed pollen abundances do not agree, there is a sub-stantial level of agreement over the North American continent for the 18 000 year interval.

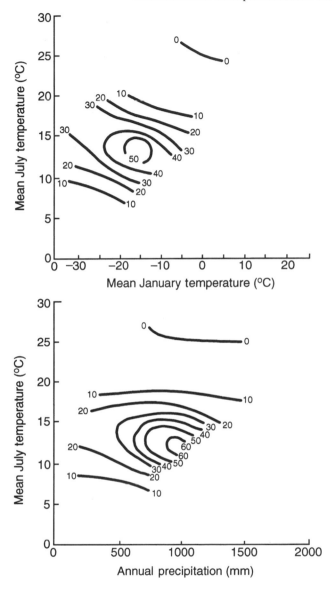

Figure 11.5. Response surfaces showing the relationship between the percentages of spruce pollen from contemporary samples in response to mean January temperature, annual precipitation and mean July temperature. From Webb *et al.* (1987).

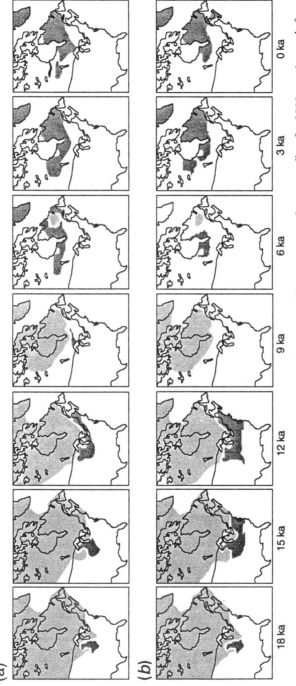

Figure 11.6. Maps of model-simulated (by CCM output) (*a*) and observed (*b*) percentages of spruce pollen for 3000 year intervals from 18,000 BP to the present (ka, thousand years). The grey region in the north represents the location of the Laurentide ice sheet as it shrinks in area from 18,000 BP and disappears at 6000 BP. Simulated values are obtained by applying the response function for spruce (see Fig. 11.3, colour plate) to responses of the NCAR 'Community Climate Model' (see COHMAP (1988) for details). Good agreement between model prediction and observed values of spruce pollen imply that the model has predicted a combination of temperature and precipitation values potentially compatible with growth of spruce trees. From Webb *et al.* (1987).

Tests of Grinnellian approaches to mapping the distributions of species under changed climates usually consist of developing reconstructions of species distributions at different times or of comparing distributions of introduced species in space. The Webb *et al.* (1987) palaeontological reconstructions are an example of the former sort of test. As examples of the latter tests, Huntley *et al.* (1989) demonstrated that the distributions of the genus *Fagus* (beech) in Europe (*F. silvatica*) and North America (*F. grandifolia*) could be fitted by the same climate-response functions. Such demonstrations at the genus level are valuable for considerations involving the interpretation of fossil pollen data, which typically are only resolved at the genus level.

Beerling *et al.* (1995) (see Chapter 6) tested the consistency of the same climate response function for the introduced species *Fallopia japonica* in Europe and in its native range in south-east Asia. They also predicted the change in the species distribution in Europe under two climate change scenarios (OSU and UKMO, see Table 11.1). Woodward (1988) also tested for consistency in factors controlling the distributions of plants introduced to North America from Europe on both continents. Palaeoecologically calibrated climate-response functions for species have been used to develop evaluations of future responses to climate change (Overpeck and Bartlein 1989; Huntley *et al.* 1995).

There are two principal difficulties in these approaches. First, because they are correlational, they do not necessarily imply that the factors used to predict the distributions of the species are primary or causal factors. In using correlations in such applications, care should be taken not to overfit the data. For example, combinations of environmental variables that vary in a north/south direction (such as various temperature-related measurements) and variables that vary east/west (moisture, in some cases) are correlated with latitude and longitude. These variables in combination represent transformed co-ordinate systems. One can map many spatial distributions in such a co-ordinate system (particularly if the variables are used in combinations with one another). For this reason, the tests of these approaches that involve plotting distributions of introduced species are important in that they allay some of the concern that attends the use of correlational data.

Second, the correlations or response surfaces are typically developed without regard to the level of CO_2 in the atmosphere. We appear to be destined to see the highest atmosphere levels of CO_2 experienced in the past several million years within the next century. The performance of plants and the degree to which they conform to correlation with environmental variables are not known. Ecologists who have applied these approaches are well aware of these difficulties and regularly identify these problems in their work. The problem of interpreting the future with elevated CO_2 from the non-elevated CO_2 past remains a problem of considerable difficulty.

The application of gap models to assess change on mosaic landscapes

Gap models have been applied to examine the response of forested systems to climate changes in both predictive and reconstructive modes. The models have been used in the reconstruction of prehistoric Quaternary forests (Solomon et al. 1980, 1981; Solomon and Shugart 1984; Solomon and Webb 1985; Bonan and Hayden 1990; Bonan et al. 1990). These reconstructions serve as a consistency test as to whether or not the species-level information in the models appears to be consistent with palaeoecological data, given a climate condition thought to have been the norm in the past. These applications have inspired model projections of possible consequences of future climate changes (Solomon et al. 1984; Solomon 1986b; Pastor and Post 1988; Urban and Shugart 1989; Bonan et al. 1990; Overpeck et al. 1990). In contrast to the modelling approaches discussed in the earlier sections of this chapter, these gap-model-based approaches to assessing the effects of climate change have more detailed resolution. Typically, these model-based predictions are for changes in species composition, vegetation structure, productivity and standing biomass.

Unfortunately, these applications are limited in spatial extent, largely because of the base information on soils, elevation and other context-related features needed to implement the model over large areas, and the information needed to initialise such models over large areas (see Chapter 10). In addition, the application of these finer resolution models is limited by computational factors. For these reasons, implementing the models for sets of conditions that sample the ranges of responses and conditions in a region has been used to provide large-area evaluations with respect to responses over environmental gradients (Solomon 1986b; Bonan 1990a,b; Smith et al. 1992b).

The models have also been applied to a wide range of environmental change-related problems that provide insight into the possible responses of forest systems to change. Before discussing the implications of gap models applied to climate change-related problems, it is useful to consider

Table 11.4. *Spectrum of increasingly more complex experiments on gap models to inspect effects of environmentally induced growth rate changes*

Level of growth alteration	Type of growth alteration	Description of model implementation
All trees in a forest	Systematic increase of growth for all species and all individuals in several different forest ecosystems. Intended as a minimal representation of the direct effects of CO_2 on trees	Increase the growth parameter ('G' in Eq. 8.7) by equivalent percentages for all species
By species of trees	Systematic decrease in growth of trees according to species at different levels according to supposed sensitivity to air pollutants	Decrease the growth parameter ('G' in Eq. 8.7) with interval percentage changes (0%, 10%, 20%) according to sensitivity to pollutants in acute situations (smelter locations, etc.)
By individuals	Change in growth rate at the individual plant level determined by implementing a plant physiology model interfaced with a biophysical model of a layered plant canopy	Interactive effects of CO_2, water and energy fluxes coupled with individual plant carbon allocation and growth model

the range of responses that can be obtained from the models that involve relatively straightforward changes in a single aspect of the models: changes in the growth rates of trees (Table 11.4).

Effects of changes in growth rates of dominant species and individuals on mosaic landscapes

Many of the verifications and validations that have been used in conjunction with gap models (see Chapter 8) involve testing the model's ability to simulate conditions along gradients and under other conditions that reflect the reliability of the model to extrapolate changes in pattern under altered environmental conditions. Of course, the fact that a given model has successfully been used to extrapolate one condition (e.g. the pattern of vegetation along a complex altitudinal gradient) provides no guarantee that the same model will successfully extrapolate some new condition (e.g. the response of a regional forest to chronic pollution). For this reason, any model extrapolations of this sort should be regarded with caution and should be augmented with primary observations as much and as soon as is possible.

One potential effect of environmental change on terrestrial ecosystems is to alter the growth rates of the species making up the terrestrial vegetation. The consequences of several types of such change have been investigated for forest ecosystems using gap models (and their derivatives). These model experiments fall into three broad classes (Table 11.4) and examples of these experiments are discussed in the sections that follow. The principal result of these model experiments is that relatively simple changes in one aspect of a plant's performance (the growth rate) can have complex consequences at the ecosystem level. Whether or not the specific simulations are accurate predictions of the future behaviour of ecosystems – the overarching result that interactions in local ecosystem processes can strongly modify the consequences of change on the individual plant processes – indicates a need to evaluate terrestrial ecosystem responses to changed conditions at multiple levels.

Uniform growth enhancement: an initial inspection of direct CO_2 effects

Measurements of CO_2 by Keeling *et al.* (1995) since 1958 at Mauna Loa Observatory, Hawaii (and elsewhere (Harris and Bodhaine 1983)) have shown a systematic increase that is attributed to the input of fossil fuel to the atmosphere (Bacastow and Keeling 1981). The complex exchanges of carbon among the oceans, land and atmosphere and the potential of the increased CO_2 to change the climate has produced a set of both significant and difficult scientific problems. One important part of this larger set of problems is the question as to whether or not the increased level of CO_2 will stimulate plant growth by increasing the supply of CO_2 for photosynthesis ('carbon dioxide fertilisation'). The increased plant growth could, in turn, cause more carbon to be stored in the terrestrial systems and, thus, reduce the rate of increase of CO_2 in the atmosphere. Kramer (1981) discusses this problem with a view to the plant physiological aspects of the problem:

> Many scientists seem to assume that the increasing carbon dioxide concentration of the atmosphere will automatically bring about an increase in global photosynthesis and dry matter (biomass) production ...

Some scientists also assert that the increase in carbon fixation by photo-synthesis will be large enough to slow down the increase in atmospheric CO_2 produced by burning fossil fuels (Bacastow and Keeling 1973). The validity of these assertions is of great importance to policy makers who need to know the global effects of the increasing use of fossil fuels. However, they are based on the assumption that the potential rate of photosynthesis is limited chiefly by CO_2 concentration and that the rate of photosynthesis is limited chiefly by a low potential rate of photosynthesis. But are these assumptions valid?

Kramer continues in his review to point out that there are potential difficulties with both of these assumptions and that 'we cannot make reliable predictions . . . until we have information based on long-term measurements.'

One aspect of this problem, which can be approached using models of the elements of a mosaic landscape, is to determine the maximum effects if both of the assumptions that Kramer identifies are met and the response of each tree in the stand is similar (see Shugart 1984; Shugart and Emanuel 1985). The question is, 'What is the biomass response expected of a forest if the growth rates of all trees are uniformly increased in a systematic fashion?' This exploration is not based on an assertion that such a response will be the case but is, instead, a determination of the magnitude of the forest response if the positive potential effects of increased CO_2 levels were realised at the individual tree level.

Shugart and Emanuel (1985) used the FORET, FORAR, FORICO, KIAMBRAM and BRIND models to investigate this response over a range of forest systems. These models simulate a considerable variety of the longer-term responses of forested landscapes and are reasonable tools with which to assess the potential consequences of perturbations on forests. A brief description of the models and a tabulation of the tests on these (and similar) models are found in Chapter 8. The models are particularly useful for inspecting the potential consequences of CO_2 fertilisation because the fundamental equation (see Eq. 8.5) used in the models to calculate the annual increase in the diameter of a tree in the simulated stand is derived from a simple growth model that balances photosynthate production against the respiratory cost of maintaining living tissue. While this growth equation is simple with respect to the complexities of the photosynthesis and growth relationships in plants, it is satisfactory to explore the effects of a uniform increase in productivity. A change in the rate of tree growth from increased photosynthate production can be simulated by adjusting the 'G' parameter in Eq. 8.7.

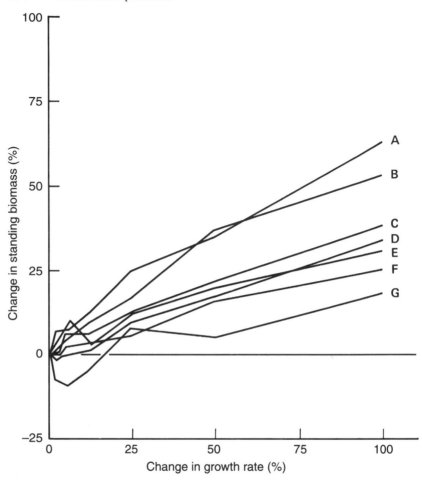

Figure 11.7. Response of seven different simulated cases of increased growth rate mimicking a maximum, non-species-specific response to CO_2 fertilisation. Lines are for the percentage of standing crop over years 300 to 500 of a 500-year simulation for systematic increases in all growth rates of 3.125, 6.25, 12.5, 25, 50 and 100% relative to a control case with no alteration in growth rate. A. FORAR model with wildfire. B. BRIND model with wildfire. C. FORAR model without wildfire. D. FORICO model. E. KIAMBRAM model. F. BRIND model without wildfire. G. FORET model. From Shugart (1984).

To inspect the potential magnitude of a fertilisation of the photosynthate production, the intrinsic growth rate (G) of each simulated tree was systematically increased by 100, 50, 22, 12.5, 6.25 and 3.125% (six levels of change) for 50 replicate computer runs of a duration of 500 simulated years each for seven different gap model cases (Fig. 11.7). The percentage change in the amount of organic matter in living plants (phytomass) was determined by averaging the phytomass on each of 50 simulated plots over the time period of 300 to 500 years and comparing these results with a control with no fertilisation of the photosynthesis production process.

The responses are complex at low levels of growth enhancement but are generally linear at higher levels. One case (the FORET model; Fig. 11.7g) actually demonstrates a diminution of phytomass at low fertilisation levels. Under inspection, this is caused by the uniform increase in growth rate allowing an increased survival of suppressed understory trees. These trees are species that tend to grow more slowly and attain a smaller mature size. Hence, there is a competition-induced change in total phytomass.

The increase in phytomass ranges from 20 to 60%. This increase is not the result of an elevated maximum phytomass at any point in the landscape. It is caused by the increased rate of filling of gaps made in the canopy by the death of large trees. For this reason, forests with a high rate of disturbance and more growth (Fig. 11.7a,b) show a larger landscape-scale response. In the present simulations, the amount of phytomass on a single simulated plot is limited by shading (for the smaller trees) or by the maximum size (for the largest trees). In actual forest systems, other limiting factors such as nutrient limitation could probably work to reduce any CO_2-induced growth enhancement.

As was pointed out in introducing this set of model experiments, these results should be regarded as the maximum expected for a uniform increase in growth rate across natural forested landscapes. A more realistic simulation using the HYBRID model in monospecies forests, but with more physiological detail at the leaf and whole plant level than the gap models used here, will be discussed below.

If the changes in growth rate were not uniform across all the species involved, then the potential expression of CO_2-induced fertilisation effects could involve changes in the community composition and could be very complex. Such responses are noted in the case of air pollutant stresses, which involve species-specific *negative* changes in growth rate, in the following section (also see Table 11.4). The responses to increased CO_2 might logically be expected to be species-specific *positive* increases in growth rate.

The effect of change in species performance with reduced growth rates

In the USA, the current pattern of generation of electricity has about 80% of total oil- or coal-fired power plants distributed in the eastern states.

This area either is currently in forests or was once forested. Increased urbanisation and industrialisation, and large fossil-fuelled electrical generating plants with tall stacks, have created regional-scale change in the air quality. There has been a widely publicised recognition that more strongly acidic rainfall (an apparent product of the long-distance transport and transformation of sulphur and nitrogen oxides from fossil fuel combustion) now falls over an increasingly larger area of the eastern USA (Cogbill and Likens 1974; Likens and Butler 1981). With these changes, the major focus in predicting the effects of pollutants on trees (and crops) has shifted from site-specific studies of the acute effects of gaseous pollutants near a pollution source to studies to understand the effects of chronic exposure to multiple pollutants from multiple sources in a regional airshed. For similar reasons, regional-scale effects of pollutants have come into prominence as a scientific problem in Europe and, recently, in developing nations.

Despite this interest, much of the information on the impact of air pollution on forests is from observations made near large smelting operations during the first half of the 20th century. In these historical cases, forests underwent decline and sometimes demonstrated increased mortality over thousands of hectares in the immediate vicinity of then unregulated smelting operations (see reviews by Scurfield 1960; Hepting 1968; Miller and McBride 1975; Smith W.H. 1981). One would like to use insights gained from these local studies (that demonstrate that pollutants such as sulphur oxides, heavy metals and hydrogen fluoride, often at some unknown level, can cause severe localised effects) to predict regional-scale effects, at presumably lower levels.

The problem in translating from acute effects to chronic effects of pollutants is formidable. It involves including the potential effects of anthropogenic stress on the growth of trees with the complexity of tree growth processes and their interactions with 'natural' environmental factors (Kozlowski 1981). This problem cannot be solved simply by the application of models, but models can be used to project the possible consequences (over long time or space scales) of inferences about the potential effects of pollutant events. A logical initial proposition to examine is, 'If the pattern of response of plants under acute pollutant stress is an index of the magnitude of their response to chronic stress, then how large a systematic reduction in the growth rate of trees is needed to cause a significant effect over a region?'

The effects of SO_2 on forest tree growth have been addressed in reviews by Miller and McBride (1975), Kozlowski (1981), and Smith

(1981). Several problems limit our ability to project the response of forests to low levels of SO_2 at a satisfactory scale of resolution, including:

1. The need for a better characterisation of atmospheric chemistry in non-urban areas and within forest stands in these areas: this lack of information characterising pollutants in rural areas stems from an allocation of air quality monitoring almost exclusively to determine exposures to urban populations
2. The lack of documentation of growth effects of chronic exposure of forest trees to SO_2 and other associated regional-scale pollutants under field conditions and over a range of pollutant concentrations
3. The need to characterise both the short- and long-term effects of SO_2 (and associated pollutants) on forest stand growth, competition and successional dynamics; it is to shed light on this part of the problem that gap models have been used.

The FORET model was used to perform model experiments on the potential effects of a chronic air pollution stress expressed as a systematic change in the growth rates of pollutant-sensitive trees (McLaughlin *et al.* 1978, West *et al.* 1980, Shugart and McLaughlin 1986; also see Kercher *et al.* 1980 for an application using a different gap model).

In these investigations, it was found that there can be profound effects on tree populations in response to relatively small (*c.* 5 to 10%) changes in annual growth rates. In some cases, tree species subjected to small decreases in their rate of growth can be totally eliminated from the simulated forest. In other cases, species that are subjected to equally small effects can actually increase as their competitors are diminished by growth reductions. The response of the ecosystem *in toto* under relatively small changes in tree growth rates can be greater than that inferred from simply averaging the reduction in growth across all the species that constitute the stand. This same result also holds if the average that is computed is a weighted average based on the composition of the stand.

The age (or stage of development) of a forest stand can have a radical effect on the nature of the pollutant response. Using the FORET model, West *et al.* (1980) found that when pollutant-sensitive trees (trees that showed extreme responses in the vicinity of point sources such as smelters) were given an annual growth-rate reduction of 20%, moderately sensitive species were given a reduction of 10% and tolerant species were given no reduction in growth rate, species showed differing responses depending on forest development at the time of addition of the stress. Trees in younger forests or in forests at a stage of development that had many suppressed trees tended to be more responsive to pollutant stress. Further, species with equivalent sensitivity (e.g. black oak, *Quercus velutina*, and yellow poplar, *Liriodendron tulipifera*: both given growth reductions of 10% in Fig. 11.8) can show different responses at different stages of forest development.

Using an extended version of the FORET model that included soil moisture effects, Shugart and McLaughlin (1986) found that the effect of stand age in altering a given species pollutant response also had a strong interaction with stand site conditions. For example, when black oak was subjected to a 5% growth reduction in a simulation of a mixed forest stand (in which sensitive species had a growth reduction of 10%, intermediate tolerance species had a reduction of 5% and insensitive species were given no reduction in growth), the resultant response varied from as much as a 22% reduction in biomass (biomass averaging 78% of control over 50 years) for young stands on dry sites to a 20% increase in some of the older stands.

Overall, there are two implications of results from simulation models for the evaluation of pollutant effects on forest systems. First, if these effects simulated by the models are actually found in natural ecosystems, it may be very difficult to project the consequences of pollutant stress from the laboratory or greenhouse to field conditions. This potential difficulty can compound the already difficult problem of determining the whole plant response to pollutants of organisms as large as trees. There is an obvious need for an understanding of forest ecosystems as an adjunct to more physiological studies of pollutant stress.

Second, there is a need to exercise care in the design of field studies on the ecosystem effects of pollutants. Shugart and McLaughlin (1986) in discussing some of these results noted the interactions among site conditions, age and species composition as a:

> complex web of interactions among site factors, age and species that could be diabolically difficult to unravel in studies that did not use considerable effort to control these interactions. In particular, indiscriminate comparisons among forests without consideration of stand age, stand structure, and site factors could well produce a tangle of seemingly conflicting results.

These difficulties indicate a need for well-designed field observations and experiments. Models will probably continue to be useful tools to evaluate potential effects and interactions involved in the evaluation of the longer-term and larger-scale projection of consequences of air pollutants on terrestrial ecosystems for some time to come.

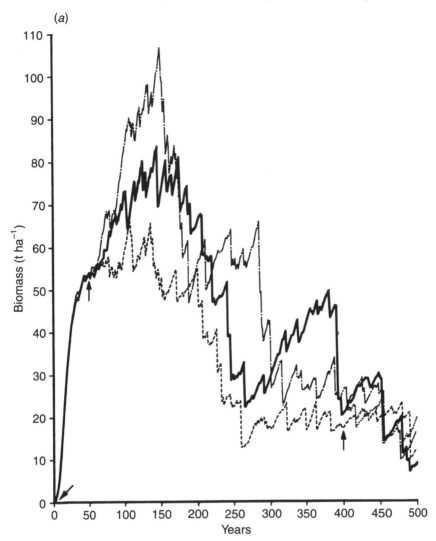

Figure 11.8. Biomass changes for (*a*) yellow poplar (*Liriodendron tulipifera*) and (*b*) black oak (*Quercus velutina*) growing in a mixed forest in which species deemed pollution-sensitive were reduced in growth by 20% and species of intermediate tolerance by 10%; tolerant species were unchanged in their rate of growth. These growth reductions were introduced at years 0 (– · – ·), 50 (– – –) and 400 (·····) during a 500-year simulation of 100 plots starting with an open plot as the initial condition. Control (—) is the same design (500 years; 100 initially open plots) with no changes in growth rates. Both yellow poplar and black oak were reduced 10% in growth (intermediate tolerance). From Shugart (1984).

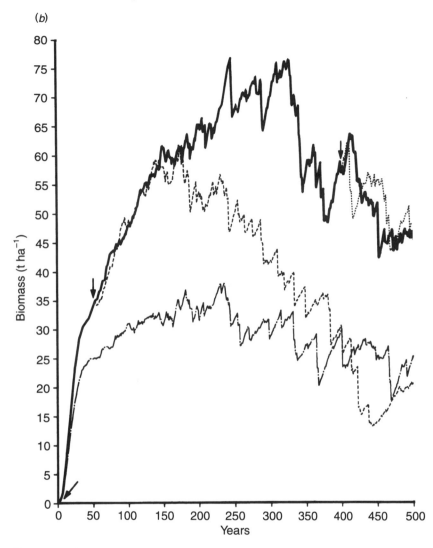

Figure 11.8 (cont.)

Simulations of direct CO_2 effects using models with increased physiological detail

Friend *et al.* (1993) initially assessed the HYBRID model by simulating the responses of forests in two cases: a lodgepole pine (*Pinus contorta*) forest under climate conditions associated with Missoula, Montana, and a white oak (*Quercus alba*) forest with a Knoxville, Tennessee climate. In this

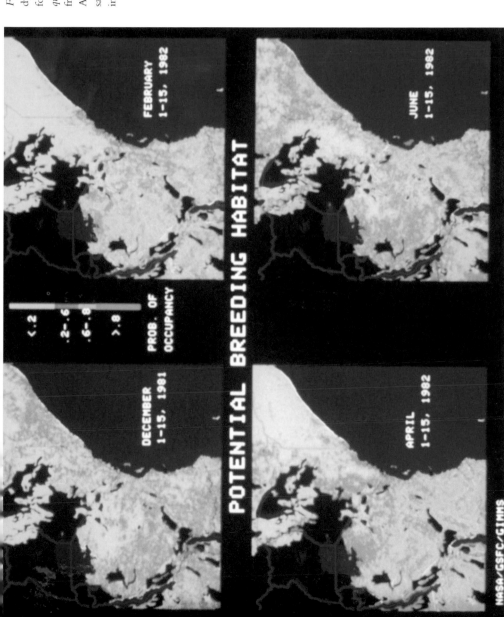

Figure 5.13. Temporal and spatial dynamics of potential breeding sites for the red-billed quelea (*Quelea quelea*) over East Africa as computed from information obtained from the AVHRR sensor on board the NOAA satellites. Season is noted on each image. Maps from Wallin (1990).

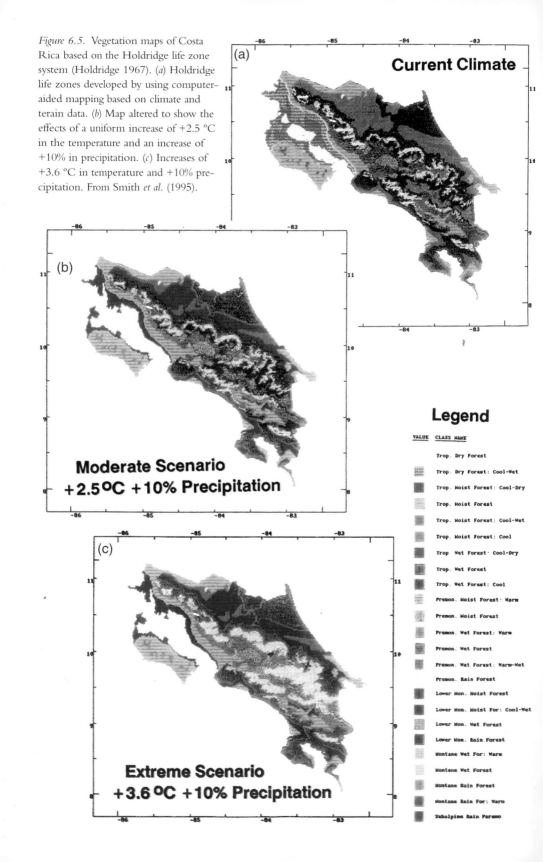

Figure 6.5. Vegetation maps of Costa Rica based on the Holdridge life zone system (Holdridge 1967). (*a*) Holdridge life zones developed by using computer-aided mapping based on climate and terain data. (*b*) Map altered to show the effects of a uniform increase of +2.5 °C in the temperature and an increase of +10% in precipitation. (*c*) Increases of +3.6 °C in temperature and +10% precipitation. From Smith *et al.* (1995).

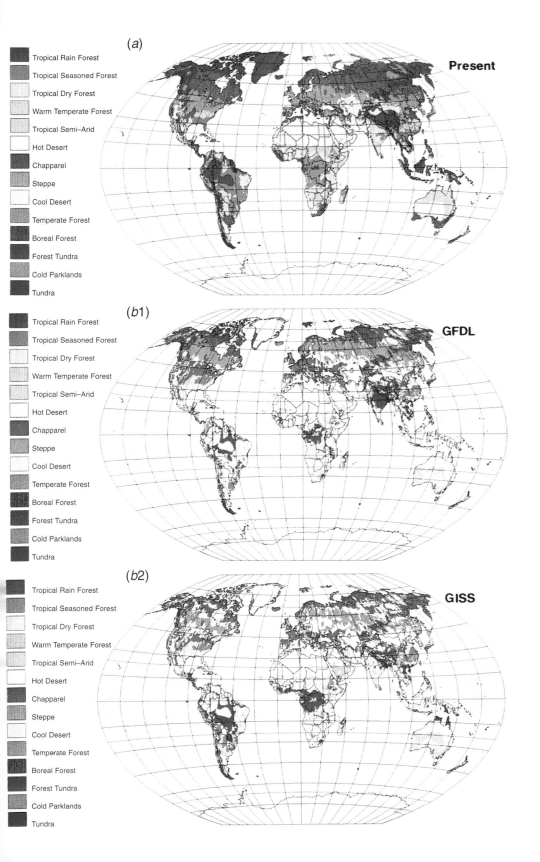

(a)

Tropical Rain Forest
Tropical Seasoned Forest
Tropical Dry Forest
Warm Temperate Forest
Tropical Semi–Arid
Hot Desert
Chapparel
Steppe
Cool Desert
Temperate Forest
Boreal Forest
Forest Tundra
Cold Parklands
Tundra

Present

(b1)

Tropical Rain Forest
Tropical Seasoned Forest
Tropical Dry Forest
Warm Temperate Forest
Tropical Semi–Arid
Hot Desert
Chapparel
Steppe
Cool Desert
Temperate Forest
Boreal Forest
Forest Tundra
Cold Parklands
Tundra

GFDL

(b2)

Tropical Rain Forest
Tropical Seasoned Forest
Tropical Dry Forest
Warm Temperate Forest
Tropical Semi–Arid
Hot Desert
Chapparel
Steppe
Cool Desert
Temperate Forest
Boreal Forest
Forest Tundra
Cold Parklands
Tundra

GISS

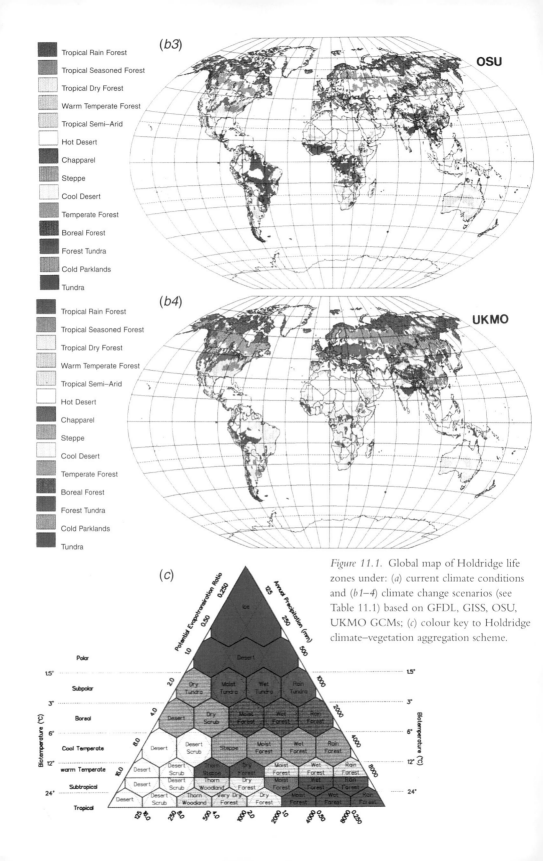

Figure 11.1. Global map of Holdridge life zones under: (*a*) current climate conditions and (*b1–4*) climate change scenarios (see Table 11.1) based on GFDL, GISS, OSU, UKMO GCMs; (*c*) colour key to Holdridge climate–vegetation aggregation scheme.

(b3) OSU

(b4) UKMO

Legend (b3):
Tropical Rain Forest
Tropical Seasoned Forest
Tropical Dry Forest
Warm Temperate Forest
Tropical Semi–Arid
Hot Desert
Chapparel
Steppe
Cool Desert
Temperate Forest
Boreal Forest
Forest Tundra
Cold Parklands
Tundra

Legend (b4):
Tropical Rain Forest
Tropical Seasoned Forest
Tropical Dry Forest
Warm Temperate Forest
Tropical Semi–Arid
Hot Desert
Chapparel
Steppe
Cool Desert
Temperate Forest
Boreal Forest
Forest Tundra
Cold Parklands
Tundra

(c)

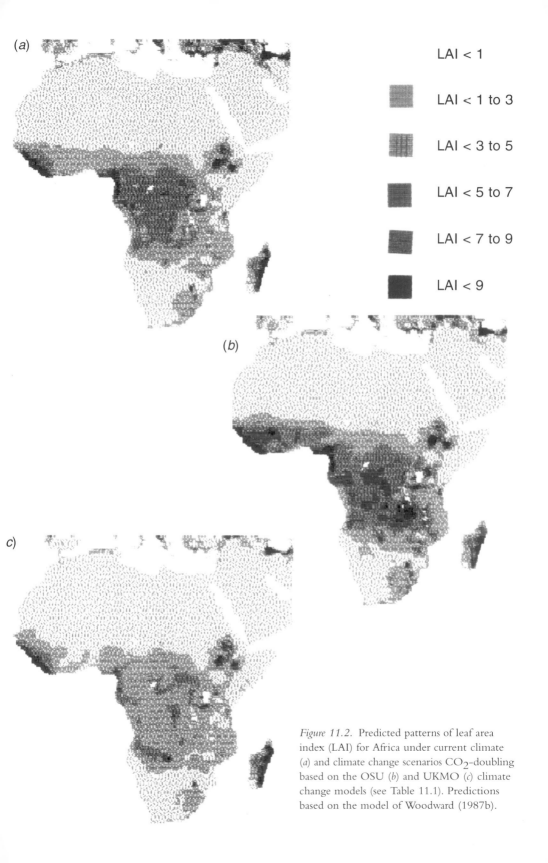

LAI < 1

LAI < 1 to 3

LAI < 3 to 5

LAI < 5 to 7

LAI < 7 to 9

LAI < 9

Figure 11.2. Predicted patterns of leaf area index (LAI) for Africa under current climate (*a*) and climate change scenarios CO_2-doubling based on the OSU (*b*) and UKMO (*c*) climate change models (see Table 11.1). Predictions based on the model of Woodward (1987b).

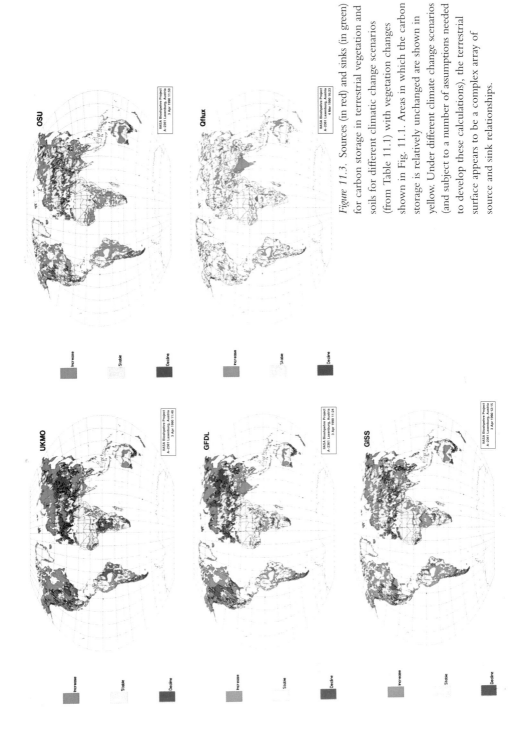

Figure 11.3. Sources (in red) and sinks (in green) for carbon storage in terrestrial vegetation and soils for different climatic change scenarios (from Table 11.1) with vegetation changes shown in Fig. 11.1. Areas in which the carbon storage is relatively unchanged are shown in yellow. Under different climate change scenarios (and subject to a number of assumptions needed to develop these calculations), the terrestrial surface appears to be a complex array of source and sink relationships.

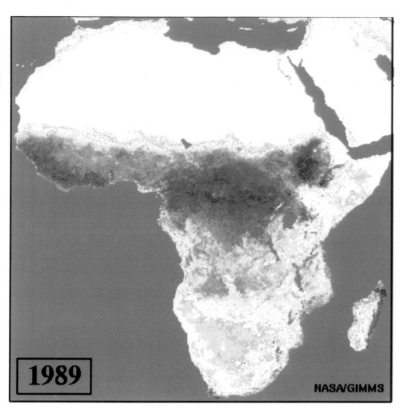

1989

NASA/GIMMS

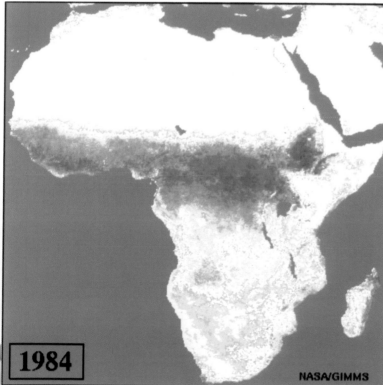

1984

NASA/GIMMS

Figure 12.5. Shifts in the vegetation greenness on the Sahelian zone in 1984 and 1989 as detected by the NDVI of 'greenness' using the visible red and the near-infrared channels of the NOAA-7 satellite. Images from C. J. Tucker and J. Kendall.

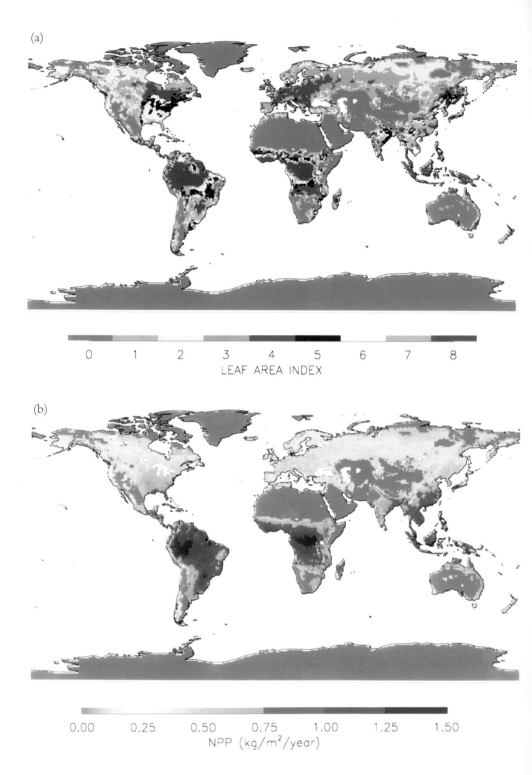

(a)

0 1 2 3 4 5 6 7 8
LEAF AREA INDEX

(b)

0.00   0.25   0.50   0.75   1.00   1.25   1.50
NPP (kg/m²/year)

Figure 13.6. (*a*) Global distribution of leaf area index as predicted by the DOLY model of Woodward *et al.* (1995). (*b*) Global pattern of net annual productivity as produced by the DOLY model.

application, the responses of the two simulated forests were compared with observations from nearby sites (leaf areas, gross primary production, water yield, etc.). Subsequently, A. K. Stevens, A. D. Friend and D. C. Mobbs (unpublished data) applied the HYBRID model to the prediction of the response of an evergreen needle-leafed plantation (*Picea rubens*) for the British Isles and the Iberian Peninsula to inspect the simultaneous and independent effects of increased atmospheric CO_2 and of climate change.

In this study, Stevens *et al.* initialised the HYBRID model with small *Picea rubens* seedlings and then simulated the forest plantation development for stands in each $0.5° \times 0.5°$ latitude/longitude block on the British Isles and Iberian Peninsula for 100 years. The resultant forests were then subjected to 100 years of altered environmental conditions (a 1% change of CO_2 per year for 100 years, a 1% per year change of climate from current conditions to those predicted for the region by Manabe and Wetherald (1987) for a doubled-CO_2 for 100 years, and both conditions together). After the initial 100 years of simulation (under 'normal' conditions), about 25% of the sites on the Iberian Peninsula were eliminated through mortality. After a second 100 years, under climate- and CO_2-changed conditions, 92% of the British Isles showed an increase in net primary productivity with an average increase of 50%.

These effects were not simply the result of the additive effects of climate and CO_2 but were a consequence of internal interactions. Figure 11.9 shows the effect of the three test conditions at the end of the second 100 simulated years for a site in southern Scotland (55.5° N, 3.0° W). As one might expect, the response to elevated CO_2 is an increase in net primary production and in the ecosystem carbon pool. The effects of climate change are to decrease both the total pool of ecosystem carbon and the net primary production (owing to the temperature effects of photorespiration). Changes at the biochemical level in response to elevated CO_2 ameliorate the photorespiration response, and the combined effects of the climate and CO_2 are for greatly increased net primary production and increased carbon storage for *Picea rubens* stands in this location.

Simulating patterns of vegetation change under altered climates with gap models

Tests based on the ability of simulation models to reconstruct palaeoecological patterns under climates of the geological past have been proposed as a necessity in testing models for their utility in global-change

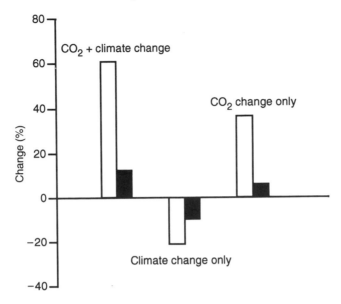

Figure 11.9. Percentage changes in net primary productivity (□)and ecosystem carbon storage (■)for a site in southern Scotland as simulated by the HYBRID model (Friend *et al.* 1993). The conditions are for the state of the forest after 100 years of climate change, CO_2 change and both changes together. From A. K. Stevens, A. D. Friend and D.C. Mobbs, unublished data.

assessments (Huntley 1990). The earliest applications of individual-based simulators to climatic change of vegetation were not in the context of future environmental change but were for testing models on their ability to reproduce the vegetation reconstructions based on palynological evidence. The first such application involved using the FORET model (Shugart and West 1977; Shugart 1984) to reconstructing a 16 000 year fossil pollen record from eastern Tennessee (Solomon *et al.* 1980, 1981). Subsequent applications have reconstructed forest communities in eastern Tennessee during the full glacial condition (Solomon and Shugart 1984). Currently, the application of simulation models to inspect the consistency of model results, inferred climates and reconstructed palaeoecological communities has become an accepted methodology (Solomon and Webb 1985) and some of the shortcomings in these applications methods have become better understood (Solomon 1986a).

The principal difficulty perceived for such applications involves the apparent circularity of the model tests. This is because the palaeoclimates, used to drive the simulation models, are often inferred from palynological evidence. Most palynological data are resolved only to the genus level. Since individual-based models predict species composition, one can further test the more specific model results against macrofossils (leaves, seeds, other tissue) and can to some degree obviate this problem. In some cases, the palaeoclimate can be inferred from other sources. For example, Bonan and Hayden's (1990) reconstruction of ice-age vegetation of Virginia was tested against palynological evidence but was based on climate reconstructions derived from marine sediments. Model applications can also provide a considerable degree of insights into theoretical responses of vegetation to climatic change and in such applications the model testing is less important than the insight provided (Davis and Botkin 1985).

Prediction of vegetation response under future climates has developed in the direction of prediction of continental responses of forests to changed conditions over large areas (Solomon 1986a; Bonan *et al.* 1990; Pastor and Post 1988). Simulated responses from the ZELIG model and the JABOWA model (Botkin *et al.* 1972) were an integral part of a recent USA assessment of potential climate change effects (Smith and Tirpak 1989). In general, this class of model has also been valuable in identifying potential complex interactions between vegetation and soils (Smith and Urban 1988) and vegetation and disturbance regimes (Overpeck *et al.* 1990) in the context of an altered climate.

Until Solomon (1986b), gap models were developed for particular sites incorporating species and processes that were important for fairly limited geographical regions. The step was taken with FORENA to include all the dominant tree species of eastern North America. The only variables left to be specified in a particular model run were climate and soil parameters. It was found that when the climate was specified for 21 locations distributed over eastern North America the model reproduced the spatial aspects of eastern deciduous, coniferous and boreal forests remarkably well. The model was physiognomically correct, simulating stunted woodland where scattered trees grow, simulating forests where forests grow and simulating growth of nothing where trees are absent on the modern landscape. Major forest types were also correctly simulated. The correct species dominants appear in the simulations for the climate at corresponding eastern North American locations.

Solomon (1986b) used FORENA to simulate the response of forests to CO_2-induced climate changes. The simulated effects of changing temperature and precipitation at a constant rate as CO_2 concentration doubles resulted in a distinctive die-back of extant trees at most locations. This is caused by climate changes, particularly temperature increases, resulting in conditions less favourable for established species and an insufficient time for individuals of other species to become established before forest decline occurs. Transient responses in species composition and carbon storage continued for as long as 300 years after simulated climate changes ceased. This is one of the earliest demonstrations that forest tree population dynamics can influence large spatial-scale carbon dynamics.

FORENA computes the effects of soil moisture on tree growth, but the simulations of Solomon (1986b) used a very deep mesic silt-loam at all sites that tended to minimise the effects of precipitation deficits. Even more drastic forest changes are predicted under CO_2-induced climate change if shallower, coarser textured soils are used in the simulations. Similar model exercises using ZELIG (Urban and Shugart 1989) with soils with less water-holding capacity predicted increased evaporative demand with increased temperatures. The resultant increase in droughtiness severely restricted tree growth in greenhouse warming climate change scenarios, even if precipitation increased. In Urban and Shugart's results, 18 of the tree species they considered would no longer grow in the south-eastern USA and much of the southern half of the south-east would not support trees. In analogous simulations for a different region, simulated biomass accumulations under CO_2-induced climate change in the Great Lakes region decreased to 23–54% of their present value (Botkin *et al.* 1989). On poor sites in both regions, forests in these simulations for the Great Lakes region could be converted to grassland or savanna with very low productivity and carbon storage in biomass.

Climate-change simulations with similar models also show changes in species composition, biomass, or both. In some cases, species composition shifts with little change in stand biomass, as is the case for simulations of forests in the Pacific Northwest (Dale and Franklin 1989). Simulations of Wisconsin and Quebec forests (Overpeck *et al.* 1990) showed that forest productivity changed with climate but resulted in small changes in species composition. Other model simulation results for the Great Lakes region (Pastor and Post 1988; Botkin *et al.* 1989) found significant changes in species composition without additional disturbances. As in simulations for the south-eastern USA, dry sites could lose the capability to grow most tree species and be converted to oak savannas or even prairies under a doubled-CO_2 climate.

Several climate model results indicate increased rates of forest disturbance as a result of weather that is more likely to cause forest fires, convective wind storms, coastal flooding and hurricanes. When increased disturbance frequency is added to the simulations, then significant changes in forest composition as well as changes in biomass are projected by the simulations.

Whether or not climate change results in species composition shifts, changes in biomass, or both, depends largely on whether or not the climate changes introduced into the model inputs are sufficient to change the climate to temperature and moisture conditions that result in dominant species either not being able to grow at all under the new conditions or losing out in the competition for light to other species that grow faster under the new climate conditions.

This section has focused on model simulations that project the effects of climate change on forest species composition and aboveground biomass and net primary production. In forest ecosystems, on average, there is nearly twice as much carbon (globally, 927 Pg, Post *et al.* 1985) stored in soil organic matter and litter as in aboveground biomass (globally, 515 Pg, Post *et al.* 1990). Most of the many essential nutrients for plant growth are supplied by the decomposition of this material. Therefore, it is critical that litter decomposition and soil organic matter dynamics be considered as well as biomass dynamics in determining the forest responses to global change.

It is possible to model decomposition and soil organic matter dynamics in a way that is consistent with aboveground population dynamics of individual-based models. Post and Pastor (1990) simulated the carbon and nitrogen changes of leaves, woody material and boles shed by individual trees and followed these litter input sources through the decomposition processes as annual cohorts. The dead parts of individuals can affect resource availability to living individuals in the same way that living leaves influence light availability in the canopy.

Decomposing litter material is the major source of organic matter and many nutrients for microbial populations in soils. The microbes convert organic forms of nutrients to inorganic forms that trees are able to take up, but the microbes also compete with the trees for these nutrients. The ratio of nutrients to carbon in leaves and other dead tree parts, especially in the form of the difficult to decompose material lignin, along with other factors such as temperature and moisture, determine the rate at which organic materials decompose and nutrients become available to trees.

The lignin:nitrogen ratio in fresh litter, a convenient index of the more general concept of litter quality (Ågren and Bosatta 1991), has been shown to be strongly correlated with the amount of nitrogen immobilisation, rate of organic matter loss and the length of time before net nitrogen mineralisation begins (Meentenmeyer 1978; Aber and Melillo 1982b) during the decomposition process. The concentration of nutrients and lignin in litter material varies among species and to a lesser extent among individuals of the same species. As a result, species composition has a strong impact on decomposition dynamics and nutrient cycling.

Individual-based models that keep track of dead material from different individuals as the material is decomposed have been developed (Aber and Melillo 1982b; Weinstein *et al.* 1982; Pastor and Post 1986). In LINK-AGES (Pastor and Post 1986), the nitrogen cycle forms a positive feedback loop. Nitrogen cycling partially controls the rate of organic matter production and species composition (through competition for available nitrogen). In turn, the rate of available nitrogen production depends on the amount and quality of organic matter produced.

The simulations that Solomon (1986b) performed were repeated using the LINKAGES model (Pastor and Post 1988), which incorporates nitrogen dynamics. The effect of the nitrogen positive feedback was to amplify further the effects of climate change. The direct effect of temperature and precipitation on forest tree growth involves both increases and decreases in productivity, largely depending on whether soil moisture becomes more or less limiting. Changes owing to direct climate effects are then accentuated by changes in species composition, altering carbon–nitrogen interactions that increase the initial effects through the positive feedback of the nitrogen cycle. Indirect species-change effects (such as alterations in the nitrogen/carbon balance of the soil) were expected to be most significant near vegetation zone boundaries such as the current boreal–cool temperate border. In Pastor and Post's (1988) application, forest simulation responses to climate change were as sensitive to the indirect effects of climate and vegetation on soil properties as they were to direct effects of temperature on tree growth.

Consistency comparisons of gap models with other approaches to modelling environmental change

GCMs typically produce climates that are warmer in the higher northern latitudes in response to increased atmospheric CO_2. Smith *et al.* (1995) developed an analysis of the possible consequences of the four different climate change cases shown in Table 11.1 that compared different modelling approaches to assess the potential effects of a warming on the high-latitude boreal forests. They used the Holdridge life zone classification to map three conditions of long-term response to climatic change:

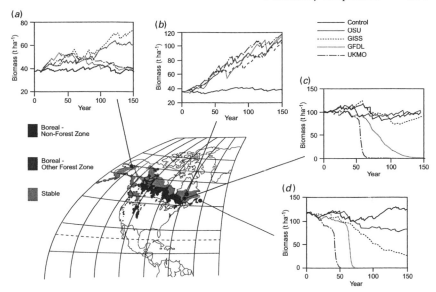

Figure 11.10. Comparison of Holdridge map for a changed climate with results from the gap model of Bonan (1990a,b) on sites that are dominated by black spruce. Gap model response for a control and four climate change scenarios for mature forests growing on (*a*) north-facing slopes in Fairbanks, Alaska; (*b*) site in northern Quebec; (*c*) site in central Quebec; and (*d*) moist site in Newfoundland.

1. Stable areas: areas of potential boreal forest under the present climatic condition that remain so after a given climate change scenario is used to alter the climate
2. Areas that change to boreal forest with the climate change
3. Areas that change from boreal forest.

A example of a resultant map from this procedure is shown in Fig. 11.10.

A forest gap model (Bonan 1990a,b; see Chapter 10) was then applied to a series of sites in the North American boreal zone (Fig. 11.10) to explore the implied transitions based on the Holdridge analysis. The model includes biophysical submodels for solar radiation, permafrost, decomposition and soil hydrology/evapotranspiration (Bonan 1989a). Several sites were selected based on their representation of different transitions based on the Holdridge analyses, and the availability of data for model verification under the current climate. Site descriptions, parameterisation and validation of the model for the locations shown in Fig. 11.10 are discussed in Bonan (1990a,b).

For each site, stand structure and composition were simulated under current climate conditions, defined by monthly statistics (mean and standard deviation) for precipitation, temperature and cloudiness. The resulting stand description after 500 years of simulation was used to initialise simulations for a control case (continued current climate statistics) and changed climate conditions based on the four GCMs in Table 11.1. Climate changes were assumed to occur over a 70-year period (years 0–70 of simulation). Annual values for each month were then determined by linearly interpolating between current and doubled-CO_2 values over this period. Climate statistics associated with the doubled-CO_2 climate were retained for the duration of the simulation. Tree species typical of the boreal forests were used in the simulation and no provision was made to allow species from other locations to 'migrate' to the simulated locations. Simulations were made for north- and south-facing slopes in central Alaska. In the latter under current climate, one would expect the site to be dominated by white spruce (*Picea glauca*), birch (*Betula papyrifera*) and aspen (*Populus tremuloides*) in a successional fire-mosaic. Under the warmer, drier conditions predicted under all the climate change scenarios in Table 11.1, the stand declines as a function of moisture stress. There are no species in the model simulation whose climatic requirements permit them to regenerate and grow on the simulated site. The differences in the timing of the decline among the scenarios reflect the different levels of warming predicted. This result is in agreement with the transition from boreal forest to cool temperate steppe (grassland) predicted for the area using the Holdridge life zone system.

In contrast to the predicted decline in forest cover on the south-facing slope, simulation of a stand in the same location on a north-facing slope (Fig. 11.10*a*) shows an increase in biomass of the dominant species, black spruce (*Picea marina*). This response appears to result from the relatively cooler conditions (and reduced heat loading) associated with north-facing slopes in this area.

Simulation of a site in the boreal forest–tundra transition zone of northern Quebec, Canada is shown in Fig. 11.10*b*. Under current conditions, one would expect the site to be an open-canopy woodland of black spruce with relatively low productivity and standing biomass. With the warming for this region predicted under all scenarios, the stand increases in biomass and is simulated to become a closed-canopy forest. This site corresponds to a northward expansion of the boreal forest in the tundra–forest boundary in the Holdridge analyses (Fig. 11.10).

In contrast to the northern Quebec site, the predicted dynamics of a black spruce stand in central Quebec (Fig. 11.10*c*) vary among the scenarios (Table 11.1). The OSU and GISS scenarios cause little change from the pattern predicted under current climate and reflect the lower warming predicted for the region by these two GCMs. In contrast, both the GFDL and UKMO scenarios predict a complete die-back of the stand, with the rates of decline directly related to the degree of warming. The area is predicted to change to a cool temperate forest under the Holdridge life zone system, and the establishment of species characteristic of that forest type would be dependent on a range of processes operating at the landscape level (e.g. rates of dispersal and establishment) not included in the model.

The predicted decline (Fig. 11.10c) in the boreal forest at its southern boundary under increased temperatures arises from the assumption (used to develop the model parameters) that the southern limits of the distributions of the component species are physiologically determined (e.g. exceed temperature tolerance for germination, establishment or growth). In contrast, if competition between boreal and northern hardwood (i.e. cool temperate forest) species is a determining factor in the southern limits of boreal forest species, then the decline in Fig. 11.10a might not occur. The increased temperatures may result in increased productivity of these stands (Bonan 1990a) and the transition to cool temperate forest would involve the eventual invasion of northern hardwood species into established boreal forest. The rate of this transition would be related to the ability of the species to invade established stands and would probably be influenced by rates of disturbances such as fire (Overpeck et al. 1990).

Similar results to those observed for central Quebec are predicted for black spruce in Newfoundland (Fig. 11.10d). Stand biomass declines under all four scenarios with the rates of decline a function of the degree of warming. As with the central Quebec site, the area is predicted to change to a cool temperate forest and the decline in black spruce is a function of the temperatures rising above those currently associated with its southern limit. Simulation of a balsam fir (*Abies balsamea*) stand on sandy soils in the same area shows slightly different results. Only the scenario with the largest temperature increase (UKMO) results in a significant decline in stand biomass. The difference between the two forest types for this location is a function of the differences between the two species in their growth response to temperature. Similar results to those observed for balsam fir are predicted for Jack pine (*Pinus banksiana*) in New Brunswick. The two warmest scenarios for the region (UKMO and GFDL) show a decline in stand biomass, while the OSU and GISS scenarios differ little from current climate conditions.

In general, Smith *et al.* (1995) found a qualitative agreement between the regional patterns predicted by the bioclimatic models and the results of the gap-model simulations for the representative sites. The results from the gap-model simulations highlight the uncertainties associated with defining the factors limiting the southern distribution of the boreal forest zone as discussed earlier.

Concluding comments

Mosaic models represent a diverse set of tools and techniques for evaluating the response of landscapes to environmental change. Importantly, many of these models have been tested against independent data and are being applied in contexts that are appropriate to prior model testing. However, the difficulty of evaluating the response of vegetation to novel circumstances is considerable. We certainly do not have the answers as to how the vegetation of landscapes, regions or continents may change, but we have made some of the initial steps toward this goal. A great value of models lies in the identification of critical factors that need to be understood and to identify research priorities.

One of the things that needs to be determined is when these models can be applied and when it is inappropriate to do so. For example, a number of scientists have met with good success in using a Grinnellian niche approach to map species to distribution patterns for past and present areas. This success lends credibility to the application of the same procedures to predict future distributions. However, we know from experience that there are species that prosper if they are moved to locations which at present are outside of their current range. We also know of species for which local distributions in no way express their ecological potential. We still need to develop the basic ecological understanding and research experience to determine when we can extrapolate species distribution and when we cannot.

In this chapter, a variety of different cases of potential responses of mosaic landscapes to change were discussed, notably climate change-related responses. Importantly, all of these investigations, using models with quite different underlying assumptions, predict that the CO_2-doubling-related climatic changes seen by meteorologists using GCMs are ecologically significant. This general result is found in the application of mapping procedures using species, genera or biomes as map units. It also arises in the application of individual-based gap models. It is not necessarily clear if these effects are 'good or bad', and the pattern of change varies depending on location. But the displacement of species and biomes from substantial portions of their current ranges seems a consistent result.

What is less clear from these explorations is whether the primary effect of CO_2 as a component of plant photosynthesis has beneficial effects sufficient to compensate for the indirect effects of greenhouse gasses (gaseous composition of the atmosphere modifying climate and climate then acting on vegetation). Results of experimental studies on ecosystems with low stature plants have produced somewhat conflicting results as to the ecosystem effects of CO_2-enrichment, and comparable experiments on forest ecosystems are only now being developed. Most experiments to date (and being designed) are too small and are conducted over too short a duration to provide insight into the effects of CO_2-enrichment on mortality rates or successional responses stemming from differential plant responses. The importance of these phenomena in vegetation dynamics will probably create a need for a strong interaction between models, field and experimental data for the foreseeable future.

12 · *Spatially interactive landscape models*

If asked whether landscape ecosystems are interactive, spatially distributed systems, most ecologists would reply, 'Yes'. Nevertheless, whilst the perception of landscape ecosystems may include (and often may emphasise) spatial interaction, most field studies of natural landscapes consider landscapes to be non-interactive mosaics. For example, after sampling landscape heterogeneity by using a random, a stratified random or a gridded sampling system, the landscape feature of interest is typically quantified as the average of the sample observations – without considering position or proximity of points in the sample scheme. Collection and recording of detailed spatial information on landscapes requires considerable effort. It is, in part, because of a general lack of information on the spatial dynamics and patterns of interactions for spatially distributed systems that the development of large-scale interactive spatial models has been limited to relatively few studies.

With respect to both computation and data storage, it is difficult to simulate, in detail, the dynamic changes of large landscapes. This is certainly the case when the computer models used for such predictions are spatially interactive models. Addition of explicit spatial considerations produces models of considerable complexity and attendant difficulty in obtaining model solutions. Even when these complex models can be solved with a relative degree of computational efficiency, the attendant complications of obtaining the model parameters and the detailed descriptions of the starting conditions (initial conditions) for such models over large areas represents an, as yet, unreachable level of detail in field data and measurement. For these reasons, the evaluation of the effects of large-scale environmental changes using interactive landscape models often involves theoretical explorations rather than predictive results.

Important issues arising from inspection of spatially interactive models involve the effects of heterogeneity in the patterns of landscapes, fragmentation of landscapes and the changes in landscape boundaries, particularly the changes in environmentally controlled boundaries. What is revealed from studies using interactive landscape models is a range of the dynamic and spatial responses that may exist in nature. These features are eliminated by the assumptions needed to develop either mosaic models (discussed in Chapter 11) or homogeneous models of landscapes (discussed in the following Chapter 13).

Effects of landscape scale in interactive landscapes

We live in a world that seems filled with fragmented landscapes, some of a 'natural' origin, and others, the products of human land-use. In Chapter 9, statistical arguments were used to posit that the attributes (such as biomass change over time following a disturbance) of a mosaic landscape of a given size can differ significantly from the same attributes for the mosaic elements or 'fragments' that make up the landscape. This can be even more the case when the elements that make up the landscape are interactive with one another (Levin 1992). One obvious such effect is the 'edge effect': the often unique assemblages of plants and animals found in the abrupt transitions of vegetation dominated by one life form to vegetation dominated by another, such as the boundary between field and forest (Leopold 1933). Edges often have significantly different microclimates, particularly with respect to light levels, wind levels and moisture conditions. A large intact landscape of forest and an equivalent area of forested woodlots differ in the amount of edge per unit forest area.

Forest pattern in edges and woodlots

Ranney (1978) simulated the expected equilibrium composition of Wisconsin woodlots of different sizes using a modified version of the spatial FOREST model (Ek and Monserud 1974a,b). Model validations in this case involved inspections of the model's ability to predict the diameter distribution of trees by species in forest edges as well as comparisons of species composition in woodlots of a particular size with equivalent model simulations. These model simulations were focused on the effects of changes in the edge-to-area ratio in forests of different sizes and the resultant changes in the relative proportions of different species. Small woodlots with a larger portion of the area in edge have a higher representation of shade-intolerant plants that prosper in the well-illuminated edge and adjacent area.

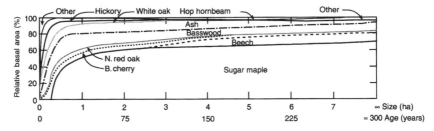

Figure 12.1. Projected equilibrium composition for square forest woodlots of differing sizes for Wisconsin. Area in the figure corresponds to the percentage of the woodlot basal area associated with each species. The species mixture changes with decreasing proportions of shade-intolerant and drought-tolerant species in increasingly larger woodlots. The stand ages are those of successional forests of a large size and are approximations based on descriptions of forest succession by regional experts. From Ranney (1978).

One intriguing result from this study was the detection of a direct correspondence between the size of the simulated woodlot and the temporal sequence (from discussions in Whitford and Salmun 1954; Ward 1956; McIntosh 1957; Curtis 1959) of ecological succession for forests in the region (Fig. 12.1). This similarity arises from the tendency for successional forests in the region to display a transition from initial dominance by shade-intolerant trees to a dominance by shade-tolerant trees in more mature stands. In small woodlots, the high light levels promote the continued success of shade-intolerant species (and the drier conditions may diminish the success of the shade-tolerant, typically more moisture requiring, mature phase species). Thus, there is a similarity between the equilibrium composition of small woodlots and the early, transient composition of successional forests. Since the FOREST model explicitly considers the seed production and seed dispersal of each tree, there also is a tendency for smaller woodlots simulated by the models to be dominated by the species that are established as mature plants in the area.

An initial ZELIG model test (Chapter 10) involved duplicating the qualitative patterns observed in a large-scale field experiment in tropical rain forests in Panama. Brokaw (1985a) artificially produced gaps of different sizes in intact rain forest and then measured the growth rate of shade-tolerant and shade-intolerant trees. With increasing gap size there is increased light (as the shading from the edge covers a progressively smaller proportion of the plot). Most trees would be expected to show growth enhancement under such improved light conditions, but this effect soon becomes asymptotic in shade-tolerant trees. The growth rate of shade-intolerant trees (which prosper in the higher light levels) continues to increase as the cleared area becomes larger. The ZELIG model was validated by reproducing these patterns of forest response (see Fig. 10.10) to experimental manipulation of landscape pattern.

While there have been relatively few model experiments using spatially explicit models of fragmented landscapes, the results from these simulations have been successfully matched to data from field experiments (e.g. Brokaw 1985a) and to patterns actually found on fragmented landscapes. The results to date imply differences in species composition (a shift to shade-intolerant species in fragmented landscapes) and function (divergent performance in forest gaps of different sizes for species with differing tolerances).

The vegetation dynamics of intact landscapes of differing sizes

The results discussed in the section above are a product of the effects of the increased presence of edges that is intrinsic to a fragmented forest landscape. There are also differences in landscapes that are a direct consequence of size, which are in addition to these edge effects. Smith and Urban (1988) used the ZELIG model to produce simulated dynamics for a 9 ha forest stand over 750 simulated years of forest dynamics. As a test, model output was disaggregated into 0.09 ha subsamples and compared with diameter frequency distributions from 0.08 ha plots collected in eastern Tennessee on the reservation of Oak Ridge National Laboratory (D. L. Urban, unpublished data). To reduce the complexity of the model output (species and size of each of the thousands of trees living on the 9 ha plot at any particular time) and provide greater ease in inspecting the simulated dynamics, a principal components analysis was developed for the vegetation structure (Fig. 12.2). Points in the principal components space (Fig. 12.2) represent stands with a greater number of smaller trees and a relatively small number of large trees on the right, and the inverse on the left. Larger-sized trees are relatively more abundant for points in the upper portion of the space and less so in the lower portion. The first principal component accounts for 31% of the variation in the data, the second for 28%.

The co-ordinate position of (0,0) at the centre of the space represents the average of all the samples. In this case, as one would expect, the average distribution for sizes of tree is an 'inverse J' distribution. There is, on average, a relatively large number of small trees and a small number of larger trees. The abundance of trees decreases rapidly and systematically as

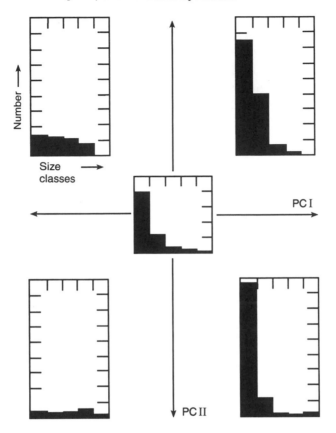

Figure 12.2. Mean diameter distributions of quadrats in four regions of the principal components space and for the pooled sample mean seen at the centre (0,0) point of the principal component axes. The histograms illustrate the qualitative differences in the abundance of different diameter size classes. From Smith and Urban 1988).

a function of their size. This distribution has been reported in a diverse array of forests and is historically associated with DeLiocourt (1898).

In 1898, DeLiocourt noted the same pattern illustrated in the centre of Fig. 12.3: the density of stems in a mixed-aged forest when plotted in equal diameter intervals results in a curve with the number of stems in each interval in constant, decreasing proportion with the next size interval (a negative exponential curve). This expected exponential decline of numbers in diameter categories (Meyer 1952) has been used as a basis for forest management (Spurr 1952; Knuchel 1953; Sammi 1969) in which forests are harvested to maintain a 'balanced' condition (Assman 1970;

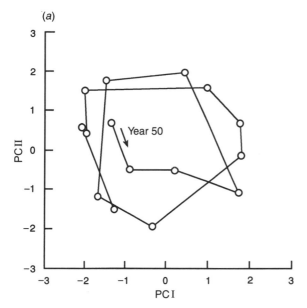

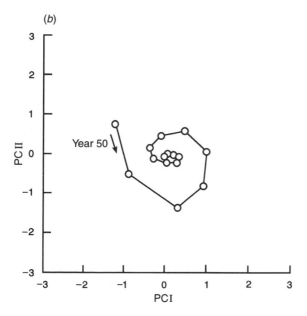

Figure 12.3. Changes in forest structure at differing spatial scales in a 9 ha forest stand simulated by the ZELIG model. (*a*) Changes in the forest structure through time for a single 0.01 ha quadrat. The initial point is for the condition of the forest quadrat in year 50 of the simulation and each subsequent point is at an interval of 50 years of elapsed time. The principal component space is as in Fig. 12.2. (*b*) Changes in forest structure for an average quadrat (mean of 900 pooled 0.01 ha quadrats). From Smith and Urban (1988).

Harper 1977) of numbers among size classes. In a sustainable harvesting schedule for forests, when the typically unimodal diameter distributions of even-aged forests are summed across ages, the resultant curve strongly resembles the curve expected for a mixed-aged forest in both shape (negative exponential) and magnitude (Assman 1970).

When the dynamics of a single small plot (0.01 ha) are plotted in the principal components space starting at year 50 and then at 50-year intervals for the 750 simulated years, the structural dynamics (Fig. 12.3*a*) of the forest consist of quasi-cyclical, non-equilibrium changes in forest structure driven by the gap-formation/gap-filling dynamics. This pattern results from the ecosystem dynamics outlined in Chapter 9 as a mosaic theory of landscape change. Importantly, the small-scale structural dynamics do not approach the average condition for the plot at co-ordinates (0,0) but appear to circle around this point. However, when the average of the 900 plots of 0.01 ha that make up the simulated forest are plotted in the same principal components space (Fig. 12.3*b*), the system dynamics consist of a regular spiral into the point representing the plot's average condition. The system behaviour is regular and equilibrium seeking.

Demography on fragmented landscapes

Populations of plants and animals typically are found in patches on landscapes of what is usually assumed to be suitable habitat. The population dynamics of the set of subpopulations of a population can often be considerably different from the expected dynamics of the population if it were homogeneous (Huston *et al.* 1988; Lomnicki 1988; DeAngelis and Gross 1992). As was discussed in Chapter 10, such populations can remain viable from recolonisation to locations of subpopulations where extinction has occurred (even when the solution to the population equations for a large homogeneous population implies extinction (Hilborn 1975)).

Wu and Levin (1994) developed a spatially interactive patch model for a well-studied annual grassland located on serpentine soils in California (Hobbs and Mooney 1985, 1991). Two species, *Lasthenia californica* (an abundant annual forb) and *Bromus mollis* (a non-native annual grass), were considered. Important spatial effects included in the model were seed dispersal, the germination and survivorship of the seedlings and the disturbance from the burrowing of the western pocket gopher (*Thomomys bottæ*). The local populations in the simulations were found to fluctuate greatly and to be subject to local extinction. The larger, landscape-level effect was for both the species to survive, but with *Bromus* populations

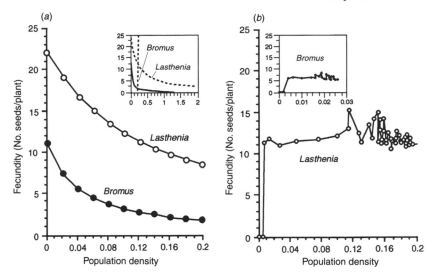

Figure 12.4. The relationship between local and landscape population fecundity from an interactive spatial model of two species (*Lasthenia californica* and *Bromus mollis*) in an annual grassland. (*a*) The relation between population density and fecundity at the local scale (the inset shows the response over a larger range of densities. (*b*) The equivalent landscape-level relationship between population density for *Lasthenia californica* (response of *Bromus mollis* is shown as an inset). From Wu and Levin (1994).

featuring an increase in abundance as a response to disturbance. One of the more important effects noted by Wu and Levin was that the relationship between the fecundity of each population as a function of population density was quite different when measured at the local level compared with that seen at the landscape level (Fig. 12.4). At the local level, simulations of the spatial responses of the species fecundity/density relate in a manner such that the fecundity systematically declines as population density increased (Fig. 12.4*a*). At the landscape level (Fig. 12.4*b*) the fecundity to density pattern for the two species had an entirely different shape: rising to a level and then fluctuating at high density. This tendency for an ecological process to differ in its fundamental nature as a function of spatial scale has been termed 'transmutation across scale' (O'Neill 1979). It is a frequently occurring feature in both modelled and observed spatial data sets.

Pulliam (1988) investigated a relatively simple model for births and deaths of populations dispersed on a fragmented landscape into sub-populations. He demonstrated that when the populations had density-dependent immigration a limited number of 'source' populations could supply immigrants that could provide 'sink' populations in areas of marginal habitat with colonists from the over-production of young. The implication of this work was that the 'sink' populations would become extinct without the re-supply from the source populations. This concept has been raised in discussions of population viability (Lidicker 1975; van Horne 1983). A body of theoretical results based on population models implies alterations in system dynamics for subdivided populations (Whittaker and Levin 1977; DeAngelis *et al.* 1986; DeAngelis 1992).

There are field observations that support these results. For example, in a 25-year study of the checkerspot butterfly (*Euphydryas editha bayensis*), the butterfly population survived as three relatively separate subpopulations. One of these subpopulations became extinct, was recolonised and again suffered local extinction in the course of the study (Wilcox and Murphy 1985). King and Mewaldt (1987) identified populations of white-crowned sparrows (*Zonotrichia albicolis*) in relatively isolated montane populations that could not exist without periodic immigration from other 'source' locations. In general, the effects of fragmentation appear to lengthen the time that a population may persist (Jenkins 1992).

Disjunct, patchy populations differ in significant ways in their dynamics from an equivalent homogeneous landscape. Metapopulations (Gilpin and Hanski 1991) have attracted considerable interest in theoretical ecology, particularly with respect to population extinction as a consequence of landscape fragmentation (Levins 1970). In many cases using relatively realistic models (such as those described above), the dynamics of patchy populations demonstrate a greater inertia in their responses. Extinction is delayed, invasions are slower and population outbreaks are more constrained. Results from theoretical studies on models and from field data appear to support one another, at least qualitatively.

Ecotone dynamics under environmental change

One situation in which one might expect environmental change to be most observable is at locations where there are sharp transitions in the vegetation from one life form to another. These ecotones have regularly been identified as locations for monitoring for response to change, and it is known that certain ecotones can change rapidly in response to environmental fluctuations (implying a sensitivity to changes of various sorts). For example, Justice *et al.* (1985, 1991) report striking shifts in the features of the boundary between the Sahara Desert and the Sahel (the seasonal rain, semiarid zone south of the Sahara (see Fig. 12.5, colour plate, for example).

Dynamic response of ecotones to external change

Noble (1993) used the vital attributes approach (see Chapter 10, particularly Tables 10.1 and 10.2) in a spatial model to simulate the changed tree-lines in the Australian Alps. He focused his investigation on the issue of whether or not ecotones represented useful (or even unique) locations to monitor the effects of global-scale changes, as has been suggested by several ecologists (Solomon 1986b; Di Castri *et al.* 1988; Dyer *et al.* 1988; Hansen *et al.* 1992). The model was developed on a 500×60 element grid. The condition simulated was for an ecotone 500 grid elements wide on an altitudinal gradient 60 elements long. At lower altitudes, the simulated trees have a tendency to invade the grassland and this tendency is decreased until a physiological limit is reached. The rate of invasion of unoccupied grid elements is at a maximum value, r, below altitude L. The rate declines linearly until it reaches a value of 0 at altitude U.

In the model (Noble 1993), the upward invasion of the trees into grassland was pushed back by the effects of wildfire. Fires were simulated to originate in the grassland at irregular intervals and to burn into the forest. The probability of a fire penetrating one cell downslope in forest is p, and the average penetration of a fire is $p/(1-p)$. In some runs of the model, the width of the fire (the number of cells along the treeline burnt) varied. A fire-frequency parameter, f, is the expected number of cells in each row burnt each year or the average number of cells that the treeline is set back by fires each years.

Under these conditions, the tendency for the treeline to 'push up' the elevational gradient could be expected to be balanced by the tendency for the fires to 'push back' the tree line and one might expect a balance between these two forces to come to an equilibrium. In the Noble (1993) model, this is the case, and the position of the equilibrium treeline is:

$$x = U - \frac{f(U-L)p}{r(1-p)} \qquad (12.1)$$

where x is the equilibrium position of the treeline.

Noble (1993) modified the basic model just described to include also other effects, notably the effects of the shape of the ecotone and the effects of dispersal beyond the treeline. In considering the shape effects on ecotone advance, he included a tendency for parts of an ecotone that extend from forest into the grassland ('convexities' *sensu* Hardt and Forman 1989) to be more/less able to migrate upslope than parts that were sheltered ('concavities' *sensu* Hardt and Forman 1989) as one simulation case. He included the possibility that trees could establish themselves beyond the tree line (with a probability of d per year) to allow the trees to advance from nuclei of small groves on sites beyond the advancing treeline.

Noble (1993) conducted a set of model experiments with different frequencies and sizes of fires, with a range of parameters for rates of spread, with alternative models with dispersal and ecotone shape effects included. Because he could solve for the expected position of the treeline using Eq. 12.1, he was able to measure both the delay in response to change and the degree to which the treeline was displaced from its eventual equilibrium position. From these model experiments (using a range of fire frequencies and appropriate rates of treeline movement for *Eucalyptus pauciflora*, the tree species that is at the treeline for much of the Australian alpine situations), he concluded that the delay in the response of treelines to a more favourable climate was so great as to limit their use in monitoring global change. The variability in the position of the dynamically determined treeline implied that the monitoring of such situations would require a large number of sample sites, or sites of unusual length on relatively homogeneous substrate.

Ecotones that might be useful for monitoring would be those with a large value of r from Eq. 12.1 (cases in which the trees were capable of rapid advances). Large values of f from Eq. 12.1 also were associated with treelines that would track the changes in environmental conditions quickly and, thus, be useful for monitoring. One could expect great variation in the value of the parameter r. Treelines have a considerable variation in the rate at which they appear to move under favourable conditions for forest expansion (Table 12.1). It is noteworthy that these rates of change vary over several orders of magnitude. In many cases, the rates of movement in Table 12.1 are the averages of highly variable patterns of treeline advance and are measured over centennial or millennial time intervals to obtain the averages listed. Noble pointed to treelines as locations at which to gain insight into the dynamics of individuals and ecosystems under extreme conditions as their most promising role in understanding global change.

Effects of alterations in cloudiness on montane landscapes with boundaries

Malanson and Cairns (1995) used a modified version of the JABOWA II model (Botkin 1993) to simulate the response of montane forests in the

Table 12.1. *Some rates of treeline movement under the influence of increasingly favourable climate conditions (also see Noble 1993). Note that these are averages in many cases of what can be highly variable processes measured over long intervals (hundreds or thousands of years). The slower rates near the top of the table are associated with alpine situations; the rapid rates near the bottom of the table are from analysis of fossil pollen occurrence following the past glaciation*

Location	Tree species	Rate (m per year)	Reference
Snowy Mountains, Australia	Australian eucalyptus (*E. pauciflora*)	0.03	Slatyer (1990)
Scandes, Sweden	*Picea abies*	0.04	Kullman (1986)
Rocky Mountains, USA	*Picea engelmannii, Abies lasiocarpa*	0.04	Marr (1977)
Cascade Range, USA	Western conifers	0.17	Franklin *et al.* (1971)
New England, USA	*Abies balsamea*	0.8–2.7	Sprugel (1976)
Hudson Bay, Canada	*Picea glabra*	5.0	Scott *et al.* (1987)
Eastern North America	Tree flora responding to deglaciation	100–400	Davis (1981a)
Europe	Tree flora responding to deglaciation	100–400	Huntley and Birks (1983)
Scotland	*Pinus sylvestris* responding to postglacial climate	350–800	Gear and Huntley (1991)

Rocky Mountains to changes in cloud cover. This model included a modification to simulate the dispersal of tree seedlings on a 20×40 grid of 30 m × 30 m cells (Hanson *et al.* 1990; Malanson *et al.* 1994). The gridded matrix was designed to mimic the conditions for a mountain slope: sub-alpine species occupied the top half of the plot and valley species conditions were found on the bottom half. The centre of the plot had an 'avalanche path' in which coniferous trees could not reproduce (acting as a partial barrier to the dispersal of conifers) and a consistently present, high-light environment. The plot was also slightly drier at the top. The top of the plot was set at 500 °C days and the bottom at 1280 °C days with regular change in degree days from the top to the bottom of the plot. The intention for arranging the model in this fashion was to reproduce the pattern of a typical spatial unit that is associated with avalanche disturbance, a regular feature of the high-elevation forests in the northern Rocky Mountains (Walsh *et al.* 1989; Cairns 1994).

In a set of simulation experiments investigating both the effects of a change in cloudiness (manifested as a reduction of 10% in ambient light) and the effects of the avalanche path, Malanson and Cairns (1995) found a decrease in the total stand basal area, the number of species and the proportion of deciduous species constituting the basal area associated with the change in cloud cover. While diversity of species was found to decrease with cloud cover, the avalanche path was actually more diverse under cloudy conditions. Further, the effect of including or omitting the effects associated with avalanches was as important as the changes in the light associated with increased cloudiness on the importance of deciduous species.

Modelling interactive landscape dynamics

Along with considerations of fragmentation and boundaries altering the dynamic responses of landscapes to change, there is an additional consideration of the effects of heterogeneity on landscape processes. Water movement, spread of fire, insect and pest outbreaks and animal movement all alter the pattern of landscapes and can produce changes in both landscape pattern and landscape processes under varying environmental conditions. The array of models that have been developed to deal with these complex interactive systems is sufficiently large and diverse that examples of the range of models in this category with a small amount of detail in the underlying formulations will be presented here.

A spatial model of tiger bush

In Chapter 7, a phenomenon called tiger bush or *brousse tigrée* was described as one of the cases in which the feedbacks between the environment and the plant processes produce regular moving patterns in the vegetation. Watt (1947) discoursed on other similar patterns in a range of different vegetations in formulating a concept of the manner in which plant processes could generate patterns.

Thiéry *et al.* (1995) developed a finite-state automata model* for *brousse tigrée* in Niger and used the resultant model to inspect the land-

---

* See Chapter 4 for an introduction to finite-state automata and Chapter 10 for other applications of finite-state automata in modelling vegetation dynamics.

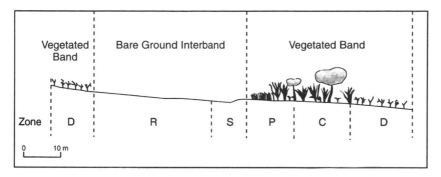

Figure 12.6. Schematic cross-section of a *brousse tigrée* showing the different zones used by Thiéry *et al.* (1995) to devise a finite-state cellular automata for this system. The zones are: D (the degraded zone); R (the run-off zone); S (the sedimentation zone); P (the pioneer front zone); C (the central zone). From Thiéry *et al.* (1995).

scape dynamics of this vegetation system. The model was developed by initially dividing the vegetation zones on a transect through a band of the vegetation into distinct zones (illustrated in Fig. 12.6):

1. The degraded zone (D) at the 'back' of the tiger stripe: the zone features a pronounced reduction in shrub cover (from 40% to 0%) and an increase in dead shrubs (to 100%)
2. The run-off zone (R) with no plant cover: smooth hard surface produces intense run-off
3. The sedimentation zone (S) also with no plant cover, but with the surface characterised by sedimentary crusts that often crack and form curled plates
4. The pioneer front zone (P), the front of the next tiger stripe: the zone may have local concavities in the boundary with the sedimentation zone that have annual grasses as the predominant cover and convexities that are covered mostly with shrubs
5. The central zone (C) in the middle of the tiger stripe: canopy cover is between 60% and 100% with shrubs toward the front of the zone and trees occurring only in the lower part.

For a more complete description of these zones and of the important role of termites in altering the surface structure of the different zones, see Thiéry *et al.* (1995). The similarity of the structure of these features and that of tiger bush in Mexico can be seen in a comparison of the two schematics (Figs. 7.4 (p. 184) and 12.6).

(a)   Upslope

−1	−2	−1
0	−1	0
1	★	1
1	1	1

Downslope

(b)   Upslope

0	−a	0
0	−a	0
0	−a	0
0	−a	0
0	−a	0
0	−a	0
b	★	b
0	3	0
0	1	0

Downslope

Figure 12.7. Convolution matrices to simulate the effect of neighbours on a focal cell (marked ★). (*a*) A simple convolution matrix. (*b*) A more complex matrix used by Thiéry *et al.* (1995) to simulate the formation of banded vegetation.

Thiéry *et al.* (1995) developed a simulation model based on relatively simple rules to change the vegetation. The model follows a cellular automata introduced by Conway (1976) called the 'Game of Life'. Dividing the landscape into cells, they assumed that the positive and negative effects of the neighbours of a given cell were expressed in the convolution matrix $C_{m,n}$. A simple example of a convolution matrix is shown in Fig. 12.7a. In this example, the asterisk indicates the focal cell. Cells directly in front of the focal cell have a negative effect of either −1 or −2 on the focal cell, reflecting the negative effects of plants in these cells having first access to water. Cells adjacent to or downslope from the focal cell have a positive effect on the plant in the focal cell by such factors as shading (reducing evapotranspiration), increased litter accumulation, reduced crusting of soil, higher water infiltration, etc.). Thiéry *et al.* actually used more complex matrices than this example and eventually used a convolution matrix of the form shown in Fig. 12.7*b*. In this formulation there are two parameters *a* and *b*, which represent upslope negative effects and lateral synergies, respectively.

There are four states possible for each simulated cell: *state 0* for a site with no tree (or a dead tree); *state 1* for a site with a seedling tree (or a senescent tree); *state 2* for a site with a small tree (or a stressed adult tree); and *state 3* for a site with a well-watered adult tree. The formula for change from one state to another over each iteration of the model was:

$$S_{i,j,k+1} = S_{i,j,k} + \mathrm{Max}[B, \mathrm{Min}(H, c\Sigma_{m,n} C_{m,n} S_{i+m,j+n,k})] \qquad (12.2)$$

where $S_{i,j,k}$ is the state (0,1,2,3) of the cell at location i,j at the kth iteration of the simulation; B is the lower limit on state changes; H is the upper limit on state changes; $\Sigma_{m,n} C_{m,n} S_{i+m,j+n,k}$ computes the value of the positive and negative effects of neighbours based on the convolution matrix given in Fig. 12.7*b*; and *c* is a calibration factor that essentially affects the rate of transition from one state to another and that can be used to match the model output to empirical data.

The model is a simple set of rules for how to change the state of the tree in a given cell depending on its present state and the state of the surrounding trees. It has three parameters, *a*, *b* and *c*, that can be varied in a model experimental or in a data-fitting mode (depending on the application).

A striking feature of this model is that it produces pattern when initialised with a random pattern (Fig. 12.8). There is also a striking similarity in the patterns generated by the model and those seen in *brousse tigrée* in the Sahel. The dynamics implied by the model have not been measured in the vegetation directly to any great degree. Indeed, Thiéry *et al.* (1995) note that they know of only one instance (noted by Leprune (1992) in Mali) where a concrete bench-mark installed in 1955 in a *brousse tigrée* stripe was found 21 years later in a barren area about 15 m from a stripe. This implies a rate of movement of about 0.75 m per year. Worrall (1959) reports rates of movement of 0.3 to 1.5 m per year for a similar grassland pattern in Sudan.

A spatial model of coupled hydrological and ecological responses to elevated CO_2

The use of the HYBRID model (Friend *et al.* 1993) to provide an initial characterisation of the direct effects of elevated CO_2 interacting with

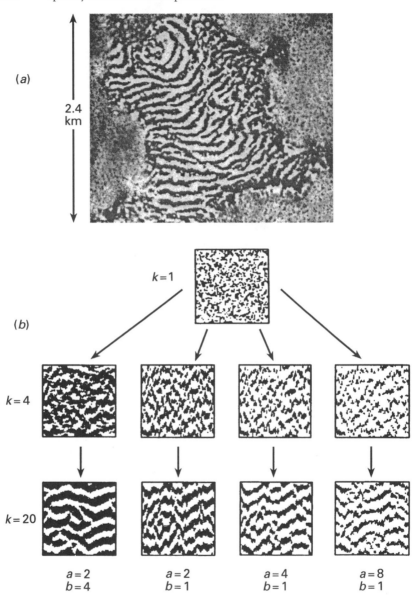

Figure 12.8. (*a*) An aerial view of a well–developed tiger bush or *brousse tigrée* pattern on a plateau in Niger (13°30′ N, 2°40′ E). The photograph is approximately 2.4 × 2.4km². (*b*) A set of simulation results from a cellular automata model by Thiéry *et al.* (1995). Parameter values for the model are at the bottoms of the columns and three different steps in the iteration of the model are shown. At iteration number $k = 1$, all the simulated landscapes are initialised with a random pattern of trees. At iterations 4 and 20, a pattern of *brousse tigrée*-like vegetation forms that is directly comparable to that in the photograph shown in (*a*).

plant canopy heat, water and CO_2 fluxes was discussed in Chapter 11. One of the possible responses to an elevated level of CO_2 in the atmosphere (at least under some circumstances) is an increased yield of water from the vegetative landscape through a reduction in transpiration demand by the plants (Aston 1984a). One potentially important spatial effect of the increased level of CO_2 in the atmosphere might be the increased amounts of water running off from elevated positions and onto lower slope positions in watersheds.

Hatton *et al.* (1992) investigated the landscape interactions of a spatially explicit watershed model on leaf area and water yield. The modelled watershed was covered with a single layer of mature spotted gum (*Eucalyptus maculata*) trees. Conditions for the model were based on those measured in an extensive study of a spotted gum forest near Kioloa, New South Wales, Australia (35°36′ S, 150°17′ E). There has been a notable study of plant–water interactions at this site, including the application of weighing lysimetry and ventilated chambers to entire, mature trees (see Aston 1984b; Dunin and Greenwood 1986; Wong and Dunin 1987; Dunin *et al.* 1988). The Hatton *et al.* (1992) study used the TOPOG model (O'Loughlin *et al.* 1989; O'Loughlin 1990), a distributed parameter catchment hydrology model. 'Distributed parameter' refers to the capability of the model to include variation in soils, vegetation type, surface, radiation and water. This basic model uses a digitised map of the three-dimensional surface of a watershed to compute water routing and the spatial distribution of conditions such as slope, aspect, incoming radiation, etc. (see O'Loughlin *et al.* 1989; O'Loughlin 1990). Coupled to TOPOG is a multi-factor plant growth model called IRM (Wu *et al.* 1993). IRM is a type of 'canopy process model' (as discussed in Chapter 13). IRM simulates the canopy conductance (based on a model by Ball *et al.* (1987)), CO_2 assimilation under different levels of nitrogen availability (from Wong and Dunin 1987; Walker *et al.* 1989) and plant response to water availability (Hatton *et al.* 1993). The model was tested by predicting the observed and predicted soil–moisture deficit over a period with rain (Fig. 12.9) and over another period without rain.

In a series of model experiments, Hatton *et al.* (1992) investigated the effects of an increase in either nitrogen or CO_2 concentration on the catchment vegetation. The effect of an increased amount of nitrogen has the effect of increasing plant growth, and subsequently plant leaf area. The increased leaf area in these simulations has the additional effect of increasing the transpiration demand and, thus, drying the watershed (compare Fig. 12.10*a* with Fig. 12.10*b*). In the case of an increased level of CO_2, the growth is also enhanced and there is an increase in the watershed leaf area.

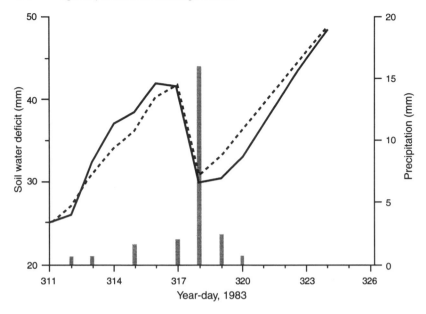

Figure 12.9. Observed (Dunin and Greenwood 1986) and predicted soil moisture deficit under an *Eucalyptus maculata* forest during a period of rain (hatched blocks). The solid line is the results from a weighing lysimeter, the dashed line is the result from TOPOG–IRM with 340 ppm CO_2 in the atmosphere. From Hatton *et al.* (1992).

However, the opposite effect occurs with respect to transpiration: there is a compensating effect of reduction in stomatal conductance (Rosenburg *et al.* 1990), and the watershed has a net increase in soil moisture (see Fig. 12.10c). The responses of the watershed to two changes, each growth enhancing and leaf-area increasing, are opposite in signs. In longer-term simulations, perhaps under more moisture-limiting conditions, the spatial responses to the changes in CO_2 in the environment could result in alterations of community composition in the areas of different moisture conditions.

Modelling the spatial pattern of herbivore grazing

In more arid zones, the effects of grazing mammals can have profound effects on the productivity and pattern of the vegetation (McNaughton *et al.* 1988). Large mammals are influenced by and, reciprocally, influence the vegetation (Laws 1970; Norton-Griffiths 1979). As was mentioned in Chapter 2, African and Asian elephant herds, as well as other large herbi-

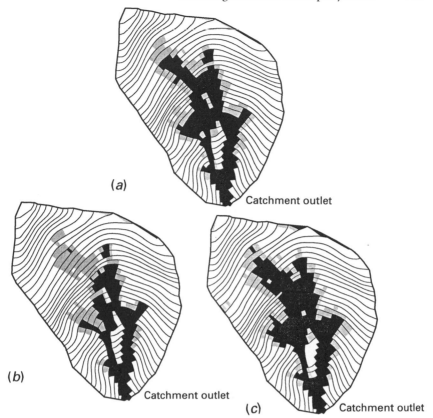

Figure 12.10. Distribution of relative volumetric water content (saturated soil has a relative volumetric water content of 100%) on a simulated forest watershed in New South Wales, Australia. Dark shading indicates areas of relative volumetric water content of 70% or greater; grey shading indicates 60–70%; white areas are less than 60%. Contour lines indicate elevation. Results of each simulation are for day 180 from year 2. (*a*) Ambient CO_2 and nitrogen conditions. (*b*) Ambient CO_2 and elevated nitrogen conditions. (*c*) Elevated CO_2 and ambient nitrogen conditions. From Hatton *et al.* (1992).

vores (Owen-Smith 1988), can alter the structure and composition of vegetation. Elephants can alter dominant tree species in the vegetation, elevate levels of tree mortality and damage in areas of relatively high elephant abundance, and maintain grasslands. Guthrie (1984) has proposed major changes in plant community organisation, habitat diversity, landscape pattern and trophic levels associated with the end of the Pleistocene Epoch and the extinction of large mammal herbivores such as the mastodon (*Mammut*) and mammoth (*Mammuthus*) (see Fig. 2.6, p. 21).

In many parts of the world, the native megaherbivore fauna has been relegated to parks and preserves. The extensive grasslands and shrublands are now grazed by cattle, sheep or goats. One of the consequences of domestic animal grazing in arid and semiarid regions has been changes in vegetation and, notably, in the rates of soil erosion. A major factor controlling the intensity of grazing and the attendant effects arises from the dependency of these domesticated animals on water. Areas surrounding water sources are subjected to heavier grazing pressure than areas further away. The spatial effects of cattle and sheep grazing (Stafford-Smith and Pickup 1993) have been simulated by a number of techniques, including one of the earlier individual-based models (Noble 1975).

Noble's approach to considering the spatial effects of grazing was framed in the problem of where best to locate watering points (from artesian sources) in large Australian sheep paddocks of tens or hundreds of square kilometres. Since they conform to landowners' property boundaries, the paddocks could have irregular shapes and the problem was to locate points for sheep to get water that would also optimise sheep production and minimise the areas of the paddock that were overgrazed. Noble approached the problem by modelling the movements of thousands of individual modelled sheep. Each sheep had six states (e.g. thirsty, hungry, extremely thirsty, etc.) and rules for sheep movement that were determined by the state of the sheep (e.g. a thirsty sheep would move toward a water source). As in reality, the modelled sheep populations tended to graze heavily in the vicinity of water sources. Depending on the shape of the paddock and on the location of water points, parts of the rangeland might not be grazed at all while other parts would receive heavy grazing.

Pickup and Chewings (1988) attacked the same problem using a different modelling approach for the dispersion of cattle grazing around each watering point. The movement of cattle from the watering point was taken as a diffusion process. The cattle were modelled in their movements following their use of the watering point as if they were particles originating at each watering point and diffusing through the vegetation. The cattle enter different vegetation types based on their preference for each vegetation and move through each vegetation type at a rate determined by the availability of forage. This process was approximated using the convection–diffusion equation:

$$\frac{\partial c}{\partial t} = \frac{1}{2}\sigma^2\left(\frac{\partial^2 c}{\partial x^2}\right) - \nu\left(\frac{\partial c}{\partial x}\right)$$

(12.3)

where: $c \equiv c(x,t)$ represents the number of cattle at time t and distance x from the watering point; σ^2 is the variance or the diffusion coefficient; and v is a drift parameter.

The solution of this equation for an impulse function (see Table 4.5 and associated discussion in Chapter 4 for use of singularity functions in linear systems) is:

$$f(t,x,v,\sigma) = \frac{x}{\sigma(2\pi t^3)^{\frac{1}{2}}} e^{\frac{(-x-vt)^2}{2\sigma^2 t}} \tag{12.4}$$

Pickup and Chewings (1988) reasoned that when presented with a mixture of vegetation types, cattle would enter each vegetation type according to their preference for the type and they introduced a preference term to the standard convection–diffusion equation:

$$c = p\frac{x}{\sigma(2\pi t^3)^{\frac{1}{2}}} e^{\frac{(-x-vt)^2}{2\sigma^2 t}} \tag{12.5}$$

where p is the preference for the vegetation types; and c is the number of cattle at distance x from the watering point.

The resultant model can be used to obtain an estimate of the number of animals within each vegetation type at a particular distance from a watering point. The grazing effect of the cattle was approximated by including features of cattle behaviour. Cattle in the summer drink at least once a day (and somewhat less frequently in the winter). They may graze when moving from the watering point to some grazing area but do not graze en route from a grazing area to the water supply. Trampling density (indicative of the negative effects of cattle walking over the vegetation) can be computed by assuming that the animals take roughly the shortest path back to water and using a computer to map these paths (a 'map analysis package'). Pickup and Chewings estimated the v, σ^2 and p parameters for Eq. 12.5 for five different types of vegetation and for wet, average and dry periods. They were then able to predict the distributions of cattle in different vegetation types under different environmental conditions (wet, average and dry) in a large paddock near Alice Springs, Northern Territory (Fig. 12.11). They were also able to show that the vegetation condition as determined using satellite remote sensing (Landsat Multispectral Scanner) corresponded to the patterns predicted by the cattle distribution model. In particular, the amount of vegetation cover in a given vegetation type decreased as a function of distance from watering points at the rate indicated by the cattle distribution model.

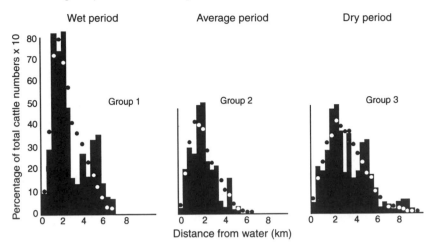

Figure 12.11. Observed and modelled (dots) cattle distributions for different vegetation groups classified by using data from the Landsat Multispectral Scanner (MSS) for wet, average and dry rainfall periods. Cattle show different preferences (*p* values from Eq. 12.5) depending on the rainfall conditions, and some discernible types of vegetation are aggregated in the figure. Using communities described by Low *et al.* (1981) for the region, Group 1 includes the cover types gilgaied plains, open woodland and eroded plains; Group 2 includes gilgaied plains and eroded plains; Group 3 includes gilgaied plains, open woodland and floodplains. Figures taken from Pickup and Chewings (1988).

Incorporating the effects of seasonal migration in animal population models

In Chapter 11, the responses in biodiversity in response to landscape fragmentation were discussed in the context of resident species existing and interacting on mosaic landscapes in which the species were resident and sedentary. The general result was that the relation between species number and available area seemed to be obtained from relatively simple models of habitat change on mosaic landscapes (see Fig. 9.18). This simple case might be expected to hold for resident populations without a great deal of migration, and as will be discussed in more detail in Chapter 14, an analogous approach has been used to predict the possible consequences to climatic change in isolated communities (McDonald and Brown 1992).

However, many communities are composed of highly mobile populations, and it is appropriate to consider the response of such species to landscape or larger-level habitat changes. Migratory birds are among the most mobile of animals and, in a sense, represent an extreme counter example to the sedentary species cases illustrated in Fig. 9.18 and discussed for mammals in McDonald and Brown (1992). In both the Old and New Worlds, there has been an apparent decline in the abundance of migratory bird species (Terborgh 1989; Baillie and Peach 1992). These continental-scale population declines are a cause of considerable concern among conservation biologists. Since both the wintering and breeding habitats of these species are subject to alteration by current human land-use changes (and potentially could also be altered by environmental change as well), it is appropriate to consider the effect of changes in habitat on the migratory birds. One issue that arises in such a case is whether the breeding habitat is limiting or if the population is primarily a consequence of changes in the wintering ground.*

Dolman and Sutherland (1994) extended a population model that can be used to predict the consequences of habitat loss for combinations of wintering and breeding habitat (Sutherland and Dolman 1994) to include the impact of loss of wintering area for bird species that migrate over long distances. In developing the model, they reasoned that the population size was determined by the balance between breeding productivity (on the breeding grounds) minus mortality and density-dependent mortality in the non-breeding season. Breeding productivity is represented as:

$$P = NF[1 + (aN)^b]^{-1} \tag{12.6}$$

where P is the productivity of the breeding population; F is the intrinsic per capita fecundity; N is the number of individuals in the population; a is a scaling constant; and b is a constant indicating the strength of density dependence (Sinclair 1989).

This model was initially developed by Maynard Smith and Slatkin (1973). The parameter b has values near 0 when density effects on breeding productivity are slight, and a value of 1 when the density dependence is perfectly compensating (Newton 1986; Galbraith 1988). In the model, the density-dependent[†] effects of competition among the birds in the population for food and other resources was represented as:

$$a'_{(s,t)} = QP_i^{-mR(s,t)_i} \tag{12.7}$$

where $a'_{(s,t)}$ is the searching efficiency of all individuals of phenotype S in patch type i; Q is the 'quest constant' and is the value of $a'_{(s,t)}$ for a solitary

---

* The idea that only a single condition might be limiting is, to some extent, a consequence of thinking of the population in the framework of homogeneous population models (Chapter 4). Heterogeneity and limiting factors are also discussed in Chapter 10, particularly in the discussion associated with Figure 10.7.

† See discussion in Chapter 4 associated with Eq. 4.7 for an example of simple density-dependent effects in a population model.

individual with no interference from other individuals; $R_{(s,t)}$ is the relative competitive ability of phenotype S in patch type i, expressed as the average of all individuals in the patch divided by the competitive ability of an individual of phenotype S (see Parker and Sutherland (1986) for more details); and m is a model parameter.

This equation for the interference of the different phenotypes of the population and the associated intraspecific competition is an extension of the more commonly used expression,

$$a_i' = QP_i^{-m} \tag{12.8}$$

with model parameters as in Eq. 12.7. Equation 12.8 is used to express the interference among individuals in which the individuals are all equal in their competitive ability (Hassell and Varley 1969). Equation 12.7 allows for differences among individuals of the same species. The parameter m (Eqs. 12.7 and 12.8) has been determined for many invertebrate species and has been estimated for oystercatchers (*Haematopus ostraegus*) to be 0.10 to 0.35 (Sutherland and Koene 1992).

Sutherland and Dolman (1994) simulated the changes in a population where the dynamics were influenced by the density- and phenotype-related effects represented in Eqs. 12.6 and 12.7 utilising different patches of habitat for breeding territories and other different areas for wintering grounds. They then experimented with the removal of wintering area, removing area of optimal habitat in one case and removing areas of poor habitat in the other. They found that in these model experiments the populations demonstrated an immediate, regular decline in the scenario in which the best winter habitat was removed first. When the poorest wintering habitat was removed first, the population densities remained relatively high initially and then dropped rather abruptly when a critical level of habitat loss was reached.

Dolman and Sutherland (1994) extended this work by considering the ability of the species to evolve (or learn new migratory pathways to and from areas of breeding and wintering) and by evaluating the effects of breeding area as well as wintering area habitat reductions. One important result was that a population with a reduction or destruction of a wintering area could produce shifts in the pattern of migration between the larger migratory network – implying that the effects of eliminating a wintering area could produce a cascade of interactions affecting subpopulations that did not even use the altered wintering area (Fig. 12.12). They also noted that the reduction of breeding habitats and wintering habitats could have a simultaneous effect on reducing the population.

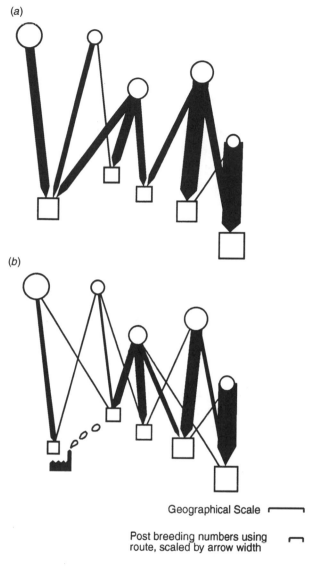

Geographical Scale ▭▬

Post breeding numbers using
route, scaled by arrow width ⌐¬

Figure 12.12. (*a*) Evolutionarily stable migration strategies and population sizes of breeding populations in which individuals have a choice of breeding sites. The distances and relative sizes of the wintering and breeding sites are indicated by their size and the number of individuals using each migration route are indicated by the width of the arrows. (*b*) New population sizes and migration routes as in (*a*) with a habitat loss of 75% in one of the breeding areas. From Dolman and Sutherland (1994).

The global carbon budget including potential spatial dynamics

The effects discussed so far have largely dealt with landscape systems. It is reasonable to ask whether these effects can be seen in the global-scale responses to changes in the environment. Smith and Shugart (1993) extended the applications for the estimation of the global vegetation and soils storage of carbon from the equilibrium cases discussed in Chapter 11 (see Fig. 11.3, colour plate and attendant discussion) to attempt an approximation of the shorter-term responses of the terrestrial surface to a climatic change. Estimates of the global carbon reserves are typically obtained from estimates for the amounts of different types of cover on the Earth's surface; these values are multiplied by other estimates for the expected amount of carbon per unit area in the vegetation and soils in each of the cover types. Estimates of changes in carbon storage in response to a climatic change apply the same carbon per unit area estimates to altered land-cover areas to obtain an equilibrium carbon storage expected for the new climate conditions.

Computation of the transient dynamics is considerably more difficult in that they require a spatial accounting for the areas that have changed. For example, if the current areal extent of grassland is $1923 \times 10^3 \, km^2$ (Smith et al. 1992a,b), and the area of grassland under a climate change is similar, one would compute no net change in the carbon storage in the long term. However, if the area of potential grassland was the same in the current and future cases but 50% of the future area comes from area that is now forest and 50% of the present area converts to desert, then the carbon change over the short terms depends, in part, on the rate that forest converts to grassland and the rate that desertification converts grassland to desert. One must also estimate the successional processes involved and one must account for the times that some of the vegetation might need to migrate to new favourable areas. The rates of pedogenesis and soil respiration must be taken into account to approximate the changes in the soil pools.

Clearly, this is a demanding problem. Based on our current understanding, ecologists are in the process of developing models and data sets for simulating the kinetics of the Earth's carbon storage under environmental change. In the Smith and Shugart (1993) study, the approach used approximations of some of these responses to obtain an initial estimate of the transient response to climatic change. The terrestrial surface was divided into $0.5° \times 0.5°$ latitude×longitude blocks. The Holdridge (1967) life zone system was used to relate vegetation to climate; the Post *et al.* (1982) mapping of soil carbon storage on Holdridge life zones was used to determine soil carbon pools for different climates; the Olson *et al.* (1983) carbon data set for carbon in the vegetation was used to determine the vegetation carbon pools.

The approach in the study was to compute the transitions of each vegetation type to another for a climate-change scenario and then to determine the rates at which the vegetation and soil carbon pools might be expected to change. For example, if a given element converted from forest to grassland, it was reasoned that such a conversion would typically be expected to occur rapidly under the influence of fire, insect outbreaks or a combination of the two. Thus, the rate of conversion would be large (0.02 per year). Conversion of tundra to forest, however, might be relatively slow if the tree species were forced to migrate into the new area here; the rate of conversion would be small (0.001 per year). Each of the transitions for life zones (vegetation and soils) for the 37×37 matrix of Holdridge types before and after a climate change was determined and assigned what were felt to be the appropriate rates. The GFDL and GISS climate scenarios from Table 11.1 with life-zone changes illustrated in Fig. 11.1 were used. A linear model results from this exercise and the model was evaluated for a step change in climate. (From Chapter 4, a step function is a singularity function. Recall that a linear system, such as this model, is superposable by definition; thus one can develop any arbitrary response from the response to a singularity function.) The rates applied to each of the transitions are shown in Table 12.2 along with the carbon fluxes associated with each transition.

Just as discussed in Chapter 11, the long-term response of the terrestrial carbon storage under both the GISS and the GFDL scenarios is to act as a carbon sink that stores more carbon. However, the balances between the rates are such that many of the transitions that increase the carbon storage under a climatic warming (boreal forests occupying what is now tundra, tundra soils forming in polar desert regions, etc.) are controlled relatively slow processes. Many of the transitions that might be sources (e.g. grasslands replacing boreal forests) could be expected to be controlled by disturbance regime and occur relatively rapidly. As a result, the terrestrial surface (which by these computations is expected to be an eventual sink for carbon) is initially a rather strong carbon source (Fig. 12.13). These calculations are only estimates of the actual responses, but the model is not particularly sensitive to changes in the parameters. Smith and Shugart (1993) found that the result illustrated in Fig. 12.13 was only slightly

Table 12.2. *Carbon pools and rates of change from an evaluation of the transient response of the global terrestrial storage of carbon under climate change conditions*

Controlling rate (per year)	Carbon losses (Pg)		Carbon gains (Pg)		Net change in carbon storage (Pg)	
	GFDL scenario	GISS scenario	GFDL scenario	GISS scenario	GFDL scenario	GISS scenario
Aboveground carbon biomass						
Successional replacement (0.004)	323.0	292.9	338.7	346.2	+15.7	+53.5
Immigration to new area (0.001)			122.4	171.8	+122.4	+158.5
Die-back/disturbance (0.02)	124.1	119.4			−124.1	−119.4
Soil carbon						
Increased decomposition (0.02)	196.8	213.3			−196.8	−213.3
Successional replacement (0.004)	87.3	111.3			+87.3	+111.3
Immigration of vegetation (0.0001)			101.9	102.0	+101.9	+102.0
Total					+37.9	+146.9

Source: From Smith and Shugart (1993).

changed by ±50% changes in each of the rates. This is because of the rather large differences in the rates in a qualitative sense. The implication of these results is that the spatial pattern of the response of terrestrial ecosystems to large-scale change, and the potential effects of such spatial phenomena as migration of vegetation, appears to matter when the global dynamics of carbon are considered. The simple model presented in Smith and Shugart (1993) is intended to identify the possible pattern of global carbon transient dynamics.

Concluding comments

Evaluation of environmental changes using spatially explicit models of landscapes represents what is for the most part a collection of case-studies.

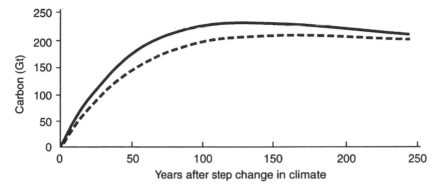

Figure 12.13. Net carbon flux to the atmosphere from the terrestrial surface in response to a step climate warming. The fluxes of carbon are derived from the estimated transient behaviour of vegetation and soils associated with climate change from two GCMs for a doubled-CO_2 condition (GFDL (—) and GISS (---) from Table 11.1 and illustrated in Fig. 11.1). From Smith and Shugart (1993).

In general, the overall result from these studies is cautionary: the effects of spatial interactions on the systems dynamics of realistic ecosystem models appear to be potentially quite significant at several different spatial scales. Indeed, the complexities of dealing with multiple spatial scales in under-standing the dynamics of ecosystems appear to be every bit as difficult a problem as dealing with multiple temporal scales.

With the exception of a few, more general, spatially explicit models (such as the ZELIG model, which can be applied to many of the same problems as the gap models discussed in Chapter 11 and can be used with parameters from the gap models), most spatial models have parameter and data-initialisation demands that limit their application to particular situations. This is likely to be the case for the foreseeable future, but progress is being made in understanding the general consequences of spatial interactions.

One of the simplest, and most regularly reported, consequences of spatial interactions is an increase in the inertia of the ecosystem response. Spatial effects such as species migration can slow the responses of landscapes to changes. Other spatial effects can reduce the rate at which species become extinct (in simple competition or predator–prey models) by providing spatial locations where species that might be eliminated in a non-spatial case can persist (at least for a time). Heterogeneity in the distributions of nutrients, or other resources, can alter the rate of processes as resources limitations occur locally.

In some senses, the spatially interactive models are an immense and complex case between the mosaic models discussed in the previous chapter and the homogeneous landscape models discussed in the chapter that follows. In the mosaic models, the interactions in the ecosystem were assumed to be non-interactive with the other surrounding elements; in the homogeneous landscape models, the interactions in space are so interactive that all parts of the landscape respond simultaneously. Because the spatial effects are important in at least some cases, the overarching consequence of the spatial modelling efforts is to emphasise the need for continued model testing.

13 · *Homogeneous landscape models*

Probably the most frequently applied approach to simulating the 'functional' response of landscapes to environmental change is to assume that the internal workings of the landscape are sufficiently well mixed to allow the system to be simulated in its entirety without consideration of the dynamics of patches or other sources of heterogeneity in the landscape. Models arising from such conceptualisation are based on the assumption that the important processes in an ecosystem can be approximated in aggregate without explicit consideration of spatial heterogeneity. While this is a model simplification, the resultant models are far from 'simple' and represent a challenge with respect to model parameter estimation and model testing.

There is a rich array of models that simulate the changes in an assumed-to-be homogeneous landscape in which the landscape processes are derived from smaller-scale observations. For example, most models of element cycling in watersheds or other ecosystems tend to view the processes as being homogeneous within the system of definition. Indeed, traditional ecosystem models were often referred to as 'point models' because they simulated the dynamics of ecosystems with no explicit references to spatial heterogeneity. Along with the traditional biogeographical or correlational approaches to understanding vegetation climate relationships (see Chapters 7 and 10), two types of homogeneous landscape model regularly used for large-scale assessments of change can be classified as canopy process models and material transfer models. As was outlined in Chapter 10, these models have been widely applied in the context of changing environmental conditions and in exploring the feedbacks between vegetation change and other global changes.

The models originate from initial attempts to formulate ecosystem models and have a rich history of development and application. Such models scale-up the response of ecological processes based on response functions developed at smaller spatial scales. For example, one might assume that the fluxes of heat, water and CO_2 associated with the functioning of a single leaf are duplicated by the sum of the responses of the billions of individual leaves making up a vegetated landscape. Models based on this assumption often are the ecological models linked to other models of ocean or atmospheric dynamics to assess the feedbacks among these major Earth systems (Ojima 1992).

A common practice in developing these models is to assume that the mathematical structure of biophysical and chemical reactions at the landscape level resembles those observed at a detailed level (perhaps with some differences in the model parameters). In an earlier example, Fig. 11.2 (colour plate) was used to illustrate the possibility that observed photosynthesis responses for shade-tolerant and shade-intolerant plants at different light levels (as measured at the leaf level) could arise from geometric considerations alone at the plant canopy level. However, there are a range of attendant difficulties in scaling-up such responses, and the dynamics of the same processes at one scale may not appear similar at larger scales. An example of this phenomena is found in the dynamics of the ZELIG model averaged over large and small scales (see Fig.12.4 (p. 389) and associated discussion in Chapter 12). Procedures for including some of the consequences of scaling-up from smaller to larger scales (particularly when there are large changes in spatial and temporal scale) are not easily generalised (Jarvis and McNaughton 1986).

The models used in large-scale or global studies often have similar internal assumptions that are a part of the rationale for scaling the models to larger space scales and longer time scales. These assumptions are also a basis for the organisation of available data on terrestrial ecosystem performance into dynamic models. There are several important assumptions.

1. *Generality*: the concept that terrestrial ecosystems function by some overarching rules that control structure and/or function. Examples might be the relation between ecosystem productivity and climatic indices (Rosenzweig 1968; Leith 1975; Box 1978), or the regularities in decomposition rates across ecosystems (Meentenmeyer 1978). These general rules can be thought to function across terrestrial ecosystems at the biome or the global level. Indeed, one of the principal rationales for using homogeneous landscape models (in the face of great perceived heterogeneity in ecological landscapes at a variety of scales) is that the operating rules apply generally over the landscape, regardless of the structure of terrestrial ecosystems.

2. *Optimality*: the concept that processes in terrestrial ecosystems will (over time) tend to optimise (or, in some cases, maximise) such attributes as leaf area index, net primary productivity or internal energy fluxes. This issue was discussed earlier (in Chapter 6) in the context of the optimality of organs in individual plants. The difficulties that were mentioned there stemmed from problems in understanding which commodities or factors were being optimised (or maximised) by plants. When the same concept is applied at the landscape level, the equivalent problem of developing a theory as to what aspects of a landscape are optimised by the diverse array of possible interactions is formidable as well. Two frequently invoked principles are *limiting* or *critical factors* – the concept that a particular internal variable, such as soil nitrogen or canopy water flux, is the keystone factor controlling the fluxes and changes in the terrestrial ecosystems – and *stoichiometry* – the concept that physical and chemical processes in terrestrial ecosystems will tend to maintain certain ratios in the concentrations of critical nutrient elements, carbon and water.

3. *Equilibrium-seeking behaviour*: the assumption, either built into the model formulations or used to estimate model parameters, that terrestrial landscapes, as represented by homogeneous landscape models, seek an equilibrium. Models based on conservation of mass, energy, etc. maintain balances of inputs and outputs, and lead naturally to an assumption of equilibrium-seeking behaviour. The formulation of many homogeneous landscape models is from the position that ecosystems may be disturbed away from their equilibrium (by disturbances or other events in their history), but their transient dynamics asymptotically seek an equilibrium condition.

These assumptions are typical of several approaches by which ecosystem data can be organised into models for the purpose of simulating large, homogeneous landscape units. Each of these assumptions is not necessarily applied to develop all homogeneous landscape models, but some subset of these assumptions is usually involved in the formulation of each of these models.

Two important categories of these models are canopy process models and material transfer models. Canopy process models simulate, from a biophysical basis, the response of a regional plant canopy that is assumed to be spatially homogeneous. Their hallmark is an explicit inclusion of a stomatal conductance term in the model and a consequent emphasis on water, CO_2 and heat fluxes. Material transfer models focus strongly on the storage and transfer of material (notably carbon, nitrogen, phosphorus and sulphur: organic matter and major nutrients). Therefore, material transfer models tend to have greater detail in their representation of soil processes and somewhat less of an emphasis on canopy processes.

Since both types of model attempt to represent all the 'important' processes involved with dynamic change or response of the vegetation, they are convergent with respect to the processes they represent. The differences lie in the emphasis on different processes in the model formulation, rather than on absolute inclusion of processes. In general, neither category of homogenous model has great ability to simulate large physical changes in the vegetation structure (outside of interpretations of the model outputs). This is considered by most practitioners to be their outstanding weakness and, consequently, is a central research challenge.

Initial results using material transfer models

Early applications of ecological models to the evaluation of the large-scale effects of environmental changes were associated with tracing the pathways of a range of different radioisotopes through the natural environment from a source (such as an emission from a nuclear reactor, radioactive fallout from testing of nuclear weapons or, potentially, thermonuclear war). The principal models used in this research were compartment models use to trace the fate of problem isotopes through the environment. The history of these models was briefly discussed in Chapter 4.

For example, Kaye and Ball (1969) produced a compartment model for the movement of radioisotopes through a tropical environment in order to evaluate the health hazards of the remarkable (certainly by today's standards) concept of using nuclear explosions to excavate a second canal across the isthmus of Panama (Martin 1969). These detonations could have produced and released into the environment over 100 different isotopes. Clearly, it was important to assess the degree of hazard of each of these. The Kaye and Ball model attempted to represent the transfer dynamics of a multiplicity of elements circulating along major environmental pathways through food, water and air to a person (depicted in the diagram as an Amerindian living in the region) The model had over 30 compartments including game animals, crops, livestock, ground water, sea turtles, fish and a banana plantation (Fig. 13.1). Movements of isotopes were computed by a compartment model representation (see Eqs. 4.18 to 4.27) of the general form:

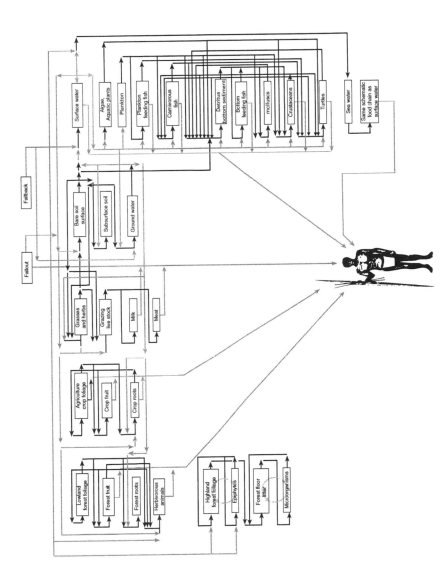

Figure 13.1. Environmental pathways moving radionuclides to a person in a tropical environment. From Kaye and Ball (1969).

$$m_j\frac{dX_j}{dt} = \sum_{i=1, i\neq j}^{n} \lambda_{ij}m_iK_{ij}X_i - \sum_{i=0, i\neq j}^{n} \lambda_{ji}m_jK_{ji}X_j - \lambda_r m_j X_j \qquad (13.1)$$

where j is a subscript designating the compartment of reference; i is a subscript designating any donor or recipient compartment other than compartment j; X is a radionuclide concentration (μCi g^{-1}); λ is an environmental transfer coefficient (days^{-1}); λ_r is the radioactive decay constant (days^{-1}); m is the biomass per unit area (g m^{-2}); and K is a selectivity factor that adjusts the environmental coefficient when it differs quantitatively from the radionuclide transfer coefficient.

The use of the parameter K in Eq. 13.1 is primarily for those radioisotopes that concentrate in foodwebs or in parts of other environmental pathways. An example might be isotopes of strontium, which tend to concentrate in the bones. Project Plowshare (Martin 1969), as the canal development project was called, did not reach a stage of application and the Kaye and Ball (1969) model was never tested. However, Jordan (1986) reports successfully using the model to predict the concentrations in plants of ^{90}Sr produced from pre-1963 nuclear weapons testing and transported as fallout to Panamanian rain forest.

This class of model is used to assess the potential health hazards from the operation of nuclear-power reactors used in electric power generation. In this context, they have been subjected to considerable scientific scrutiny, and a significant effort has been focused on the development of such models. The relative success and degree of internal testing of these simple material transfer models (Jordan 1986) is probably a strong contributing factor to the willingness to use homogeneous landscape models for other classes of large-scale model evaluations. Over time, the application of material transfer models has progressed through a spectrum of increasingly difficult applications (Table 13.1).

In some cases, the more recent material transfer models use the ratios of essential nutrients observed in a variety of ecosystems to compute the expected balances of carbon and other nutrients. Thus, if the ratio of carbon to nitrogen (C/N-ratio) is low relative to that expected, then one might expect increased productivity in the presence of ample nitrogen to increase the carbon content of the system and restore the ratio to some nominal value. Constancy in ratios among carbon and other essential nutrients is often found in aquatic systems and is a logical consequence of a chemistry-oriented view of living protoplasm functioning in a well-mixed liquid. The use of stoichiometric ratios to control the rates at which ecosystem processes occur derives from a similar philosophy and is logically consistent with assuming the terrestrial system to be well mixed or homogeneous.

Table 13.1. *Large-scale predictions from material transfer models. The degree of complexity, internal feedback, and non-linearity in the model formulation tends to increase from top to bottom of the table*

Environmental problem	Aspects of the problem that influence formulation of material transfer model	Typical models used
Determine the rate of transfer and fate of radionuclides in the environment	Radioactive isotopes in very small amounts move passively through a number of physical, chemical and biologically mediated pathways. The levels of radioactivity are below those that could be expected to produce direct effects	Linear compartment models (Chapter 4) often solved for the equilibrium condition
Determine the build-up of pesticides through different levels of foodchains	Pesticides are organic compounds that tend to increase in their concentration as they are transported through from the prey to the predator (biomagnification)	Linear compartment models. Non-linear models to imitate the predator feeding dynamics
Assess the consequences of direct and indirect effects of atmospheric pollutants on ecosystem productivity	Need to extrapolate effects observed over a short period of time with models that include the direct effects of a pollutant. Indirect effects (pollutant alters soil chemistry, which affects plant growth) add complexity	Simulation models with sections of models that represent key processes
Assess the effects of climatic change on natural ecosystems	Response of vegetation to climatic change operates at multiple temporal scales. Feedback from a climatically altered vegetation to the atmosphere to further change the climate is potentially important	Correlational approaches using the vegetation response to the spatial variation in climate to interpret climate change. Simulation models with sections of models that represent key processes
Assess the direct effects of elevated atmospheric CO_2 on natural ecosystems	Potential changes in the CO_2, water and heat budgets of plants can produce changes in vegetation at a multiplicity of scales	Simulation models with sections of models that represent key processes. Basis of models often involves theories as to controlling variables of ecosystem functions

Material transfer models applied at continental scales

As was discussed in Chapter 10, the development of material transfer models was related to radionuclide models such as the one discussed above and the models developed during the International Biological Programme in the 1960s and 1970s. Over time, the developers of these models focused on understanding how the rates of transfer of material from different compartments changed under the influence of different environmental conditions (particularly temperature and moisture) and on interactions with other compartments (notably interactions between carbon and nitrogen compartments). With the incorporation of additional mechanism in the growth and production processes, the models have become convergent on the biophysical models used in canopy process models that will be discussed in the following section.

Comparing material transfer models at test sites

Table 13.2 outlines the mechanisms used in a suite of material transfer models used to examine the response of forests to climate change and the effects of CO_2 enrichment of the atmosphere. These models (BIOMASS (McMurtrie et al. 1990), BIOME-BGC (Running and Coughlan 1988; Running and Gower 1991; Hunt and Running 1992), CENTURY (Parton et al. 1987, 1988; Sanford et al. 1991), HYBRID (Friend et al. 1993), MBL-GEM (Rastetter et al. 1991), PnET-CN (Aber and Federer 1992), Q (Bosatta and Ågren 1991a,b)) all attempt to simulate the productivity of a coniferous forest by considering processes involving photosynthesis, respiration, nitrogen dynamics and water use. They differ as to their overall structure, in their time steps and as to which of several different available options are used to represent key processes (Table 13.2). The models were all tested with respect to their ability to reproduce carbon dynamics at two locations: a 10-year-old *Pinus radiata* plantation located 20 miles from Canberra, Australia (35°21′ S, 148°56′ E), and a *Pinus sylvestris* plantation located at Ivantjärnsheden near Jädraås, Sweden (60°49′ N, 16°30′ E). Subsequent to this model testing, the suite of models was inspected in their prediction of the effects of a 4°C warming, an increase in the concentration of the atmospheric CO_2 (700 p.p.m.), and both the warming and the CO_2 change together.

The formulation of this suite of models by different scientists for different objectives offers several comparisons of the choices in the formulation of material transfer models. The models are at different time steps (Table

Table 13.2. *Approaches to modelling processes included in different ecosystem material transfer models (see text for model descriptions)*

Model attribute		BIOMASS	BIOME-BGC	CENTURY	Models MBL-GEM	HYBRID	PnET-CN	Q
Time step[a]		D	D/A	M	M/A	D/A	M	A
Transpiration model[b]		PM	PM	PM	I	PM	WUE	–
Photosynthesis model[c]		F	F/P	I	P	F	P	I
Photosynthesis scalar[d]		B	B/N	C	B/N	B/N	N	B/N
Respiration model[e]		G/M	G/M	M	G/M	G/M	G/M	–
Respiration scalar[f]		N	B	B	N	B	N	–
Carbon allocation controlled by[g]		A/V	V	C	V	V	C	V
Turnover of leaves[h]		C	C	C	C	C	C	C
Leaf nitrogen[i]		I	V	V	V	I	V	V
Litter nitrogen[j]		–	P	P	P	–	P	C
Decomposition rate modelled by[k]		–	AET	–	LCI	–	R	M
Nitrogen mineralisation[l]		–	P	–	C/N	–	C/N	P

Notes:

[a] D, daily; M, monthly; A, annually.

[b] PM, Penman–Monteith equation; I, soil moisture is an input variable; WUE, water-use efficiency determines water use.

[c] F, Farquhar *et al.* (1980) biochemical model; P, empirical model; I, empirical model with net primary production only.

[d] B, biomass; N, foliar nitrogen; C, climate.

[e] G, growth respiration modelled; M, maintenance respiration modelled.

[f] B, biomass; N, tissue nitrogen.

[g] A, allometric equations; V, variable percentage; C, constant percentage.

[h] C, constant percentage.

[i] I, input; V, variable.

[j] P, constant percentage of leaf nitrogen; C, constant value.

[k] AET, actual evapotranspiration and lignin; LCI, ligno-cellulose index; R, residence time for a given stage; M, microbial growth rate.

[l] P, proportional to decomposition; C/N, maintains a constant carbon to nitrogen ratio.

Source: From Ryan *et al.* (1997a).

13.2) and consequently represent the various processes with differing resolution (recall the issue of scale discussed in Chapter 3).

Production models that include the direct effects of CO_2 concentration often use a model that duplicates the diffusion processes that bring CO_2 into the leaf and account for the biochemical processes involved with photosynthesis. Variants of the Farquhar et al. (1980) model of the photosynthesis process are often used as such models. This approach will be presented in more detail in the section below. If the production model is formulated at a longer time scale (or if the parameters needed for the Farquhar et al. (1980) model are not available), then the effects of CO_2 concentration may be included from empirical calibrations for either gross or net productivity.

Often water flux is computed using an idealised plant canopy that performs somewhat like a leaf, and the Penman–Monteith (e.g. Penman 1948; Monteith 1973) evapotranspiration equation is applied to compute transpiration for the vegetation canopy. Since the detailed photosynthesis models and the detailed transpiration models all involve the functioning of the leaf stomata and the resultant effects on the diffusion of water into and CO_2 out of the leaf, it is natural to include both of these equations to simulate the important coupling of CO_2 and water. A somewhat less mechanistic, but nonetheless quite valid, approach to the problem is to include a water-use efficiency function to specify how much carbon is fixed for a given amount of water lost.

The development of these sorts of model is a bit like a chess game. One move in choosing a certain formulation for a given process implies a counter move of choosing a related formulation for another process. For example, important decisions to be made in developing such models involve the degree to which the CO_2 and water diffusion processes are represented. Since the stomatal opening in leaves can change very rapidly with changing microenvironmental conditions, the data demand and time resolution increase greatly as the processes are considered in detail. Other important factors such as the decomposition of dead leaves and the release of nitrogen and other nutrients occur at slower rates, and the representation of these as controlling factors calls for longer time step models and consequently less detail in the production processes.

Given our level of understanding of ecological processes and the modelling of whole ecosystems, the diversity of modelling approaches is probably healthy for our science. In the case of the seven models listed in Table 13.2, the models responded quite differently to the cases involving changes in temperature or CO_2 concentration (Ryan et al., 1997b). The models disagreed in their predictions of the amount of carbon stored over the longer term by the forest, the direction of the response to the combined effect of CO_2 and temperature, and the magnitudes of the responses under control conditions. However, a consistent prediction of forests as strong sinks for storing carbon held over all the models.

Applications of material transfer models at continental scales

Some of the initial applications of material transfer models involved computation of regional or global transfers of carbon, nitrogen, sulphur and phosphorus (e.g. Bolin and Cook 1983). These models have been aimed at understanding the relative magnitudes of the different elements and have generally been developed as linear compartment models.

Until relatively recently, the detailed data sets required to run the more mechanistic material transfer models over continental and global areas have constituted a major logistic problem. Parton *et al.* (1995) applied the CENTURY model to investigate the effects of climatic change on grassland ecosystems globally, and Melillo *et al.* (1993) have applied the TEM model to investigate global productivity patterns.

In the Melillo *et al.* (1993) study, the TEM model (Fig. 13.2) was calibrated for 18 different vegetation types representing the terrestrial cover (polar desert/alpine tundra, wet/moist tundra, desert, tropical savanna, etc.). The parameter estimation for the model is described in Raich *et al.* (1991) and McGuire *et al.* (1992) and consists of the application of empirical functions for observed rate process, and estimations of rates of transfer taken from literature sources. Several of the functions in the model have components involving monthly temperature and/or moisture conditions, so that the model is capable of predicting responses to climatic change. The model is applied by specifying monthly temperature and moisture conditions to a grid cell with a given soil and vegetation cover. The soil-moisture computations are developed by another model of large-scale hydrology as a separate calculation from the model run. The TEM model does not change the vegetation cover with climatic changes, and the interpretations are all with respect to the current vegetation. Using climate-change scenarios from four different GCMs, the TEM model predicted relatively little change in global net primary production (ranging from a decrease of 2.4% to no change), but this was a consequence of increases in net primary production in some ecosystems being offset by decreases in others.

Parton *et al.* (1995) used the CENTURY model (See Chapter 10,

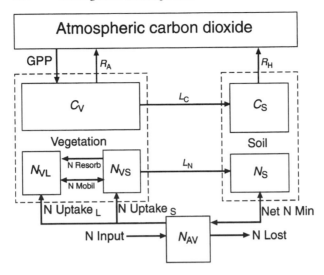

Figure 13.2. Schematic diagram for the TEM model of Melillo *et al.* (1993). Of the compartments in the model: C_V represents the carbon in the vegetation, N_{VL} is labial nitrogen in the vegetation, N_{VS} is the structurally incorporated nitrogen in the vegetation, C_S is the carbon in the soil, N_S is the nitrogen in the soil and N_{AV} is the inorganic nitrogen pool. Of the fluxes or transfers of carbon in the model: GPP is gross primary production, R_A is plant respiration, R_H is soil (hetereotrophic) respiration, and L_C is the carbon transferred to the soil by litter production. Of the transfers of nitrogen (L_N), N Resorb is the nitrogen that is resorbed from senescing tissue, N Mobil is nitrogen mobilised between structural and labile nitrogen, N Uptake$_L$ is the nitrogen uptake by vegetation into the labile pool and N Uptake$_S$ is the nitrogen uptake by vegetation into the structure; Net N Min is the nitrogen exchanged between the organic and inorganic fractions of the soil. Inputs of nitrogen from the system are N Input and losses are N Lost. From Melillo *et al.* (1993).

particularly Fig. 10.15 (p. 334) for details of the model) to evaluate the potential effects of climatic change on the world's grasslands by calibrating and then running the model at 31 temperate and tropical grassland sites. They then compared the performance of the models with climate-change scenarios produced by two different GCMs, one from the Canadian Climate Center (CCC) and the other from the Geophysical Fluid Dynamics Laboratory High Scenario (GFHI). Both of the climate-change scenarios are similar in temperature changes (2–51°C depending on location) but differ in precipitation changes. The GFHI scenario shows increases in precipitation for all regions, while the CCC scenario has decreases in areas of temperate steppe and humid savanna. Under climate change, the grasslands are simulated as being a source of carbon to the atmosphere globally, with the effect being ameliorated somewhat by including the potential positive effects of having more CO_2 in the atmosphere. In their model experiments, most of the detectable changes in plant production and soil organic matter occur during the first 25 years of the dynamic response.

These (Melillo *et al*. 1993; Parton *et al*. 1995) and other large-scale applications of material transfer models are in their initial stages of development. The functions used in the models are a mixture of mechanistic models and empirical functions. The mechanistic formulations almost inevitably involve scaling-up from relatively short-term and small-area experiments; the empirical formulations are limited by a lack of systematic large-scale data on ecological processes for calibration and testing. The model developers are well aware of these problems. Indeed, one of the more refreshing aspects of many of these model descriptions is the openness in discussing the shortcomings of the models. What is clear is that the interactions between the vegetation and the environment at global scales are important in determining the dynamics of global recycling of important elements. Melillo *et al*. (1993) note, 'Our results indicate that the simultaneous interactions among the dynamics of carbon, nitrogen, and water affect the ability of vegetation to incorporate elevated CO_2 into production'. This conclusion is true with respect to other material transfer models, based on different assumptions and developed by independent groups. The need for more and better information on the processes controlling ecosystem function is essential.

Canopy process models at continental scales

As was discussed in Chapter 6, there are relationships between climate variables and measured features of plant tissues, plants, physiognomy and composition that have been noted since the time of Theophrastus. Ecosystem functional characteristics also have related empirically to climate for such variables as net primary productivity (Rosenzweig 1968; Leith 1975; Box 1978) or rate of decomposition (Meentenmeyer 1978; Raich and Schlesinger 1992). These empirical relationships have been used in material transfer models for the global carbon cycle (Esser 1984; Dai and Fung 1993) and in climate models (Henderson-Sellers 1990). Nonetheless, relations between plant processes and CO_2 concentrations are potentially so important that one expects that empirical relations, based on large-scale observations under present

CO_2 concentrations, may not hold under future higher CO_2 conditions.

This latter consideration has been a strong motivation for the development of more mechanistic material transfer models (e.g. Melillo *et al.* 1993; Parton *et al.* 1993, see discussion above) as well as models of land-surface process within climate models (Dickinson 1986, 1988; Sellers *et al.* 1996). These models all have certain empirical relationships embedded in them: often concerned with the pattern of biomes across the Earth's surface. Woodward (1987b) related leaf area to climate variables using a mechanistic model involving temperature and moisture conditions. Bonan (1993a,b) demonstrated that basic plant physiological models can be used to explain observed relationships between ecosystem variables and climate at continental scales. Woodward *et al.* (1995) have developed a model of canopy processes that also simulates the changes in some of the features of the vegetation considered along with producing estimates of canopy processes.

Models of land surface interactions with the atmosphere: estimating the productivity of the Earth

The plant leaf can be thought of as balancing a set of complex interactions involving heat (including light), water and CO_2. Light is an essential component of the photosynthesis process, but structures that capture light often tend to capture heat. Leaves can radiate some of this heat, but much of the cooling for leaves is accomplished by the evaporation of water from inside the leaf. The stomatal openings that allow water to pass out of the leaf also are the pathways for CO_2 into the leaf. Canopy process models are essentially based on representing these mostly biophysical processes (often measured at the leaf level) at the plant canopy level.

Under ideal conditions, primary production has been related to the absorption of photosynthetically active radiation (PAR) (Monteith 1977; Linder 1985; Landsberg 1986). Perhaps since crops are often growing in good conditions (well-watered, nutrient-augmented systems), this has led to the development of crop production models that relate absorbed PAR (APAR) to net primary production. Such models are of the form:

$$P_n = \epsilon_n \sum_t N_i S_i \qquad (13.2)$$

where P_n is net primary productivit; ϵ_n is the dry matter yield of APAR (an efficiency measured in $kg J^{-1}$); $N_i S_i$ is APAR, where N_i is the proportion of incident PAR absorbed by the canopy over the time interval i. The value of N_i is derived from satellite measurements of the reflectance of red and near-infrared radiation (computed as a corrected ratio of red (0.58–0.69 μm wavelengths) and infrared (0.725–1.1 μm wavelengths) (see Sellers (1985, 1987) and Tucker and Sellers (1986) for more details) using NDVI = (IR − R)/(IR + R), where IR and R are the near-infrared and red measured radiances, respectively. Red radiation is photosynthetically active and near-infrared is not, so the NDVI index provides a measure of APAR.

Since the development of this equation (Monteith 1977), the critical parameter ϵ_n has been measured for a range of systems.

Recent studies have found the ϵ_n efficiency (thought to be relatively constant) to vary across vegetation types and under different environmental conditions (Jarvis and Leverenz 1983; Russell *et al.* 1989; Prince 1991). Variation in the value of ϵ_n was viewed as being caused by departure from the ideal conditions that underlie the formulation of Eq. 13.2 and by the effects of respiration (Fig. 13.3). Accordingly, Prince (1991) developed an alternative model of the form:

$$P_n = \epsilon_g \sum_t N_i S_i \qquad (13.3)$$

and

$$P_n = f(\epsilon_g) Y_g Y_m d \sum_t N_t S_t \qquad (13.4)$$

where f is a function relating the effects of stress on the unstressed value of ϵ_n operating to close the stomata or on the coupling of light and dark reactions in photosynthesis; d is the proportion of biomass lost in death, decay and grazing; ϵ_g is the gross energy fixation measured in dry matter equivalent per unit APAR (kg J^{-1}); N_t is the proportion of incident PAR absorbed by the canopy over time interval t and S_t is the incident PAR over that time interval. Y_m is the proportion of assimilate not used in maintenance respiration, where:

$$Y_m = \left(1 - \frac{R_m}{P_g}\right)$$

for R_m as maintenance respiration and P_g as gross photosynthesis; Y_g is the efficiency of conversion of assimilate into biomass, including growth respiration, R_g and is derived from:

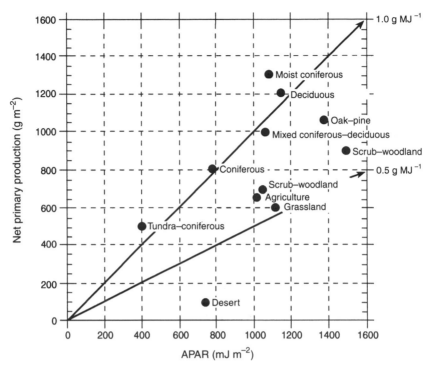

Figure 13.3. Relation of primary production to the APAR absorbed by plant canopies. The net primary production data (see Table 3.1) are from Whittaker and Likens (1975). The APAR is determined from an eight year average of monthly AVHRR values for NDVI measurements (see Goward and Huemmrich 1992). The incident PAR is estimated from a 10 year average of ultraviolet-reflectance measurements from the Total Ozone Mapping Spectrometer (TOMS) satellite instrument (see Eck and Dye 1990). The diagonal lines represent constant biomass–energy quotients (ϵ_n, see Eq. 13.2) of $1.0\,\mathrm{gMJ}^{-1}$ and $0.5\,\mathrm{gMJ}^{-1}$. Well-watered forest systems tend to lie on the $\epsilon_n = 1.0\,\mathrm{gMJ}^{-1}$ diagonal and ecosystems associated with more arid conditions appear at lower values of ϵ_n. From Prince *et al.* (1994).

$$Y_g = \left(1 - \frac{R_g}{Y_m P_g}\right)$$

The parameter ϵ_n in Eq. 13.2 is equivalent to $f(\epsilon_g) Y_g Y_m d$ from Eqs. 13.3 and 13.4.

Equations 13.2 and 13.4 are important because they allow investigators to relate the absorption of PAR to net primary productivity at global or continental spatial scales. These large-scale surveys typically are developed using the AVHRR, an instrument carried on-board meteorological satellites. One application of data from these satellites was discussed in Chapter 5 with predicting the breeding habitat distribution at different times over Africa for the pest bird *Quelea quelea*. Typical spatial resolution for these data is 16 km^2. Therefore, the equations produce a net productivity estimate by treating what is often complex, heterogeneous terrain as if it behaved as a leaf. It is not clear why this necessarily should be the case. Field (1991) proposes a 'functional convergence hypothesis' – the concept that the investment in additional light capture machinery (leaves, chloroplasts, chlorophyll) should stop when a limitation owing to a shortage of any resource prevents the exploitation of the additional light capture machinery.

These equations have been used to allow satellite remote sensing to attempt to measure net primary productivity. Initial evaluations of net primary productivity based on satellite-based observations include studies across some of the major ecotones of the Earth, such as the African Sahel (Tucker *et al.* 1985a; Prince *et al.* 1991) or the boreal forest (Justice *et al.* 1985).

One of the most important results from this work has been to relate the total global changes in the seasonal pattern of NDVI (Eq. 13.3) to the variation in the CO_2 concentration of the atmosphere. Initial observations by Tucker *et al.* (1986) demonstrated that the patterns of NDVI were strongly correlated with the changes in the latitudinal pattern of CO_2 in the atmosphere (Fig. 13.4). The implication is that the annual variation in global atmospheric CO_2 is a consequence of the balance of net primary productivity of the terrestrial surface pulling down the CO_2 during the growing season and decomposition processes producing CO_2 (sometimes poetically referred to as the 'breathing' of the Earth's terrestrial surface). Tucker *et al.* (1986) state in their conclusions that, 'Our analysis demonstrates the measurable link between atmospheric CO_2 drawdowns and terrestrial NDVI dynamics and suggest that there may be quantitative relationships between multi-temporal satellite data and atmospheric CO_2 drawdowns'. This interpretation has been reinforced by the work of Fung *et al.* (1987), who demonstrated that the productivity and decomposition rates for different biomes were of the appropriate magnitudes to produce such effects.

Models of land–surface interactions with the atmosphere: simulating the structure and the function of the terrestrial surface

The models treated in this chapter focus on the function of landscapes (quantified by measurements of productivity, nutrient cycling, etc.).

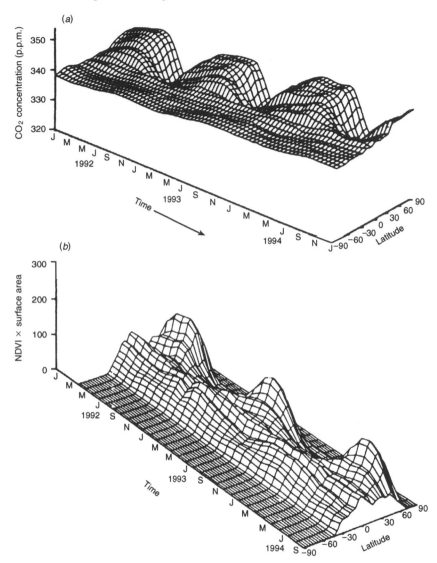

Figure 13.4. (*a*) Time (three years) and latitude plotted for atmospheric CO_2 concentration (from Harris and Bodhaine 1983). (*b*) The equivalent three years of NDVI from the visible red and the near-infrared channels of the NOAA-7 satellite plotted as a three-dimensional surface against time and latitude zone (from Tucker *et al.* 1986). The seasonal effects seen in the northern latitudes in the dynamics of CO_2 are also seen in the northern latitudes for NDVI dynamics. Note also the relatively constant values of NDVI and CO_2 for the equatorial values, and the influence of deserts in depressing NDVI in the 20° to 10° N latitude zone.

However, a central problem in understanding the feedback between climate and vegetation at large scales is predicting the response of the Earth's terrestrial vegetation to global change using models that can simultaneously produce predictions of the changes in *both* the structure and the function of ecosystems in response to environmental changes. The structural aspect of this problem arose in the previous two chapters with models that could represent the changes in the structure (species composition, species physiognomy, spatial heterogeneity) of the terrestrial surface resulting from environmental changes.

Woodward *et al.* (1995) have developed a model that dynamically changes the leaf area of locations across the Earth's surface. This model, called DOLY in an early description (Baskin 1993) but unnamed in its documenting publication (Woodward *et al.* 1995), combines features of canopy process models with material transfer models. In particular, the model provides both the net primary productivity and the leaf area of locations across the Earth as output variables. To provide an impression of the formulation of this class of model and of the manner in which filed and laboratory observations are incorporated into models, the photosynthesis section of this model will be treated in some detail.

The DOLY model bridges from cellular biochemical processes to leaf-, canopy-, plant- and landscape-level processes. The communication among these levels is through internal parameters and assumptions about the manner in which productivity-related responses scale-up.

At the cellular level, the DOLY model represents the biochemical processes of photosynthesis and net CO_2 assimilation, using the expression:

$$A_b = V_c \left(1 - \frac{0.5P_o}{\tau P_c} \right) - R_d \tag{13.5}$$

where A_b is the net rate of CO_2 assimilation; V_c is the rate of carboxylation; P_o is the internal partial pressure of O_2; P_c is the internal partial pressure of CO_2; R_d is the dark respiration; and τ is the specificity factor for Rubisco for CO_2 relative to O_2 (Jordan and Ogren 1984).

Oxygen competes with CO_2 and ties up Rubisco. This reduces the yield of Rubisco with respect to CO_2 carboxylation. The attendant loss in photosynthate production rate is called photorespiration. The value of τ increases as a function of temperature (Harley *et al.* 1992), such that:

$$\tau(T_k) = e^{-\left(3.949 - \frac{28.99}{0.00831 T_k} \right)} \tag{13.6}$$

where T_k is temperature in kelvins.

Farquhar *et al.* (1980) developed a biochemical model of photo-synthesis designed to simulate the net photosynthetic effects of pho-torespiration rate with changes in CO_2 or light. The model determines the carboxylation rate (V_c in Eq. 13.5), which is taken to be the minimum rate of at least two processes (one rate is limited by Rubisco (denoted W_c in Eq. 13.7, below); the other is limited by the rate of ribulose bisphos-phate regeneration in the Calvin–Benson cycle (denoted W_j in Eq. 13.7, below)). An addition to the Farquhar *et al.* (1980) representation was developed by Sharkey (1985) and Harley and Sharkey (1991), who noted that under conditions of high internal CO_2 concentration photosynthesis rate could be controlled by the triose phosphate utilisation and a corre-sponding carboxylation rate, W_p. The logic is that whichever of the three potential rates of carboxylation was the minimum, that rate would ulti-mately be the controlling factor, or:

$$V_c = \min\{W_c, W_j, W_p\} \tag{13.7}$$

where W_c is the Rubisco-limited rate of carboxylation; W_j is the ribulose bisphosphate-limited rate of carboxylation; and W_p is the triose phos-phate-limited rate of carboxylation.

Each of these potentially limiting rates is computed in the DOLY model, and the rate that is expected to be limiting (the minimum from Eq. 13.7) in a given leaf layer with the given environmental conditions is used.

When Rubisco limits the carboxylation rate:

$$W_c = \frac{V_c^{\max} P_c}{P_c + K_c \left(1 + \dfrac{P_o}{K_o}\right)} \tag{13.8}$$

where K_c and K_o are the Michaelis rates for carboxylation and the com-peting process of oxygenation by Rubisco (Farquhar *et al.* 1980; Harley *et al.* 1992). Both of these coefficients depend on temperature. The maximum rate of carboxylation by Rubisco, $V_c^{\max}$, is related to tempera-ture by:

$$V_c^{\max}(T) = V_c' k_V(T) \tag{13.9}$$

where V_c is an estimate of the carboxylation rate under maximum, light-saturated conditions (Friend (1991) and Reynolds *et al.* (1992) provide procedures to estimate V_c'.); $k_V(T)$ is a temperature response function from McMurtrie and Wang (1993).

If the rate of ribulose bisphosphate regeneration limits carboxylation:

$$W_j = \frac{JP_c}{4\left(P_c + \frac{P_o}{\tau}\right)} \tag{13.10}$$

where J is the light-saturated rate of electron transport:

$$W_j = \frac{\alpha I}{\left(1 + \frac{\alpha^2 I^2}{J_{max}^2}\right)^{0.5}} \tag{13.11}$$

in which I is the irradiance; α is the efficiency of light conversion (0.24 mol electron mol^{-1} photons) (Harley et al. 1992); and J_{max} is linearly dependent on V_c^{max}.

Based on data for 106 plant species derived by Wullschleger (1993):

$$J_{max} = 29.1 + 1.64\,V_c^{max}. \tag{13.12}$$

The maximum rate of electron transport is related to temperature. The value of V_c^{max} is assumed to be equal to that of V_c' (in Eq. 13.9) when the temperature is 25 °C, and it is further assumed that:

$$J_{max} = (29.1 + 1.64\,V_c')k_j(T) \tag{13.13}$$

In Eq. 13.13, the function that describes the temperature response is:

$$k_j(T) = 1 + 0.04(T-25) - 1.54 \times 10^{-3}(T-25)^2 \\ -9.42 \times 10^{-5}(T-25)^3 \tag{13.14}.$$

Finally, if triose phosphate utilisation limits photosynthesis, then the carboxylation rate is:

$$W_p = 3U + \frac{0.5\,W_{min}P_o}{\tau P_c} \tag{13.15}$$

where W_{min} is the minimum of W_c and W_j and $U = 5.79 \times 10^{-7} + 0.0569\,J_{max}$ (Wullschleger, 1993).

The equations for the carbon assimilation process in the DOLY model are drawn from functions intended to define the expected relations of the photosynthesis process with respect to CO_2 concentration, light levels and temperature. Several of the response functions in this section of the model are based on comparative studies across a large number of different plants (e.g. Reynolds et al. 1992; Wullschleger, 1993). The fundamental concept of identifying the critical limiting factors under

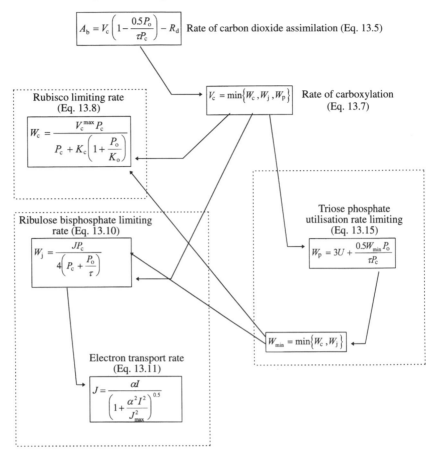

$$A_b = V_c\left(1 - \frac{0.5P_o}{\tau P_c}\right) - R_d \quad \text{Rate of carbon dioxide assimilation (Eq. 13.5)}$$

$$V_c = \min\{W_c, W_j, W_p\} \quad \text{Rate of carboxylation (Eq. 13.7)}$$

Rubisco limiting rate (Eq. 13.8)

$$W_c = \frac{V_c^{max} P_c}{P_c + K_c\left(1 + \frac{P_o}{K_o}\right)}$$

Ribulose bisphosphate limiting rate (Eq. 13.10)

$$W_j = \frac{J P_c}{4\left(P_c + \frac{P_o}{\tau}\right)}$$

Triose phosphate utilisation rate limiting (Eq. 13.15)

$$W_p = 3U + \frac{0.5 W_{min} P_o}{\tau P_c}$$

$$W_{min} = \min\{W_c, W_j\}$$

Electron transport rate (Eq. 13.11)

$$J = \frac{\alpha I}{\left(1 + \frac{\alpha^2 I^2}{J_{max}^2}\right)^{0.5}}$$

Figure 13.5. Interactions of the principal equations for photosynthesis rate limitation in the DOLY model (Woodward *et al.* 1995). Equation numbers correspond to notation in the text.

different conditions originates with Farquhar *et al.* (1980) and is shown in Fig. 13.5.

These photosynthesis calculations (Eqs. 13.5 to 13.15) are the cellular level computations in DOLY and are concerned with the biochemical processes in the model. At the leaf level, the model treats stomatal conductance and the supply of CO_2 to the inside of the leaf. The canopy section of the model deals with light extinction, respiration and the interaction between nitrogen content of leaves and the maximum assimilation of CO_2. The photosynthesis calculations are done for each leaf layer in the canopy so the different rate-limiting processes can act at different depths in the canopy. Plant-level responses are involved with the computation of evapotranspiration. At a landscape level, the model computes nitrogen uptake and the soil moisture dynamics. Throughout the model, the approach is to use basic mechanisms for different chemical and biophysical processes with response functions derived from more detailed studies. The DOLY model is unique in that along with computing a mechanistic hierarchical interaction among fundamental plant processes, it also computes structural features of the vegetation – notably leaf area. Woodward *et al.* (1995) provide the details of these other parts of the DOLY model.

The DOLY model has been tested with respect to its ability to reproduce maximum assimilation rates of CO_2 under a range of environmental conditions and for a diverse array of ecosystems ranging from tundra to arid vegetation to rain forests (from field observations collected by Raich *et al.* 1991 and McGuire *et al.* 1992, see Fig 10.17 and related discussion in Chapter 10). Other tests include:

1. Photosynthetic responses of cotton plants (from Harley *et al.* 1992) under different temperature conditions
2. Functions relating stomatal conductance to soil moisture tested on wheat (Gollan *et al.* 1986) and sycamore, *Acer pseudoplatanus* (Khalil and Grace 1993)
3. Simulation of canopy photosynthesis and evapotranspiration from the *Eucalyptus maculata* forests (collected by Wong and Dunin (1987)) and discussed in the context of estimating parameters for spatial models in the section of Chapter 12 relating to Figs. 12.9 and 12.10 (p. 399)
4. Leaf area predictions successfully tested for 35 globally distributed sites (Woodward 1987b).

When the models are applied to estimate global leaf area (Fig. 13.6*a*, colour plate) the resultant pattern matches the distribution expected from the pattern of biomes (Woodward 1987b; Prentice *et al.* 1993) and obtained by satellite remote sensing. Geographical patterns of annual net primary productivity (Fig. 13.6*b*) generally match those produced by material transfer models (Melillio *et al.* 1993) and remote-sensing-based canopy process models (Potter *et al.* 1993).

Continental-scale changes in terrestrial ecosystems: a performance comparison among homogeneous landscape models

VEMAP (1995), a consortium of ecosystem scientists, compared six different models (Table 13.3) for conditions of doubled CO_2 and several different climate change scenarios for the conterminous USA (i.e. the US

Table 13.3. *Models used in the VEMAP (1995) comparison of vegetation change in response to climatic changes and CO_2 increases*

Model (citation)	Type of model	Principal internal variables responding to climatic change	CO_2-response is obtained from model by
BIOME2 (Haxeltine and Prentice, VEMAP (1995))	Biogeographical	Biome type (from plant types) based on calibrations to climate variables Linear relation between GPP and absorbed photo-synthetically active radiation (Monteith 1972, 1981a,b). Ratio of actual to potential evapotranspiration and temperature also considered	Change in GPP Change in competitive differences in C_3 and C_4 plants
MAPPS (Neilson et al. 1992)	Biogeographical	Biome type (from plant types) from calibrations to climate variables Leaf area calculated from water balance considerations	Change in stomatal conductance alters water balance calculations
CENTURY (Parton et al. 1987, 1988)	Material transfer	Carbon, phosphorus, nitrogen and sulphur dynamics with internal transfers among compartments controlled by calibrations to climatic variables	Reduction in nitrogen content in vegetation Changes in actual evapotranspiration
TEM (Raich et al. 1991)	Material transfer	Calibration of rates of transfers of carbon and nitrogen among compart-ments to existing data with a strong emphasis on $C : N$ ratios	Modification of GPP Actual and potential evapotranspiration not changed by elevated CO_2
BIOME-BGC (Running and Coughlan 1988)	Canopy process	Biophysical responses used as a basis to simulate daily photosynthesis and evapotranspiration	Reduction in nitrogen content in vegetation Modification of biophysical model parameters (canopy conductance)
DOLY (Woodward et al. 1995)	Canopy process	Biophysical responses used as a basis to simulate daily photosynthesis and evapotranspiration Statistical calibrations produce biome types as an output option	Modification of biophysical model parameters (stomatal conductance)

Note:
GPP, gross primary productivity.

states sharing common boundaries). The models were subjected to an initial calibration to soils, climate and vegetation data sets and then subsequently used to simulate the responses of the vegetation (at equilibrium) for climatic changes associated with a doubled atmospheric CO_2 condition simulated by three different GCMs, to the direct effects of an increase in CO_2 concentration and to both the climate and CO_2 effects acting in concert. The orchestration of the amount of information needed to implement this rather diverse array of models, the differences and richness of the model outputs, the different assumptive structures in the models all made this initial exercise in large-scale model comparison a considerable challenge in logistics, as well as a scientific challenge.

The three climate-change scenarios were based on the GCMs developed by the Geophysical Fluid Dynamics Laboratory (GFDL R30 2.22°×3.75° grid, Manabe and Wetherald 1987; Wetherald and Manabe 1990), Oregon State University (OSU, Schlesinger and Zhao 1989) and the United Kingdom Meteorological Office (UKMO, Wilson and Mitchell 1987). Of these climate models, the OSU model produced the smallest increase in average annual temperature (+3.0 °C) and the smallest increase in annual precipitation (+4%). Comparable numbers for the GFDL R30 case were +4.3°C increase in average annual temperature and +21% increase in annual precipitation, and for the UKMO model they were +6.7°C increase in average annual temperature and +12% increase in annual precipitation.

The six ecological models vary significantly in their input requirements (Table 13.4). These differences reflect differences in model formulation and differences in the spatio-temporal resolution of functions in the models. Perhaps not unexpectedly given the differences in model formulation and resolution, the six models produced rather different results when subjected to large changes in the environment. For example, three of the models capable of simulating change in vegetation pattern (Fig. 13.7) produce, in one case a decrease in forest area under climatic warming scenarios (MAPPS model), in another case a relatively slight change in forest cover (DOLY model), and in a third case significant increase in forest area (the BIOME2 model). There is a similar variation in the net primary productivity and carbon storage predicted by other models (BIOME-BGC, CENTURY and TEM) (Fig. 13.8). Depending on the model and climate change scenario considered, the climate-change effects produce a range of net primary production changes from −6.5% to +17.0% of the baseline and the total carbon storage is changed from between −37.6 and +4.3%. The simultaneous effects of climate change and direct CO_2 effects ranged from +1.7% to +34.6% for net

Table 13.4. *Input requirements for models used in VEMAP (1995) simulation of ecosystem change in responses to climate and CO_2 changes. Required values are indicated by X except for climate variables where the models differ by requiring daily (D) or monthly (M) resolution data*

Input variable	BIOME2	DOLY	MAPPS	BIOME-BGC	CENTURY	TEM
Surface climate						
Air temperature						
Mean	M	D	M			M
Minimum		D[a]		D	M	
Maximum	D					
Precipitation	M	D	M	D	M	M
Humidity		D[b]	M[c]	D[d]		
Solar radiation	M[e]	D[f]		D[g]		M[h]
Wind speed		M	M			
Vegetation type				X	X	X
Soil						
Texture	X[i]	X[j]	X[k]		X[l]	X[m]
Depth		X		X	X	
Water-holding capacity		X		X		
Soil carbon and nitrogen	X					
Location						
Elevation				X	X	X
Latitude	X			X	X	X

Notes:

[a] Also uses the absolute daily minimum temperature for the length of the climate record.

[b] Average daytime relative humidity.

[c] Vapour pressure deficit.

[d] Average daytime relative humidity.

[e] Percentage possible sunshine hours.

[f] Daily mean irradiance.

[g] Total incident solar radiation.

[h] Percentage cloudiness.

[i] Categorical soil type.

[j] Percentage sand, silt and clay.

[k] Percentage rock, sand, silt, organic matter and clay.

[l] Percentage sand, silt and clay.

[m] Percentage sand, silt and clay.

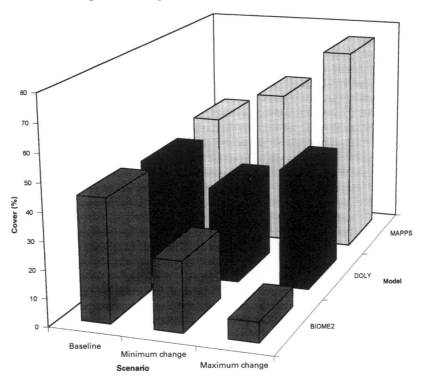

Figure 13.7. Responses from three different models simulating the change in potential forest cover of the conterminous USA. The baseline condition is the percentage of the area potentially covered by forests. The minimum change is the smallest deviation from this baseline condition under three different climate-change scenarios and the maximum change is the largest such deviation. Data from VEMAP (1995).

primary productivity and -32.7% to $+14.6\%$ for total carbon storage (again depending on model and scenario).

Three of the models do not simulate the changes in the structure of the vegetation in response to climate and require that vegetation types and other aspects of the vegetation be specified external to the simulation (Table 13.4). The three models that do predict variation in vegetation structure do so quite comparably with respect to the baseline (current climate) condition but depart from one another to a great degree under altered conditions (Fig. 13.7). These latter differences arise logically on consideration of the formulations for the dominant functions in the models (particularly with regard to the degree either thermal or moisture conditions are taken as the controlling factors on vegetation distribution). VEMAP (1995) also experimented by

(a)

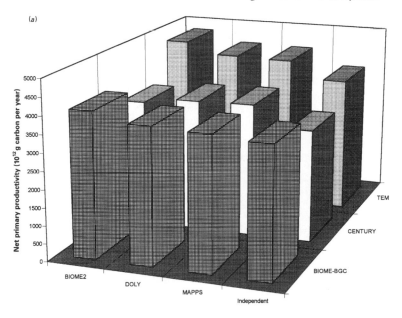

(b)

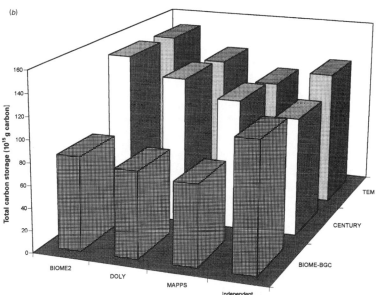

Figure 13.8. Changes in net primary productivity (*a*) and total carbon storage (*b*) for the conterminous USA using all combinations of three material transfer models (BIOME-BGC, CENTURY and TEM) coupled with three models capable of simulating biome changes in response to climatic change for the climate changes implied by the UKMO GCM and a doubling (to 710 p.p.m.) of atmospheric CO_2. 'Independent' indicates the response of the three material transfer models without the vegetation changes indicated by the three models simulating biome changes.

interacting the models to allow those models that could specify vegeta-tion pattern (BIOME2, DOLY and MAPPS) to feed modified condi-tions into the three models that did not have this capability (BIOME-BGC, CENTURY and TEM). These latter three models were then used to compute changes in net primary productivity and carbon storage under the combined effects of climate change and the direct effects of CO_2. Figure 13.8 shows these results for the UKMO climate-change scenario. The other cases are similar in pattern. These results indicate that there is relatively greater consistency in the model predictions from all possible combinations of models for the expected net primary productivity of the 48 conterminous states (Fig. 13.8a) than for the expected carbon pools (Fig.13.8b). It appears that the effects pro-duced by adding a capability to change ecosystem structure to homo-geneous landscape models emphasising the simulation of ecological processes has a significant effect on their performance. The variation in the performance of the models in all cases indicates a need for further model development and for the development of the critical test data sets to distinguish appropriate from inappropriate performance of these large-scale models.

Concluding comments

The concept that one can make any progress in predicting landscape ecosystem change by assuming landscapes to be homogeneous runs counter to the intuition of some ecologists. It is important to emphasise that system abstraction is a standard procedure in developing scientific theory (particularly theory with a strong mathematical basis). Demographic models, population genetics models, models of natural selection all feature underlying assumptions of spatial homogeneity. In the case of homogeneous landscape models, the model formulation typically involves positing general ecosystem theory and utilising assumptions regarding tendencies of ecosystems to seek equilibria or optima.

Homogeneous landscape models are, in most cases, point models that are limited in the amount of detail considered by the spatial detail of the environmental data they represent. Most material transfer models require an underlying biome map (along with large spatial data sets of climate and soils data) and are typically resolved over relatively large areas. Canopy process models have this same limitation unless they are tied to direct observations from satellites to represent critical features of the terrestrial surface (often APAR). The satellites typically used for these applications often have spatial resolutions of 1 km to 4 km on a side. The base data sets for the underlying climatology are usually interpolated from the coarse global web of meteorological stations to 0.5° latitude × 0.5° longitude blocks based on data sets of global terrain. This is a resolution of ~50 km on a side at the equator. Therefore, virtually any model that attempts to represent global pattern based on such data sets must map large areas of the surface as if they were homogeneous.

The models treated in this chapter have been applied successfully to reproduce important features of the Earth's vegetation as a response to the environment. Some of the models are calibrated to available global data sets but have mechanistic functions. Others are less calibrated and scale-up representations of underlying processes. Most of the models are primarily syntheses of existing knowledge about the functioning of leaves, plants and vegetation to larger scales. In this, they represent general, hierarchically structured theory on the way one of the major Earth systems works.

Comparisons of the performance of these models either at points or over continental areas are an important further step in the development of these models. Many of the comparisons to date have found that homogeneous landscape models often agree relatively well for test data under current conditions but depart in their predictions of cover, production or water-use when required to predict novel conditions (particularly involving altered CO_2 concentrations in the atmosphere). This is to be expected at this stage of model development. What is being identified is a need for further model development and, an almost uniform plaint from the model developers, more data and experiments on critical processes.

However, there are problems in the fundamental assumptions underlying homogeneous landscape models that will be difficult to unravel. One of the most important involves the allocation of carbon fixed by the photosynthesis process. The allocation of photosynthate is a whole-plant-level process that is highly dependent on the status of the individual plants involved. Therefore, a forest with a leaf area index of four could be a senescing forest with a broken canopy and considerable amounts of the photosynthate allocated to plant respiration. A younger forest with the same leaf area would have less tissue respiration demand and a surfeit of photosynthate to allocate to plant growth, to root growth, to subsidising mycorrhizae and obtaining nitrogen and phosphorus, to insect chemical defence compounds, etc. With only the state variables of leaf area, biomass and other aggregated variables as system state variables, homogeneous landscape models have to make assumptions about allocation that infer a constancy in allocation patterns from the underlying dynamic structure of the vegetation. Allocation as a function of vegetation structure is in the domain of individual-based models, and one would expect interactions among different models with differing underlying assumptions to continue as a proper procedure to understand the terrestrial surface.

14 · *Global change*

In their application to larger spatial-scale problems, the different land-scape models discussed in the previous four chapters have strengths and weaknesses that, to a degree, are complementary. Many of the models that simulate the landscape elements for mosaic landscapes (Chapter 11) have demonstrated a capability to predict changes in the structure of vegetation associated with environmental change but are limited in con-tinental- and global-scale applications by a lack of the species-specific or location-specific information that they require. Interactive landscape models (Chapter 12) have provided cautionary results that point to the potential importance of spatial interactions in altering the rates of pro-cesses to a significant degree. These spatially interactive models are data and information demanding to a degree that currently limits their applications to case studies. Models of ecosystem material flows (Chapter 13) scale-up to larger scales relatively easily (given the appropriate base data sets) but have little in their internal mechanisms to change the ecosystem structure. This prevents the material-flow models from exten-sive applications involving the feedbacks of structural changes onto the underlying processes. An obvious 'quick fix' to this problem is to calibrate the appropriate parameter changes to reflect modification of ecological processes in response to structural changes in ecosystems. However, this approach does not provide any dynamic interaction between ecosystem pattern and process.

Ecologists have a considerable amount to learn before the large-scale dynamics of the terrestrial surface under environmental change condi-tions can be predicted with confidence. Nevertheless, considerable progress has been made. The predictive state of the science has progressed greatly over a few decades, particularly since the mid-1980s. We are beginning to apply a variety of landscape models to the problem of under-standing the ways that natural ecosystems might respond to large-scale environmental change. Some examples of these findings will be the topic of this chapter (as well as earlier chapters) and are part of an ongoing research agenda. What are some of the larger features of this research programme? What are the principal challenges? One important issue is to understand what large-scale environmental change might do to the Earth's biota. What are the potential effects of environmental change on the biodiversity? A second issue concerns the way the terrestrial ecosystems may interact with the environment to ameliorate or amplify changes in either the global carbon budget or the atmosphere. A third regards the interactions of these natural systems with the state of humankind on our planet.

Effects of global environmental change on the Earth's terrestrial biota

Environmentally induced changes in the vegetation, such as those discussed in Chapter 11, represent an adverse condition for the diversity of terrestrial species. Both the climate and the biota of the terrestrial surface have had changes, in some cases of great magnitude, through geological history. Nevertheless, the potential reduction in the number of species on the terrestrial surface in response to environmental change and human alteration of the surface is significant and substantial.

One of the more obvious potential consequences of the shifts in the potential vegetation cover, such as those outlined in Chapter 11, is a pronounced alteration of the habitats of plants and animals. Identification of potential changes in ecoclimatic zones has been conducted for natural areas designated as major nature reserve systems (Leemans and Halpin 1992). This analysis provides information on the numbers of natural sites experiencing a change in established ecoclimatic zone for each protected area and identifies the direction of climatic change suggested under each scenario for these sites. The percentage of over 2600 nature reserve sites impacted under four GCM scenarios (see Table 11.1 (p. 345) for citations to these four scenarios). is listed in Table 14.1 (Leemans and Halpin 1992). This type of provisional analysis may be useful in providing initial information for the development of ecosystem monitoring and modelling networks at a global scale by describing expected directional changes at a large distribution of well-studied natural areas. Clearly, global changes in natural vegetation and in nature reserves implied by Fig. 11.1 (colour plate) and recorded in Table 14.1 could have a profound effect on the biotic diversity of terrestrial species.

Table 14.1. *Proportion of 2600+ nature reserves listed by the World Conservation Monitoring Centre and Man and the Biosphere (MAB) reserves (243 sites) that undergo a shift in ecoclimatic zone based on the shifts in aggregated Holdridge life zones presented in Fig. 11.1*

GCM	World Conservation Monitoring Centre nature reserves (% changed)	Man and the Biosphere reserves (% changed)
OSU	53.3	48.5
GISS	63.3	59.2
GFDL	66.4	62.9
UKMO	80.5	82.7

Source: From Leemans and Halpin (1992).

Effects of shifts in species ranges on biodiversity

Peters (1992) documents several types of change that might be expected as responses to climatic changes. For example, species distributions along altitudinal gradients might be expected to move up in altitude as the climate warms (Fig. 14.1). Species on the very tops of mountains have no options to move upward and would be subjected to higher likelihood of local extinction. Species from lower elevations would be restricted to smaller areas and into fragmented subpopulations as they were displaced up mountains (a topic that will be discussed in the section that follows). One might also expect the lowest altitudes to be colonised by species from locations further south (Fig. 14.1*b*).

On flatter terrain, one might expect reserve systems for a species to shift outside of its range, if the changes in the range distributions of the species are sufficiently great (Fig. 14.2). On the modern landscape, which in many instances is highly fragmented, species populations in locations with suitable habitat in the process of degrading as a result of environmental change might be unable to migrate across the urban and agricultural matrix in which their habitat occurred to new sites. Corridors of natural habitat between biodiversity reserves have been proposed as a potential solution to this problem, but whether such corridors are effective is an open question. It seems likely that active programmes of moving species from one location to another may also be a potential solution. Some species are more movable than others and such a programme could be expected to have differential success across a range of species.

One would expect the species that would be most able to survive and

(a)

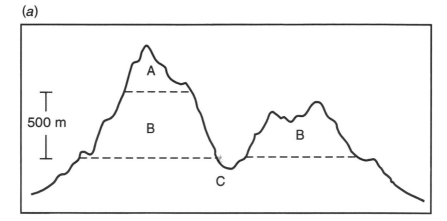

(b)

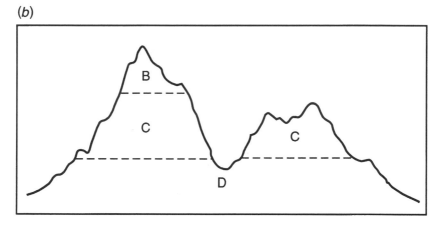

Figure 14.1. (*a*) Hypothetical present distributions of three species (A, B and C) distributed at different elevations along a mountain; (*b*) Same species distributions when displaced upward by a climatic warming. Species A becomes extinct. Species B is shifted up the mountain and the total habitat area for the species is reduced. Species C is fragmented into two smaller subpopulations. A new species D can colonise the lowest habitats. From Peters (1992).

perhaps prosper in a region with a changed environment and fragmented landscape to have such attributes as high dispersal rates, high reproductive success in novel situations, the ability to adapt quickly to new environmental regimes, and rapid growth and reproduction rates. These are attributes often associated with weeds and early successional plant species. The attributes of species that one currently associates with successful invasion would be the attributes that might ensure survival in a changing world. More sedentary species, highly adapted to a particular situation, would be likely losers in a changed terrestrial environment.

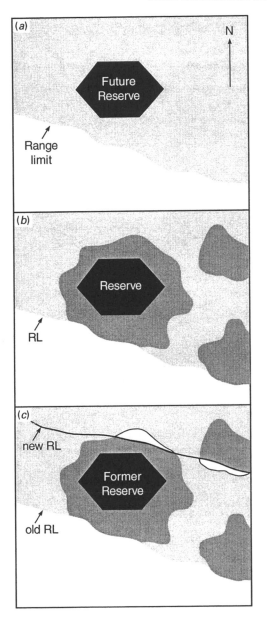

Figure 14.2. Climatic warming may cause species within biotic reserves to disappear as the ranges of the species cease to overlap the reserves. (*a*) Species distribution shown in crosshatched area before either human fragmentation or climatic change. (*b*) Habitat fragmentation decreases the species and a reserve is declared in a large region of suitable habitat. (*c*) Climate change displaces the species range from the reserve area. RL indicates the southern boundary of the species range. From Peters (1992).

An example of the application of community theory to biodiversity change

There has been a considerable body of theoretical work developed on the structure of plant and animal communities (Chapter 5). This work has been applied in several instances to attempt to understand how assemblages of interactive populations might respond to climatic change. For example, McDonald and Brown (1992) assessed the possible effects of climatic change on the diversity and composition of mammal faunas in the isolated mountain ranges in the Great Basin region of the American south-west. The Great Basin is an extensive, primarily desert region that lies between the Rocky and the Sierra Nevada Mountains. There are numerous mountain ranges in the basin inhabited by species that remain as relics from the Pleistocene Epoch when the conditions in the basin were moister and cooler. Presumably, at this earlier time, the now montane and isolated species distributions were more continuous across the Great Basin.

McDonald and Brown (1992) investigated the possible effects of a 3°C warming on the montane mammal fauna. The problem was one of the shifts in the vegetation types or mammal habitats up the mountains in response to a climatic warming.* McDonald and Brown reasoned that the area of each vegetation would shrink in extent in this process and computed these area changes from topographic maps. Indeed, some of the higher elevation types could be eliminated completely if the elevational range with which they were associated became higher than the mountains when adjusted for a warmer climate. By determining the change of different habitat areas, McDonald and Brown determined the expected number of species for the adjusted habitat area following a climatic change. The reduction in the number of species expected in each habitat was determined and the fauna was inspected to determine the species most likely to become extinct. The resultant predictions of change indicated a 9 to 62% loss of species for the 19 isolated mountain ranges studied. Of the 14 boreal mammals considered in the study, only two are predicted to survive on all the mountain ranges. Of the remaining 12 species, three are expected to become extinct and nine suffer reduction in their ranges.

Brown (1995) uses this evaluation as an example of what he calls a

---

* The potential for non-linear changes in vegetation pattern and the potential for re-arrangement in the composition and character of vegetation zones in mountains that have been associated with past climatic changes have been discussed in Chapter 2 (see Fig. 2.5, p. 19). Such changes would present a difficulty for this particular mode of analysis of change.

'macroscopic' approach to ecology and cautions that there are a number of critical assumptions involved in making such determinations. The approach is analogous to the others grouped in Chapter 13 as 'homogeneous landscape' approaches. In this case, the processes that relate habitat area with species richness (and manifested as a species versus area curve) are investigated for the manner in which they might work at a whole system level to produce a response to a climatic change.

Lessons from the past: change in the palaeoecological record

The palaeoecological record of changes in ecosystems associated with climatic change often produces a more complex set of potential responses than those just discussed. Patterns of species distributions have changed in surprising ways in response to climatic changes in the past. Species that are currently associated with one another often have migrated and colonised new areas in response to past changes in patterns that are independent of one another. This complexity in response may not necessarily be deleterious to some of the species inhabiting the changing habitats. Using the tundra biome as an example, Graham (1992) points out that the re-organisation of communities in the face of climatic change which has been seen in the past suggests that 'climatic warming may not necessarily push the tundra biome into the Arctic Ocean *en masse*, but this biome may be reorganised in way similar to, although not identical to, the tundra-like environments of the late Pleistocene'. (This response is seen when biogeographical correlations between biome patterns and climate data are used to interpret the potential magnitudes of global climatic warming (see Chapter 11).) If species respond to habitat structure rather than the species composition of the habitat, there are species that might survive quite well in habitats that are quite altered in their species composition but have the appropriate structure.

Protection of biotic diversity at the regional level in the face of habitat shifts will likely need to involve an understanding of species management, the knowledge of how to introduce species populations to new suitable locations and a greatly improved understanding of at least a marginal level of information on the habitat requirements and needs of a wider range of species. It was pointed out in Chapter 5 that in 1927 J. Grinnell, the developer of the term niche, recommended that descriptions of birds include the climatic limits of the ecological niche to provide a 'phylogenetically significant system for designating bird's ranges'. Such information, not just for birds but for a wide array of plant and animal species, is essential to manage species populations today and is a prerequisite to any future plans to attempt to manage species in the face of habitat change in response to future changes in the environment.

The changes in biotic diversity associated with a global environmental change should be moderated by the heterogeneity of the landscapes. One might expect the shifts of species ranges that might accompany a climatic change to leave behind relict populations at suitable sites in the landscape. This certainly appears to have been the case in the warming seen since the ending of the Pleistocene Epoch. Also, over the last 10 000 or more years, the palaeoecological record often features large, and often inexplicable, increases in some species and decreases in others. One might expect the same sorts of long-term response in the future.

The terrestrial surface and its interactions with the atmosphere

The relation of relatively simple models of the large-scale dynamics of the Earth's surface productivity and atmospheric composition (discussed in Chapter 13) also anticipate another application of canopy process models, namely representing the terrestrial surface in models of the general circulation of the atmosphere. There has been initial experimentation with adding a dynamic terrestrial surface to physically based models of the atmospheric circulation. While the models used in these exercises may lack the richness of biological and ecological detail desired by ecologists, the initial results portray important feedbacks between terrestrial surface and atmosphere and imply a more important terrestrial surface role in the ultimate functioning of the entire Earth system than was anticipated by many oceanographers and atmospheric scientists.

Models of land-surface interactions with the atmosphere: coupling terrestrial surface conditions to climate models

According to Dickinson (1992), the first inclusion of dynamic interaction between the land-surface and atmospheric processes in a GCM was in Manabe's (1969) inclusion of a simple 'bucket' hydrology model (Budyko 1974). In this conceptualisation, points on the surface of the Earth have a water-holding capacity (the 'bucket' size). The balance between precipitation and evapotranspiration determines the degree to which this capacity is filled and any excess precipitation beyond this capacity is considered to be runoff.

There have been two different efforts on the part of the climate-modelling community to understand the effect of changes in the surface on the climate. The first type of effort involves sensitivity analyses of GCM responses to changes in the parameters that are used to represent the surface. The second is focused on the development of models of the processes that must be included to better represent the land. Results from the sensitivity analyses motivate development and determine priorities in the modelling effort.

One of the earliest and most interesting sensitivity studies was developed by Dickinson and Henderson-Sellers (1988) who used a representation of the interaction between the terrestrial surface called BATS (Biosphere–Atmosphere Transfer Scheme, Dickinson et al. 1986) to investigate the effects of forest clearing in Amazonia on climate. (The first experiment was developed by Henderson-Sellers and Gornitz (1984) using a relatively coarse spatial resolution GCM. This model experiment used a two-layered representation of the hydrology but did not include a plant canopy.) The model experiment was produced by converting the land cover over a large area of Amazonia from a surface with parameters for roughness, albedo and moisture-transfer properties appropriate to forest to a surface with parameters appropriate for grassland. They found that the effect of removing forest cover was to increase the temperature by $3°C$ in the deforested region and pointed to the need to characterise better the terrestrial surface in terms of its influence on the atmosphere. Subsequently, several other studies have been made of the problem of evaluating the consequences to climate of Amazonian clearing. Henderson-Sellers et al. (1993) have subsequently developed additional model experiments based on their initial work and reviewed other efforts (Table 14.2).

Multiple comparisons of the effects of deforestation over a large area of the Amazon Basin tend to agree in predicting a response with hotter and drier conditions (Table 14.2). Temperature increases from the various models range from near zero to as much as $3°C$. Water supplied from evaporation is reduced by as much as 500 mm, and precipitation is reduced by as much as 640 mm. Using virtually any of the biogeographical techniques used in climate-change effects assessment (see Chapter 13), these would be seen as climatic changes of sufficient significance to generate a long-term shift in the expected vegetation. Given that tropical forest clearing in Amazonia (as elsewhere) is proceeding at a substantial rate (Malingreau and Tucker 1988), these possible feedbacks between clearing of forests and climatic conditions are a source of concern.

Table 14.2. *Simulated responses of Amazonia to rain forest clearing and conversion based on five different GCMs of the atmosphere*

	GISS[a]	CCM0B[b]	UKMO[c]	NMC[d]	CCM1-Oz[e]
Climate model attributes					
Resolution (° latitude/ ° longitude)	8×10	4.5×7.5	2.5×2.75	1.8×2.8	4.5×7.5
Terrestrial surface model	2-layered hydrology model	BATS[f]	Canopy Interception Model	SiB[g]	BATS[f]
Clouds represented by	55% cloud cover	30% cloud cover	Not given	Fixed means by latitudinal zones	Cloud cover from Slingo (1989)
Oceans represented by	Fixed heat transport	Fixed sea surface temperatures	Fixed sea surface temperatures	December conditions	Mixed layer ocean model
Length of control simulation (years)	20	3	3	1	6
Length of deforested simulation (years)	10	1	3	1	6
Climate change results					
Temperature change (°C)	0	+3	+2.4	+2	+0.6
Precipitation change (mm)	−220	0	−490	–640	−588
Evaporation change (mm)	−164	−200	−310	–500	−232

Notes:
[a] Henderson-Sellers and Gornitz (1984).
[b] Dickinson and Henderson-Sellers (1988).
[c] Lean and Warrilow (1989).
[d] Shukla *et al.* (1990), Nobre *et al.* (1991).
[e] Henderson–Sellers *et al.* (1993).
[f] Dickinson *et al.* (1986).
[g] Sellers (1985, 1987).
Source: From Henderson-Sellers *et al.* (1993).

Just as model experiments in Amazonia have indicated significant feedbacks between the vegetation and climate, vegetation alteration in other regions has also been shown to have potentially disruptive effects on climate. The conversion of sparsely vegetated semiarid shrublands to desert represents a substantial change in the features of the terrestrial surface vis-à-vis the atmosphere. Xue and Shukla (1993) used the SiB model of Sellers (1985, 1987; also Sellers *et al.* 1986) to investigate the effects of drought in desertification for a sub-Saharan zone, the Sahel. They conducted several model experiments that resulted in reduced rainfall to the Sahel when it was converted to desert. The onset of the rainy season was delayed by one month and the zone of maximum rainfall in the Sahel was shifted toward the equator. The overall implication of the results was to shift the Sahel toward more desert-like conditions. Schlesinger *et al.* (1990) noted that internal ecosystem feedbacks also had a tendency to reinforce the desertification process.

Bonan *et al.* (1992) investigated the effects of deforestation of the boreal forest worldwide on global climate using the National Center for Atmospheric Research climate model CCM1 (Williamson *et al.* 1987). In this simulation, the soil properties were based on the BATS model and the vegetation properties on the SiB model. The model was run under two conditions: one with a dynamically changing sea-surface temperature and ice-sheet formation, the other with these features prescribed. The removal of forest increased the albedo of the land surface and resulted in colder air temperatures with respect to the control conditions. The resultant cooling was substantial and affected the temperatures of the entire northern hemisphere (Fig. 14.3). The effects were more pronounced in the case in which the atmosphere was more interactive with the ocean. Bonan *et al.* (1992) summarised these results as, '. . . the summer cooling caused by deforestation is sufficient to prevent forest regrowth in much of the deforested area. Thus, boreal deforestation may initiate a long-term irreversible feedback in which the forest does not recover and the treeline moves progressively farther south'.

The broad implications of results from investigations of forest clearing in tropical and boreal regions, and desertification in the subtropical zones, are similar in pointing to a re-inforcing feedback between the extant vegetation and the climate conditions. These initial conclusions match more anecdotal and smaller-scale studies between climatic conditions and vegetation (Table 14.3). Historical figures, such as Columbus and Noah Webster, observed that changes in the land cover seemed manifest in changes in the climate. Clearly, we need to understand the atmospheric–

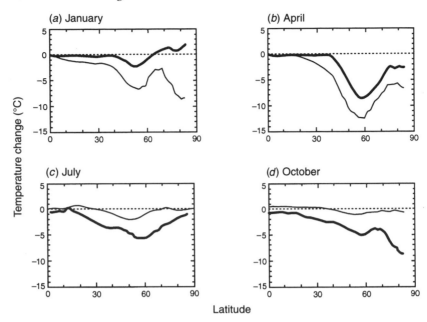

Figure 14.3. Average air temperature of experimental departures from control conditions at a height of 2 m above the land surface averaged by latitudinal zone for the northern hemisphere. Thicker lines are simulations with an interactive sea surface temperature and sea ice dynamics. Thinner lines are for simulations with prescribed sea surface temperatures and sea ice. From Bonan *et al.* (1992).

terrestrial surface interactions much better, given the implications of model results to date.

Anthropogenic changes in the global carbon budget

Given the importance of CO_2 as a component of the atmosphere and its role as a greenhouse gas, along with its independent importance as a major constituent of the photosynthesis process, it may be surprising that there are major aspects of the global budget of CO_2 that are not known. Pre-eminent among the missing information regarding the global carbon budget is our inability to account completely for the fate of all of the carbon produced by the burning of fossil fuels. To appreciate the implications of this situation, one must first understand the broad features of the global carbon budget.

The Earth can be thought of as having three active carbon reservoirs (Post *et al.* 1990):

Table 14.3. *Anecdotal observations of effects of vegetation on the local climate and comparable results from model or observational studies*

Historical observation		Recent measurement or model result	
Response	Citation	Response	Citation
Forests cause the West Indies to have more rainfall than the deforested Azores and Canary Islands (Columbus)	*vide* Keen (1959)	Forested islands in the West Indies receive three times the rainfall compared with deforested islands	Anthes (1984)
Winters became more variable after Europeans deforested North America (Noah Webster 1799 (in Kittredge, 1948); also Becquerel 1853)	Noah Webster, *vide* Kittredge (1948)	Significant amounts of water transpired retard night-time cooling by reducing outgoing terrestrial radiation and by elevating the dewpoint temperature	Schwartz and Karl (1990); Hayden (1994)
'How do they [forests] modify the temperature of the country?'	Becquerel (1853)	Simulations of global conditions with GCMs show increased temperatures with evapotranspiration 'turned-off'	Shukla and Mintz (1982); Fennessy *et al.* (1994)
Discussion of water-related effects of vegetation on climate	Becquerel (1853)	Biogenic hydrocarbons from plants alter the emissivity of the atmosphere and increase low night-time temperatures, particularly in areas with dry air	Hayden (1994) for comparison of several deserts

Source: Information for the table is abstracted from Hayden (1994) which provides more detail.

1. The *atmosphere*, containing 748 Pg carbon mostly as gas CO_2
2. The *terrestrial surface*, containing >2000 Pg carbon of which 430–830 Pg is estimated to be in the living vegetation and 1200–1600 Pg is in the soils. The variation arises from different methods used to characterise the terrestrial carbon budget (see Post *et al.* 1990; also Chapter 11).

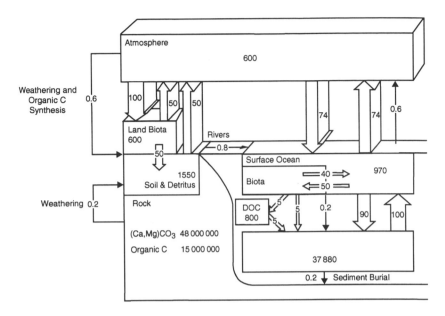

Figure 14.4. The global carbon budget for the pre-Industrial Revolution condition. Diagram is based on the IPCC (Watson *et al.* 1990) calculations with modifications by Sarmiento and Siegenthaler (1992). The numbers inside the boxes are in Pg of carbon and the arrows are fluxes of carbon in Pg per year.

3. The *oceans*, containing 38 000 Pg carbon (37 000 Pg dissolved inorganic carbon consisting of dissolved CO_2 and the ions, HCO_3^- and CO_3^{2-}, 1000 Pg dissolved organic carbon, and 30 Pg particulate organic carbon).

By comparison, the reservoir of recoverable fossil fuels is thought to be of the order of 4000–5000 Pg carbon.

Along with the difficulty in measuring the carbon storage in the heterogeneous terrestrial surface (as evidenced in the variability of the estimates, above), the global carbon budget presents a second measurement challenge. The magnitudes of the annual fluxes of materials among these major components of the global carbon cycle are large. Roughly, 100 Pg moves from the atmosphere to the ocean and another 100 Pg fluxes from the atmosphere into the terrestrial surface. Each of these carbon fluxes from the atmosphere is balanced by return fluxes to the atmosphere of similar magnitude ($\approx$100, each). Figure 14.4 illustrates the global carbon budget and associated fluxes of material based on an estimate of global carbon budget before the industrial revolution and the

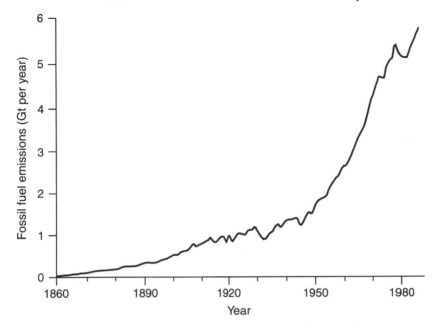

Figure 14.5. Releases of carbon to the atmosphere from fossil fuel burning, cement production and flaring of natural gas since the Industrial Revolution. Events such as the Great Depression and world wars appear to produce short drops in the rate of release. In general, the rate of increase has been about 4.3% annually. In the past few decades, the 1973 oil embargo and the sharp oil price increases in 1980 have produced declines in fossil fuel use. From Marland *et al.* (1989).

associated anthropogenic release of CO_2 into the atmosphere from the burning of fossil fuels.

The human contribution of CO_2 to the atmosphere can be estimated from a variety of historical and economic sources. Since the 1860s and the attendant industrialisation until the present, the release of carbon to the atmosphere from fossil fuel combustion has increased regularly (with occasional declines associated with historical events such as the Great Depression and World Wars I and II) at a rate of about 4.3% per year. In 1989, the release was 5.9 Gt per year (Fig. 14.5). There is a large body of evidence that this input of carbon to the atmosphere is responsible for the systematic increase in the level of CO_2 in the atmosphere record since the late 1950s at Mauna Lao Observatory in Hawaii and elsewhere (see Fig. 2.10 (p. 25) and attendant discussion in Chapter 2). Indeed, the roughly 27% increase in the amount of CO_2 in the atmosphere (Fig. 14.6) that has occurred since the Industrial Revolution is more than accounted for by

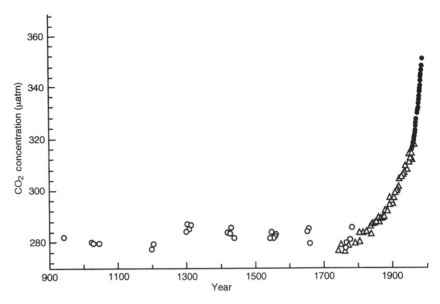

Figure 14.6. Atmospheric CO_2 over the last millennium as measured at Mauna Loa (●) by C. D. Keeling (Boden *et al.* 1991), in bubbles in glacial ice at Siple Station (▲) by Neftel *et al.* (1985) and Friedli *et al.* (1986, see also Fig. 2.11 p. 26), and at the South Pole (○) by Siegenthaler *et al.* (1988). From Sarmiento and Bender (1994).

the release of fossil fuels. The CO_2 that can be accounted for by increases in the atmospheric concentration of CO_2 is called 'the air-borne fraction'. Between 1959 and 1979, this fraction was 55.9% (Keeling *et al.* 1995).

The CO_2 released by human activities and not remaining in the atmosphere must go to either the oceans or the terrestrial ecosystems. For the oceans, this uptake is strongly controlled by physical processes involving diffusion and mixing of ocean waters, chemical processes involving carbon chemistry and biological processes of production and decomposition. For the terrestrial surface, the uptake of CO_2 involves a spectrum of biological processes with differing time responses, a variety of natural disturbances (notably fires) and effects from human land-use. The heterogeneity of the land surface poses a challenge to the measuring of carbon pools and fluxes.

In the oceans, CO_2 is exchanged rapidly between the atmosphere and surface waters (through mixing by the winds) and the atmosphere and surface ocean are in approximate equilibrium with respect to the partial pressures of CO_2 in each. This effectively means that for the ocean to take up additional net amount of CO_2 there must be some processes that transfer carbon to the deeper ocean waters (Post et al. 1990). This latter transfer is difficult to measure, but since the 1960s and 1970s, tritium (^3H) and ^{14}C produced by the testing of nuclear weapons have been used as tracers for understanding ocean mixing. Compartment models (called a 'box-diffusion' model, see Chapter 4) with internal transfers that match the movement of tritium (a tracer for water movement) and ^{14}C (a tracer for transport of carbon) suggest that the carbon uptake by the oceans was between 23 and 30 Pg between the years 1958 and 1980 (26% to 34% of the fossil fuel carbon put into the atmosphere in that period).

If the amount of carbon produced by burning fossil fuels and remaining in the atmosphere (the air-borne fraction) is ~60% and the amount in the oceans is $27 \pm 3\%$, then some increase in the degree of carbon storage on the land surface is implied as a consequence of our understanding of the global carbon cycle. However, many terrestrial ecologists are inclined to think of the terrestrial surface as a net source of carbon – not as a sink for carbon. Several studies in the 1980s estimated the clearing of land in specific regions to imply transfers of 0.4 to as much as 2.6 Pg carbon annually (Post et al. 1990). Houghton et al. (1983) and Houghton (1989) have drawn together a large range of sources to produce a historical accounting of the terrestrial surface as a source of carbon to the atmosphere.

If one uses the estimates of carbon produced by burning fossil fuel, the uptake of the oceans and the air-borne fraction and then computes the global carbon budget of the Earth starting at the time of the Industrial Revolution (Fig. 14.4) to the present, the result is that there is a substantial amount of carbon that is unaccounted for by our current understanding of the dynamic functioning of the Earth (Fig. 14.5). The size of this missing sink is currently of the order of 2 Pg carbon per year. Given the magnitudes of the fluxes of carbon among the earth–ocean–atmosphere components (shown for a pre-industrial condition in Fig. 14.4), this missing sink is relatively small and, therefore, is difficult to measure directly (Fig. 14.7). Computations from ocean models imply that this sink for carbon must be on the terrestrial surface. Computations by Keeling et al. (1989) and Tans et al. (1990) using the subannual variation in the concentration of CO_2 with latitude (see Fig. 13.4b, p. 430) strongly imply that this missing sink for carbon is in the Northern Hemisphere temperate latitudes.

Work to improve our understanding of the global carbon cycle and the

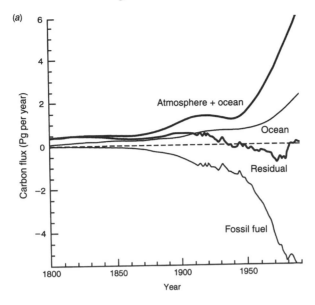

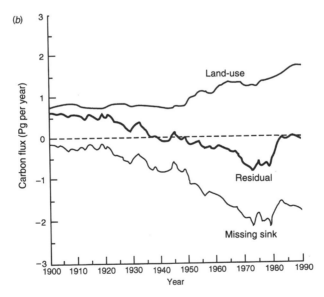

Figure 14.7. Fluxes of carbon from earth, ocean and atmospheric systems. (*a*)
Time history of the anthropogenic CO_2 fluxes. Ocean uptake is computed by an
ocean model by Sarmiento *et al.* (1992), fossil fuel production from Marland *et al.*
1989 (see Fig. 14.5), atmospheric concentrations from Fig. 14.6. The residual curve
is the carbon unaccounted for in the balance of these data. (*b*) When the residual
term from (*a*) is combined with estimates of carbon sources from land clearing and
land-use (Houghton 1992), one obtains an estimate of a 'missing sink' for carbon
of the order of 2 Pg per year. From Sarmiento and Bender (1994).

magnitudes of the fluxes of carbon is ongoing. There is evidence that some of the processes that influence the carbon budget can change relatively abruptly and in unpredictable ways (Keeling *et al.* 1995). Analyses of variations in the stable ^{13}C isotope of carbon in the atmosphere and oceans (Broecker and Peng 1993; Tans *et al.* 1993) as well as small variations in the oxygen:nitrogen ratio in air (Keeling and Shertz 1992) are providing valuable insights as to the sizes of some of the major fluxes associated with the global carbon cycle. Nevertheless, we have much to learn about the global carbon budget of the Earth, and even more to learn about the dynamics of the coupled earth–ocean–atmosphere system.

Human society's adaptability to global change

Understanding how the biosphere functions as a mechanism; knowing whether the feedbacks among ocean, atmosphere and land are mutually stabilising; predicting how the terrestrial system might change in response to changes in other Earth systems are all challenges for this generation of scientists. The results we have to date show how much more there is to be learned. We do have very good evidence that technological human society has modified components of the earth, air and ocean systems, and our lack of knowledge in the face of ongoing environmental modification sounds an alarm. So does our relative lack of ability to predict the effects of environmental change on human society itself.

The effects of large-scale environmental changes on past civilisations appear in many cases to have been dire. Bryson and Murray (1979) in an early review of climate impacts on society note, for example, declines in Mycenaean society in Greece and the Mill Creek society in what is now Iowa, both in 1200 BC, and both owing to drought. They also cite times of famine in Bergthørsson's (1969) reconstruction of Icelandic temperature (based largely on the long record of drift ice and historical observations). The list of past civilisations where collapses are totally or partially attributable to environmental change is long and, as historical reconstructions, subject to considerable debate.

Of course, all environmental changes are not necessarily bad. Sage (1995) speculates that the independent and more or less contemporaneous development of agriculture in several different locations about 10 000 years ago (Cohen 1977; Blumler and Byrne 1991) is attributable to a CO_2 increase from below 200 μmol mol^{-1} to about 270 μmol mol^{-1} in the atmosphere between 15 000 and 12 000 years BP. The 200 μmol mol^{-1} atmospheric concentration before this time may have been too low to support enough productivity to establish agriculture. Sage points to the potential productivity increase of 25% to 50% in C_3 plants (observed under controlled conditions) associated with a 200 to 270 μmol mol^{-1} increase in CO_2. Other effects of increasing CO_2 in this range would be a pronounced increased in the performance of C_3 crops over C_4 weeds, increased levels of biological nitrogen fixation and increased plant capacity to obtain nutrients and water.

The difficulty in assessing the direct effect of CO_2 on plant performance is most apparent in natural systems in which the factors limiting plant growth may not be CO_2 levels. Rangeland managers, foresters and some third-world farmers may not notice appreciable increases in plant growth from increased CO_2 because nutrient supply from the rate of litter decomposition cannot match the potential increases in photosynthesis rate. Further, one of the potential benefits from increased CO_2, an increase in the per unit area leaf efficiency in using water, may be offset by plants growing larger leaf areas. In agricultural systems, particularly for high-intensity, high-technology agriculture with optimal spacing, augmentation of nutrients and irrigation, one might expect the responses of the agricultural systems to be the greatest to the direct effects of CO_2 (Strain 1992). However, these responses are not even across all crops (Table 14.4). Further, the positive direct effects of increased CO_2 are also positive effects for some weeds (Parry 1992). When the direct and indirect effects of increased CO_2 are considered, the agricultural response is even more complex for understanding the effects of crop pests and pathogens (Parry 1992; Cammell and Knight 1992).

What does the current assessment of the consequences of global change portend? Will the disruptions of climate change on vegetation and soil systems, and on the agricultural systems upon which we depend, be as disruptive on our society as climate fluctuations have been on other cultures in the past? Will the increased presence of CO_2 in the atmosphere provide a boon for crop systems and agricultural surplus? Which regions or countries will be winners or losers in a changing world? Probably the most straightforward answer is that we simply do not know. We surely do not know with any certainty.

Indeed, the greatest difficulty in dealing with issues in the area of global change is the challenge of the potentially large effects of global change on human society combined with the high level of uncertainty as to what will happen. This is conpounded by the time scale of the potential changes and the time scales at which economic and political systems work. Gore (1992) summarises his book on this topic with:

Table 14.4. *Average predicted growth and yield for different C$_3$ photosynthesis pathway plants for a doubling of CO$_2$ concentration from 330 to 660 p.p.m.*

Crop	Immature crops		Mature crops	
	Number of records	Increase in biomass (%)	Number of records	Increase in marketable yield (%)
Fibre crops[a]	5	124	2	104
Fruit crops[b]	15	40	12	21
Grain crops[c]	6	20	15	36
Leaf crops[d]	5	37	9	19
Pulses[e]	18	43	13	17
Root crops[f]	10	49	–	–
C$_3$ weeds[g]	10	34	–	–
Trees[h]	14	26	–	–
Average (total)	(83)	40 ± 7[i]	(51)	26 ± 9

Notes:

[a] Cotton (*Gossypium hirsutum*).

[b] Cucumber (*Cumumis sativus*); eggplant (*Solanum melongena*); okra (*Abelmoschus esculentus*); pepper (*Capsicum annum*); tomato (*Lycopersicum esculentum*).

[c] Barley (*Hordeum vulgare*); rice (*Orza sativa*); sunflower (*Helianthus annuus*); wheat (*Triticum aestivum*).

[d] Cabbage (*Brassica oleoracea*); white clover (*Trifolium repens*); fescue (*Festuca elatior*); lettuce (*Lactuca sativa*); Swiss chard (*Beta vulgaris*).

[e] Bean (*Phaseolus vulgaris*); pea (*Pisum sativum*); soybean (*Glycine max*).

[f] Sugar beet (*Beta vulgaris*); radish (*Raphanus sativus*).

[g] *Crotalaria spectabilis*; *Desmodium paniculatum*; Jimson weed (*Datura stramonium*); pigweed (*Amaranthus retroflexus*); ragweed (*Ambrosia artemisiifolia*); sticklepod (*Cassia obstusifolia*); velvet leaf (*Abutilon theophasti*).

[h] Cotton (*Gossypium deltoides*).

[i] Mean ± 95% confidence limits.

Source: From Parry (1992) and Warrick *et al.* (1986).

> For civilization as a whole, the faith that is so essential to restore the balance now missing in our relationship to the earth is the faith that we do have a future. We can believe in that future and work to achieve it and preserve it, or we can whirl blindly on, behaving as if one day there will be no children to inherit our legacy.

Scientists studying global change hopefully will continue to give us an ability increasingly to understand what we are doing to our planet and how the biosphere functions. This information will be a necessary element of an understanding of the consequences of our actions. We may through this gain an insight as to the acceptable human population density and level of use of the world's resources. It will remain to develop the mutual will to act upon this increased knowledge.

Concluding comments

The capabilities for predicting the responses of terrestrial ecosystems to global change and to the 'natural fluctuations' in the Earth's climate have developed to a remarkable degree since the mid-1980s. Our ability to predict the consequences of change derives from an array of models, each with different underlying assumptions and with different data requirements. There are three very significant problems that present themselves as scientific challenges for the terrestrial ecologist attempting to understand the state of the biosphere following a large-scale environmental change: the effect of change on the biota, on the life support system of the Earth and on human society.

Biodiversity
The consequences of global environmental change on the diversity of the terrestrial biota are potentially dire. As in the palaeontological record, one would expect environmental change to affect some species positively and others negatively. Because of the fragmented nature now of much of the terrestrial landscape (and the rather desperate situation for certain endangered species, species on islands and populations isolated on mountain tops), a higher level of extinction of species will occur than might be expected from comparable past climatic changes operating on a flora and fauna inhabiting intact and continuous natural landscapes. If we are to preserve the biotic diversity of the terrestrial surface and simultaneously preserve a valuable resource of genetic, aesthetic and commercial importance, we must aggressively engage ourselves in collecting the necessary information to do so. Ecology is rich with quantitative methods for assessing and managing biodiversity that have been used with success in assessing habitat change and in predicting patterns of species response to such changes. Chapter 5 provides examples of some of these and references to scores of others. What is now needed is information on the requirements and ecological tolerances of a wide array of plant and animal species. With this, we have a reasonable start on what is needed to make enlightened decisions about the management of the Earth's diversity. Importantly, the information that would allow us better to preserve biodiversity under conditions of climatic or global change is the same information that we

already need to conserve biodiversity in the face of widespread human alteration of the Earth.

Climate–vegetation feedbacks

In this chapter, a brief overview of the potential interactions between atmospheric systems and the terrestrial surface has been provided. In synopsis, examples have been given at the local, regional, continental and global levels of potentially significant feedbacks between the terrestrial vegetated surface and the atmosphere. In many cases these feedbacks seem self-reinforcing and stabilising: the presence of deserts may reinforce the hot arid conditions that are associated with deserts or desertification; the presence of rain forests interacting with the atmosphere may produce precipitation regimes that favour rain forests. Perhaps this is to be expected. In a changing environment, one would expect terrestrial-surface–atmosphere feedbacks that were destabilising to have been excited and, therefore, dynamically eliminated by the recent environmental fluctuations (see Chapter 2). The feedbacks between surface and atmosphere raise the question of reversibility of changes on the terrestrial surface induced by human land-use and land conversion. We are accustomed to think of land as a renewable resource with respect to human use. If, for example, overgrazed dry grasslands turn into deserts but do not revert to grasslands when grazing is reduced, then the abuse of such land systems becomes not land conversion but land consumption. Proper stewardship of the land requires a more fundamental understanding of the multi-scale interactions between the atmosphere and terrestrial ecosystems.

The global carbon cycle

The global carbon cycle involves not only the feedbacks between the atmosphere and the terrestrial surface at a global scale but also includes the additional complexity of interactions with the oceans in determining the amount of CO_2 in the atmosphere. Sciences sometimes develop compelling, central problems that can be a focus of intellectual, observational and experimental endeavour and the understanding of the global carbon cycle is such a problem. It involves processes at the biochemical level (photosynthesis, respiration) and at the physiological level (stomatal function, allocation of photosynthate in plants of different forms in different environmental conditions) right through to understanding continental dynamics of vegetation and the circulation of the atmosphere and oceans. It is a rich and non-trivial area in which to advance knowledge. The elucidation of the global carbon budget may be one of the richest problems to be presented to ecologists. It is also an extremely central applied problem in that it implies the level of energy utilisation for development that can occur on our planet. For the importance of the problem area, the understanding of the heterogeneous terrestrial ecosystems and their role in the global carbon cycle is in its infancy. Our current difficulty in knowing the fate of about one-half of the carbon produced by human consumption of fossil fuels and released into the atmosphere is a symptom of how much work remains to be done.

References

Aber, J. D. and C. A. Federer. 1992. A generalized, lumped parameter model of photosynthesis, evapotranspiration and net primary production in temperate and boreal forest ecosystems. *Oecologia* 92: 463–474.

Aber, J. D. and J. M. Melillo. 1982a. *FORTNITE: A Computer Model of Organic Matter and Nitrogen Dynamics in Forest Ecosystems.* University of Wisconsin Research Bulletin, No. R3130.

Aber, J. D. and J. M. Melillo. 1982b. Nitrogen immobilization in decaying hardwood leaf litter as a function of initial nitrogen and lignin content. *Canadian Journal of Botany* 58: 416–421.

Aber, J. D., D. B. Botkin and J.M. Melillo. 1978. Predicting the effects of differing harvest regimes on forest floor dynamics in northern hardwoods. *Canadian Journal of Forest Research* 8: 306–315.

Acevedo, M. F. 1980. Tropical rain forest dynamics: a simple mathematical model (pp. 219–227). In: J.I. Furtado (ed.). *Tropical Ecology and Development.* International Society for Tropical Ecology.

Acevedo, M. F. 1981. On Horn's Markovian model of forest dynamics with particular reference to tropical forests. *Theoretical Population Biology* 19: 230–250.

Acevedo, M. F., D. L. Urban and H. H. Shugart. 1995. Models of forest dynamics based on roles of tree species. *Ecological Modelling* 87: 267–284.

Adams, S. M. and D. L. DeAngelis. 1987. Indirect effects of early bass–shad interactions on predator population structure and food web dynamics (pp. 103–117). In: W. C. Kerfoot and A. Sih (eds.). *Predation in Aquatic Ecosystems.* University Press of New England, Hanover, NH.

Agenbroad, L. D. 1984. New World mammoth distribution (pp. 90–108). In: P.S. Martin and R.G. Klein (eds.). *Quaternary Extinctions: A Prehistoric Revolution.* University of Arizona Press, Tucson, AZ.

Ågren, G. I. and E. Bosatta. 1991. Dynamics of carbon and nitrogen in the organic matter of the soil: a generic theory. *American Naturalist* 138: 227–245.

Aikman, D. P. and A. R. Watkinson. 1980. A model for growth and self thinning in even-aged monocultures. *Annals in Botany (London)* 45: 419–427.

Alard, P. G. 1974. Development of an empirical competition model for individual trees within a stand (pp. 22–37). In: J. Fries (ed.). *Growth Models for Tree and Stand Simulation.* Royal College of Forestry, Stockholm, Sweden.

Allee, W. C. 1934. Concerning the organization of marine coastal communities. *Ecological Monographs* 5: 541–554.

Allen, T. F. H. and T. W. Hoekstra. 1992. *Toward a Unified Ecology.* Columbia University Press, New York.

Allen, T. F. H and T. B. Starr. 1982. *Hierarchy: Perspectives for Ecological Complexity.* University of Chicago Press, Chicago.

Allison, T. D., R. E. Moeller and M. B. Davis. 1986. Pollen in laminate sediments provides evidence for a mid-Holocene forest pathogen outbreak. *Ecology* 67: 1101–1105.

Alvarez, W. W., F. Asaro and H. V. Michel. 1980. Extraterrestrial cause for the Cretaceous–Tertiary extinction. *Science* 208: 1095–1108.

Anderson, B. W., R. D. Ohmart and J. Rice. 1983. Avian and vegetation community structure and their seasonal relationships in the lower Colorado River valley. *Condor* 85: 392–405.

Anderson, E. 1984. Who's who in the Pleistocene: a mammalian beastiary (pp. 5–39). In: P.S. Martin and R.G. Klein (eds.). *Quaternary Extinctions: A Prehistoric Revolution.* University of Arizona Press, Tucson, AZ.

Anderson, M. C. 1964. Light relations of terrestrial plant communities and their measurement. *Biological Review* 39: 425–486.

Anderson, S. H. and H. H. Shugart. 1974. Habitat selection of breeding birds in an East Tennessee deciduous forest. *Ecology* 55: 828–837.

Anthes, R.A. 1984. Enhancement of convective precipitation by mesoscale variations in vegetative covering in semiarid regions. *Journal of Climate and Applied Meteorology* 23: 540–553.

Anway, J. C., E. G. Brittain, H. W. Hunt, G.S. Innis, W. J. Parton, C.F. Rodell and R. H. Sauer. 1972. *ELM: Version 1.0. U.S.I.B.P Grassland Biome Technical Report No. 156.* Colorado State University, Fort Collins, CO.

Armstrong, J. T. 1965. Breeding home range in the nighthawk and other birds: its evolutionary and ecological significance. *Ecology* 46: 619–626.

Arney, J. D. 1974. An individual tree model for stand simulation in Douglas-fir (pp. 38–46). In: J. Fries (ed.). *Growth Models for Tree and Stand Simulation. Research Notes 30.* Department of Forest Yield Research, Royal College of Forestry, Stockholm, Sweden.

Arthur, W. 1987. *The Niche in Competition and Evolution.* Wiley, New York.

Ash, J. B. 1981. *The Nothofagus Blume (Fagaceae) on New Guinea* (pp. 355–397). In: J. L. Gressitt (ed.). *Ecology and Biogeography of New Guinea.* W. Junk, The Hague.

Assmann, E. 1970. *The Principles of Forest Yield Study.* Pergamon Press, Oxford.

Aston, A. R. 1984a. The effect of doubling atmospheric CO_2 on streamflow: a simulation. *Journal of Hydrology* 67: 273–280.

Aston, A. R. 1984b. Evaporation from eucalypts growing in a weighing lysimeter: a test of the combination equations. *Agricultural and Forest Meteorology* 31: 241–249.

Atjay, G. L., P. Ketner and P. Duvigneaud. 1979. Terrestrial primary production and phytomass (pp. 129–181). In: B. Bolin, E. T. Degens, S. Kemp and P. Ketner (eds.). *The Global Carbon Cycle. Scope 13.* Wiley, New York.

Atkinson, I. A. E. 1970. Successional trends in the coastal and lowland forest of Manna Loa and Kilauea volcanoes, Hawaii. *Pacific Science* 24: 387–400.

Atkinson, P. M. and F. M. Danson. 1988. Spatial resolution for remote sensing of forest plantations (pp. 221–223). In: *Proceedings of the IGARSS '88 Symposium,* Edinburgh, Scotland (13–16 Sept., 1988). European Space Agency Publication Division, Geneva.

Aubréville, A. 1933. La forêt de la Côte d'Ivoire. *Bulletin de la Comite de Afrique Occidentale Française* 15: 205–261.

Aubréville, A. 1938. La forêt colonaile: les forêts de l'Afrique occidentale française. *Annales Academie Sciences Colonaile* 9: 1–245. (Translated by S. R. Eyre. 1991. Regeneration patterns in the closed forest of Ivory Coast, pp. 41–55. In: S. R. Eyre (ed.). *World Vegetation Types*. MacMillan, London).

Austin, M. B. 1980. an exploratory analysis of grassland dynamics: an example of a lawn succession. *Vegetatio* 43: 87–94.

Austin, M. B. and L. Belbin. 1981. An analysis of succession along an environmental gradient using data from a lawn. *Vegetatio* 46: 19–30.

Austin, M. P. 1985. Continuum concept, ordination methods and niche theory. *Annual Review Ecological Systems* 16: 39–61.

Austin, M. P. and T. M. Smith. 1989. A new model of the continuum concept. *Vegetatio* 83: 35–47.

Austin, M. P., R. B. Cunningham and R. B. Good. 1983. Altitudinal distribution in relation to other environmental factors of several eucalypt species in southern New South Wales. *Australian Journal of Ecology* 8: 169–180.

Austin, M. P., R. B. Cunningham and P. M. Flemming. 1984. New approaches to direct gradient analysis using environmental scalars and statistical curve-fitting procedures. *Vegetatio* 55: 11–27.

Austin, M. P., A. O. Nicholls and C. P. Margules. 1990. Measurement of the realized quantitative niche: environmental niches of five *Eucalyptus* species. *Ecological Monographs* 60: 161–177.

Axelrod, D. I. 1966. Origin of deciduous and evergreen habits in temperate trees. *Evolution* 20: 1–15.

Ayala, F. J., M. E. Gilpin and J. G. Ehrenfeld. 1973. Competition between species: theoretical models and experimental tests. *Theoretical Population Biology* 4: 331–355.

Bacastow, R. B. and C. D. Keeling. 1973. Atmospheric carbon dioxide and radiocarbon in the natural carbon cycle: II. Changes from A.D. 1700 to 2070 as deduced from a geochemical model (pp. 186–135). In: G. M. Woodwell and E. V. Pecan (eds.). *Carbon and the Biosphere*. CONF-720510. National Technical Information Service, Springfield, VA.

Bacastow, R. B. and C. D. Keeling. 1981. Atmospheric carbon dioxide concentration and the observed airborne fraction (pp. 103–112). In: B. Bolin (ed.). *Carbon Cycle Modeling. SCOPE 16*. Wiley, New York.

Baes, C. F., Jr, H. E. Goeller, J. S. Olson and R. M. Rotty. 1977. Carbon dioxide and climate: the uncontrolled experiment. *American Scientist* 65: 310–320.

Baillie, G. and W. Peach. 1992. Population limitation in Palearctic–African migrant passerines. *Ibis* 13: S120-S132.

Baker, J. A. and S. T. Ross. 1981. Spatial and temporal resource utilization by southeastern cyprinids. *Copeia* 1981: 178–179.

Ball, J. T., I. E. Woodrow and J. A. Berry. 1987. A model predicting stomatal conductances and its contribution to the control of photosynthesis under different environmental conditions (pp. 221–224). In: J. Biggins (ed.). *Progress in Photosynthesis Research*. Martinus Nijhoff, Dordrecht.

Barker, G. 1985. *Prehistoric Farming in Europe*. Cambridge University Press, Cambridge.

Bartlein, P. J., I. C. Prentice and T. Webb, III. 1986. Climate response surfaces from pollen data for some eastern North American taxa. *Journal of Biogeography* 13: 35–57.

Bartlett, A. and E. S. Barghoom. 1973. Phytogeographic history of the isthmus of Panama during the past 12,000 years (a history of vegetation, climate, and sea-level change) (pp. 203–299). In: A. Graham (ed.). *Vegetation and Vegetational History in Northern Latin America*. Elsevier, New York.

Barton, A. M., N. Fetcher and S. Redhead. 1989. The relationships between treefall gap size and light flux in a Neotropical rain forest in Costa Rica. *Journal of Tropical Ecology* 5: 437–439.

Baskin, Y. 1993. Ecologists put some life into models of a changing world. *Science* 259: 1694–1696.

Bazzaz, F. A. 1979. The physiological ecology of plant succession. *Annual Reviews of Ecology and Systematics* 10: 351–371.

Bazzaz, F. A. and Pickett, S. T. A. 1980. Physiological ecology of tropical succession: a comparative review. *Annual Reviews in Ecology and Systematics* 11: 287–310.

Beard, J. S. 1980. The physiognomic approach (pp. 33–64). In: R. H. Whittaker (ed.) *Classification of Plant Communities*, 2nd edn. W. Junk, The Hague.

Becquerel, A. 1853. Connection between forestry and climate. In: F. B. Hough, 1877. *Report Upon Forestry* 1: 221–283.

Beerling, D. J., B. Huntley and J. P. Bailey. 1995. Climate and the distribution of *Fallopia japonica*: use of an introduced species to test the predictive capacity of response surfaces. *Journal of Vegetation Science* 6: 269–282.

Bega, V. 1974. *Phytophthora cinnamomi*: its distribution and possible role in 'ohi'a decline on the island of Hawaii. *Plant Disease Reporter* 58: 1069–1073.

Bella, I. E. 1971. A new competition model for individual trees. *Forest Science* 17: 364–372.

Beltrami, E. 1987. *Mathematics for Dynamic Modeling*. Academic Press, Orlando, FL.

Berger, A. 1984. Accuracy and frequency stability of the earth's orbital elements during the Quaternary (pp. 3–126). In: A. Berger, J. Imbrie, J. Hays, G. Kukla and B. Saltzman (eds.). *Milankovitch and Climate Part 1. NATO ASI Series: Series C, Mathematics and Physical Science*, Vol. 126. D. Reidel, Dordrecht.

Bergthørsson, P. 1969. An estimate of drift ice and temperature in Iceland for 1000 years. *Jökull* 19: 94–101.

Bernabo, J. C. and T. Webb III. 1977. Changing patterns in the Holocene pollen record of northeastern North America: a mapped summary. *Quaternary Research* 8: 64–96.

Betancourt, J. L., T. R. Van Devender and P. S. Martin (eds.). 1990. *Packrat Middens: The Last 40 000 Years of Change*. University of Arizona Press, Tucson, AZ.

Birks, H. J. B. 1989. Holocene isochrone maps and patterns of tree-spreading in the British Isles. *Journal of Biogeography* 16: 503–540.

Bledsoe, L. J., R. C. Francis, G. L. Swartzman and J. D. Gustafson. 1971. PWNEE: a grassland ecosystem model. *USIBP Grassland Biome Technical Report No. 64*. Colorado State University, Fort Collins, CO.

Blumler, M. A. and R. Byrne. 1991. The ecological genetics of domestication and origins of agriculture. *Current Anthropology* 32: 23–53.

Boden, T. A., R. J. Sepanski and F. W. Stoss (eds.). 1991. *Trends '91: A Compendium of Data on Global Change*. ORNL/CDIAC–46. Carbon Dioxide Information Analysis Center, Oak Ridge National Laboratory, Oak Ridge, Tennessee, TN.

Bolin, B. and R. B. Cook (eds.). 1983. *The Major Biogeochemical Cycles and their Interactions. Scope 21.* Wiley, New York.

Bolin, B., B. R. Döös, J. Jager and R. A. Warwick (eds.). 1986. *The Greenhouse Effect, Climatic Change, and Ecosystems. SCOPE 29.* Wiley, Chichester.

Bonan, G. B. 1989a. A computer model of the solar radiation, soil moisture, and soil thermal regimes in boreal forests. *Ecological Modelling* 45: 275–306.

Bonan, G. B. 1989b. Environmental factors and ecological processes controlling vegetation patterns in boreal forests. *Landscape Ecology* 3: 111–130.

Bonan, G. B. 1990a. Carbon and nitrogen cycling in North American boreal forests. I. Litter quality and soil thermal effects in interior Alaska. *Biogeochemistry* 10: 1–28.

Bonan, G. B. 1990b. Carbon and nitrogen cycling in North American boreal forests. II. Biogeographic patterns. *Canadian Journal of Forest Research* 20: 1077–1088.

Bonan, G. B. 1993a. Do biophysics and physiology matter in ecosystem models? *Climatic Change* 24: 281–285.

Bonan, G. B. 1993b. Physiological derivation of the observed relationship between net primary production and mean annual temperature. *Tellus* 45B: 397–408.

Bonan, G. B. and B. P. Hayden. 1990. Using a forest stand simulation model to examine the ecological and climatic significance of the late-Quaternary pine/spruce pollen zone in eastern Virginia. *Quaternary Research* 33: 204–218.

Bonan, G. B. and H. H. Shugart. 1989. Environmental factors and ecological processes in boreal forests. *Annual Reviews in Ecology and Systematics* 20: 1–28.

Bonan, G. B., H. H. Shugart, and D. L. Urban. 1990. The sensitivity of some high-latitude boreal forests to climatic parameters. *Climatic Change* 16: 9–29.

Bonan, G. B., D. Pollard and S. L. Thompson. 1992. Effects of boreal forest vegetation on global climate. *Nature* 359: 716–718.

Boorstin, D. J. 1983. *The Discoverers.* Random House, New York.

Booth, T. H. 1985. A new method for assisting species selection. *Commonwealth Forestry Review* 64: 241–250.

Booth, T. H. 1990. Mapping regions climatically suitable for particular tree species at the global scale. *Forest Ecology and Management* 36: 47–60.

Booth, T. H. 1991. A climatic/edaphic database and plant index prediction system for Africa. *Ecological Modelling* 56: 127–134.

Booth, T. H., H. A. Nix, M. F. Hutchinson and T. Jovanovic. 1988. Niche analysis and tree species information. *Forest Ecology and Management* 23: 47–59.

Bormann, F. H. and G. E. Likens. 1979a. *Pattern and Process in a Forested Ecosystem.* Springer-Verlag, New York.

Bormann, F. H. and G. E. Likens. 1979b. Catastrophic disturbance and the steady state in northern hardwood forests. *American Scientist* 67: 660–669.

Borough, C. J., W. D. Incoll J. R. May and T. Bird. 1978. Yield Statistics (pp. 201–205). In: W. Hillis and A. G. Brown (eds.). *Eucalypts for Wood Production.* CSIRO, Melbourne, Australia.

Bosatta, E. and G. I. Ågren. 1991a. Dynamics of carbon and nitrogen in the organic matter of the soil: a generic theory. *American Naturalist* 138: 227–245.

Bosatta, E. and G. I. Ågren. 1991b. Theoretical analysis of carbon and nutrient interactions in soils under energy limited conditions. *Soil Society of America* 55: 728–733.

Botkin, D. A., R. A. Nisbet and T. E. Reynales. 1989. Effects of climate change on forests of the Great Lake states (pp. 2–1 to 2–31). In: J. B. Smith and D. A. Tirpak (eds.). *The Potential Effects of Global Climate Change on the United States: Appendix D*

– *Forests*. Office of Policy, Planning, and Evaluation, US Environmental Protection Agency, Washington, DC.

Botkin, D. B. 1993. *Forest Dynamics: An Ecological Model*. Oxford University Press, Oxford.

Botkin, D. B., J. F. Janak and J. R. Wallis. 1972. Some ecological consequences of a computer model of forest growth. *Journal of Ecology* 60: 849–872.

Boudet, G. 1972. Désertification de l'Afrique tropicale sèche. *Adansonia Series 2* 4: 505–524.

Bourlière, F. 1973. The comparative ecology of rain forest mammals in Africa and tropical America: some introductory remarks (pp. 279–292). In: B. J. Meggers, E. S. Ayensu and W. D. Duckworth (eds.). *Tropical Forest Ecosystems in Africa and South America: A Comparative Review*. Smithsonian Institution Press, Washington, DC.

Box, E. O. 1978. Geographical dimensions of terrestrial net and gross primary productivity. *Radiation and Environmental Biophysics* 15: 305–322.

Box, E. O. 1981. *Macroclimate and Plant Forms: An Introduction to Predictive Modeling in Phytogeography*. W. Junk, The Hague.

Brandani, A., G. S. Hartshorn and G. H. Orians. 1988. Internal heterogeneity of gaps and tropical tree species richness. *Journal of Tropical Ecology* 4: 99–119.

Broaler, S. B. and C. A. H. Hodge. 1964. Observations on vegetation arcs in the northern region, Somali Republic. *Journal of Ecology* 52: 511–544.

Broecker, W. S. and T. -H. Peng. 1993. Evaluation of the ^{13}C constraint on the uptake of fossil fuel CO_2 by the ocean. *Global Biogeochemical Cycles* 7: 619–626.

Brokaw, N. V. L. 1985a. Gap-phase regeneration in a tropical forest. *Ecology* 66: 682–687.

Brokaw, N. V. L. 1985b. Treefalls, regrowth, and community structure in tropical forests (pp. 101–108). In: S. T. A. Pickett & P. S. White (eds.). *The Ecology of Natural Disturbance and Patch Dynamics*. Academic Press, New York.

Brown, G. S. 1965. Point density in stems per acre. *Forest Research Notes No. 38*. Forest Research Institute. New Zealand Forest Service.

Brown, J. H. 1995. *Macroecology*. University of Chicago Press, Chicago.

Brown, J. L. 1975. *The Evolution of Behavior*. Norton, New York.

Brown, W. L. and E. O. Wilson. 1956. Character displacement. *Systematic Zoology* 5: 48–64.

Brylinski, M. 1972. Steady-state sensitivity analysis of energy flow in a marine ecosystem (pp. 81–101). In: B. C. Patten (ed.). *Systems Analysis and Simulation in Ecology*, Vol. II. Academic Press, New York.

Bryson, R. A. and T. J. Murray. 1979. *Climates of Hunger: Mankind and the World's Changing Weather*. University of Wisconsin Press, Madison, WI.

Budowski, G. 1965. Distribution of tropical American trees in the light of successional process. *Turrialba* 15: 40–42.

Budowski, G. 1970. The distinction between old secondary and climax species in tropical Central American lowland forests. *Tropical Ecology* 11: 44–48.

Budyko, M. I. 1974. *Climate and Life*. Academic Press, New York.

Bugmann, H. K. M. 1994. On the Ecology of Mountainous Forests in a Changing Climate: A Simulation Study. PhD Thesis Number 10638. Swiss Federal Institute of Technology, Zurich, Switzerland.

Bugmann, H. K. M. and A. M. Solomon. 1995. The use of a European forest model in

North America: a study of ecosystem response to climate gradients. *Journal of Biogeography* 22: 477–484.

Bugmann, H. K. M., X. Yan, M. T. Sykes, P. Martin, M. Lindner, P. V. Desanker and S. G. Cumming. 1995. A comparison of forest gap models: model structure and behaviour. *Climatic Change* 34: 289–313.

Burgman, M. A. 1989. The habitat volume of scarce and ubiquitous plants. *American Naturalist* 133: 228–239.

Burrows, C. J. 1990. *Processes of Vegetation Change.* Unwin Hyman, London.

Burton, P. J. 1980. *Light Regimes and Metrosideros Regeneration in a Hawaiian Montane Rain Forest.* MSc Thesis. Botany Department, University of Hawaii, Honolulu.

Busby, J. R. 1986. A bioclimatic analysis of *Nothofagus cunninghamii* (Hook.) Oerst in south-eastern Australia. *Australian Journal of Ecology* 11: 1–7.

Busing, R. T. 1991. A spatial model of forest dynamics. *Vegetatio* 92: 167–179.

Busing, R. T. and E. E. C. Clebsch. 1987. Application of a spruce–fir forest canopy gap model. *Forest Ecology and Management* 20: 151–169.

Cain, M. L. 1985. Random search by herbivorous insects: a simulation model. *Ecology* 66: 876–888.

Cain S. A. and G. M. de Oliveira Castro. 1959. *Manual of Vegetation Analysis.* Harper, New York.

Cairns, D. M. 1994. Development of a physiologically mechanistic model for use at the alpine treeline ecotone. *Physical Geography* 15: 104–124.

Cammell, M. E. and J. D. Knight. 1992. Effects of climate change on the population dynamics of crop pests. *Advances in Ecological Research* 22: 117–162.

Canham. C. D. 1988. An index for understory light levels in and around canopy gaps. *Ecology* 69: 1634–1638.

Cannell, M. G. R. 1982. *World Forest Biomass Data and Primary Production Data.* Academic Press, New York.

Capen, D. E. (ed.). 1981. *The Use of Multivariate Statistics in Studies of Wildlife Habitat. General Technical Report RM–87.* Rocky Mountain Forest and Range Experimental Station, US Department of Agriculture Forest Service. Fort Collins, CO.

Caswell, H. 1978. Predator-mediated coexistence: A non-equilibrium model. *American Naturalist* 112: 127–154.

Caswell, H., H. E. Koenig, J. A. Resh and Q. E. Ross. 1972. An introduction to systems science for ecologists (pp. 3–80). In: B. C. Patten (ed.). *Systems Analysis and Simulation in Ecology,* Vol. II. Academic Press, New York.

Cattelino, P. J., I. R. Noble, R. O. Slatyer and S. R. Kessell. 1979. Predicting the multiple pathways of plant succession. *Environmental Management* 3: 41–50.

CDAC. 1983. Changing Climate. *Report of the Carbon Dioxide Assessment Committee.* National Academy Press, Washington, DC.

Chapin, F. S., P. M. Vitousek and K. Van Cleve. 1986. The nature of nutrient limitation in plant communities. *American Naturalist* 127: 48–58.

Chapin, F. S., A. J. Bloom, C. B. Field and R. H. Waring. 1987. Plant responses to multiple environmental factors. *BioScience* 37: 49–57.

Chen, D. 1982. Evaluation on the development of Korean pine/broadleaved forest ecosystems. *Journal of the Northern Forestry University* Vol. Monograph on Korean Pine Forests: 1–17 (in Chinese).

Cheng B., G. Ding, G. Xu and Y. Zhang. 1986. Biological nutrient cycling in the broadleaved-*Pinus koraiensis* (Korean pine) forest of Changbaishan Mountain (pp. 110–114). In: H. Yang, Z. Wang, J. N. R. Jeffers and P. A. Ward (eds.). *The Temperate Forest Ecosystem. Proceedings of the ITE Symposium No. 20*, Antu, China. 5–11 July 1986. Institute of Terrestrial Ecology, Grange-over-Sands, UK.

China's Vegetation Editing Committee. 1980. *China's Vegetation.* Scientific Publishing House of China, Beijing (in Chinese).

Christensen, N. L. 1985. Shrubland fire regimes and their evolutionary consequences (pp. 86–100). In: S. T. A. Pickett and P. S. White (eds.). *The Ecology of Natural Disturbance and Patch Dynamics.* Academic Press, New York.

Clarke, F. L. 1875. Decadence of Hawaiian forest. *All About Hawaii* 1: 19–20.

Clarke, G. L. 1954. *Elements of Ecology.* Wiley, New York.

Clarke, W. C., K. H. Cook, G. Marland, A. M. Weinberg, R. M. Rotty, P. R. Bell, L. J. Allison, C. L. Cooper. 1982. The carbon dioxide question: perspectives for 1982 (pp. 3–43). In: W. C. Clarke (ed.). *Carbon Dioxide Review: 1982.* Clarendon Press, Oxford.

Clements, F. E. 1916. *Plant succession: An analysis of the development of vegetation.* Carnegie Inst. Pub. 242. Washington, DC.

Clements, F. E. 1928. *Plant Succession and Indicators.* Wilson, New York.

Clements, F. E. 1936. Nature and structure of the climax. *Journal of Ecology* 24: 252–284.

Clutter, J. L. 1963. Compatible growth and yield models for loblolly pine. *Forestry Science* 9: 345–371.

Cody, M. L. 1968. Habitat selection and interspecific territoriality among silvid warblers in England and Sweden. *American Naturalist* 102: 107–147.

Cody, M. L. 1974. *Competition and the Structure of Bird Communities.* Princeton University Press, Princeton, NJ.

Coffin, D. P. and W. K. Lauenroth. 1989a. A gap dynamics simulation model of succession in a semiarid grassland. *Ecological Modelling* 49: 229–266.

Coffin, D. P. and W. K. Lauenroth. 1989b. Disturbances and gap dynamics in a semiarid grassland: a landscape-level approach. *Landscape Ecology* 3: 19–27.

Cogbill, C. V. and G. E. Likens. 1974. Acid precipitation in the Northeastern United States. *Water Resource Research* 10: 1133–1137.

Cohen, M. N. 1977. *The Food Crisis in Prehistory: Overpopulation and the Origin of Agriculture.* Yale University Press. New Haven, CT.

Cohen, W. B. and T. A. Spies. 1992. Estimating attributes of Douglas fir/western hemlock forest stands from satellite imagery. *Remote Sensing of the Environment* 41: 1–18.

Cohen, W. B., T. A. Spies and G. A. Bradshaw. 1990. Semivariograms of digital imagery for analysis of conifer canopy structure. *Remote Sensing of the Environment* 34: 167–178.

COHMAP (Cooperative Holocene Mapping Project). 1988. Climatic changes of the last 18,000 years: observations and model simulations. *Science* 241: 1043–1052.

Cole, C. V., G. S. Innis and J. W. B. Stewart. 1977. Simulation of phosphorus cycling in semiarid grassland. *Ecology* 58: 1–15.

Cole, K. 1982. Late Quaternary zonation of the vegetation in the eastern Grand Canyon. *Science* 217: 1142–1145.

Cole, K. 1985. Past rates of change, species richness, and a model of vegetational inertia in the Grand Canyon, Arizona. *American Naturalist* 125: 289–303.

Colgan, M. W. 1983. Succession and recovery of a coral reef after predation by *Acanthaster planci* (L.). *Proceedings of the International Coral Reef Symposium* 2: 333–338.

Connell, J. H. 1978. Diversity in rain forests and coral reefs. *Science* 199: 1302–1310.

Connell, J. H. 1983. On the prevalence and relative importance of interspecific competition: evidence from field experiments. *American Naturalist* 122: 661–696.

Connell, J. H. 1990. Apparent versus 'real' competition in plants (pp. 9–26). In: J. B. Grace and D. Tilman (eds.). *Perspectives on Plant Competition*. Academic Press, San Diego, CA.

Connell, J. H., and R. O. Slatyer. 1977. Mechanisms of succession in natural communities and their role in community stability and organization. *American Naturalist* 111: 1119–1144.

Connor, E. F. and E. D. McCoy. 1979. The statistics and biology of the species–area relationship. *American Naturalist* 113: 791–833.

Conway, J. H. 1976. *On Numbers and Games.* Academic Press, London.

Cook, R. E. 1977. Raymond Lindeman and the trophic-dynamic concept in ecology. *Science* 198: 109–122.

Cornell, H. V. and J. H. Lawton. 1992. Species interactions, local and regional processes, and limits to the richness of ecological communities – a theoretical perspective. *Journal of Animal Ecology* 61: 1–12.

Cornet, A. F., C. Montaña, J. P. Delhoume and L. Lopez-Portillo. 1992. Water flows and the dynamics of desert vegetation stripes (pp. 327–345). In: A. J. Hansen and F. DiCastri (eds.). *Landscape Boundaries: Consequences for Biotic Diversity and Ecological Flows. Ecological Studies 92.* Springer Verlag, Berlin.

Cottam, G. and R. P. McIntosh. 1966. Vegetation continuum. *Science* 152: 546–547.

Cowan, I. R. 1968. The interception and absorption of radiation in plants. *Journal of Applied Ecology* 5: 367–379.

Cowan, I. R. 1971. Light in plant stands with horizontal foliage. *Journal of Applied Ecology* 8: 579–580.

Cowan, I. R. 1982. Regulation of water use in relation to carbon gain in higher plants (pp. 549–587). In: O. L. Lange, P. S. Noble, C. B. Osmond and H. Ziegler (eds.). *Physiological Plant Ecology, Encyclopedia of Plant Physiology (NS),* Vol. 12B. Springer Verlag, Berlin.

Cowan, I. R. 1986. Economics of carbon fixation in higher plants (pp. 133–170). In: T. J. Givnish (ed.). *On the Economy of Plant Form and Function*. Cambridge University Press, Cambridge.

Cowles, H. C. 1899. The ecological relations of the vegetation on the sand dunes of Lake Michigan. *Botanical Gazette* 27: 95–117, 176–202, 281–308, 361–369.

Cox, D. L. 1980. A note on the queer history of 'niche'. *Bulletin of the Ecology Society of America* 61: 201–202.

Craighead, J. J., J. S. Summer and G. B. Scaggs. 1982. *A definitive system for analysis of grizzly bear habitat and other wilderness resources using LANDSAT multispectral imagery and computer technology.* Wildlife-Wetlands Institute Monographs Number 1. University of Montana Foundation, Missoula, MT.

Cramer, W. P. and R. Leemans (1993). Assessing impacts of climate change on vegetation using climate classification systems (pp. 190–217). In: A. M. Solomon and H. H. Shugart (eds.). *Vegetation Dynamics and Global Change*. Chapman & Hall, New York.

Cramers, J. A. 1766. *Anleitung zum Forst-Wesen*. Waisenhaus-Buchhandlung, Braunschweig.

Croll, J. 1867. On the eccentricity of the Earth's orbit, and its physical relations to the glacial epoch. *Philosophical Magazine* 33: 119–131.

Crowley, T. J., D. A. Short, J. G. Mengel and G. R. North. 1986. Role of seasonality in the evolution of climate during the past 100 million years. *Science* 231: 579–584.

Curran, P. J. 1980. Multispectral remote sensing of vegetation amount. *Progress in Physical Geography* 4: 315–341.

Curran, P. J. 1988. The semivariogram in remote sensing. *Remote Sensing of the Environment* 24: 493–507.

Curtis, J. T. 1959. *The Vegetation of Wisconsin: An Ordination of Plant Communities*. The University of Wisconsin Press, Madison, WI.

Dai, A. and I. Y. Fung. 1993. Can climate variability contribute to the 'missing' CO_2 sink. *Global Biogeochemical Cycles* 7: 599–609.

Dale, V. H. and J. F. Franklin. 1989. Potential effects of climate change on stand development in the Pacific Northwest. *Canadian Journal of Forest Research* 19: 1581–1590.

Dale, V. H. and M. A. Hemstrom. 1984. *CLIMACS: A Computer Model of Forest Stand Development for Western Oregon and Washington*. *Research Paper. PNW–327*. US Department of Agriculture/Forest Service, Washington DC.

Daniels, R. F. 1976. Simple competition indices and their correlation with annual loblolly pine growth. *Forest Science* 22: 454–456.

Dansereau, P. 1968. The continuum concept of vegetation responses. *Botanical Review*. 34: 253–332.

Dansgaard, W., H. B. Clausen, N. Gendestrap, C. Hammer, S. J. Johnsen, P. M. Kristinsdottir and N. Reeh. 1982. A new Greenland deep ice core. *Science* 218: 579–584.

D'Arrigo, R., G. C. Jacoby and I. Y. Fung. 1987. Boreal forest and atmosphere–biosphere exchange of carbon dioxide. *Nature* 329: 321–323.

Daubenmire, R. 1966. Vegetation: identification of typal communities. *Science* 151: 291–398.

Davis, M. B. 1976. Pleistocene biography of temperate deciduous forests. *Geoscience and Man* 13: 13–26.

Davis, M. B. 1981a. Quaternary history and the stability of forest communities (pp. 132–153). In: D. C. West, H. H. Shugart and D. B. Botkin (eds.). *Forest Succession: Concepts and Application*. Springer Verlag, New York.

Davis, M. B. 1981b. Outbreaks of forest pathogens in Quaternary history. *Proceedings of the IV International Palynology. Conference* (Lucknow) 3: 216–227.

Davis, M. B. 1983. Quaternary history of deciduous trees of eastern North America and Europe. *Annals of the Missouri Botanical Garden* 70: 550–563.

Davis, M. B. 1986. Climatic instability, time lags, and community disequilibrium (pp. 269–284). In: J. Diamond and T. J. Case (eds.). *Community Ecology*. Harper and Row, New York.

Davis, M. B. and D. B. Botkin. 1985. Sensitivity of cool-temperate forests and their fossil pollen record to rapid temperature change. *Quaternary Research* 23: 327–340.

DeAngelis, D. L. 1992. *Dynamics of Nutrient Cycling and Food Webs.* Chapman & Hall, London.

DeAngelis, D. L. and L. J. Gross (eds.) 1992. *Individual-based Models and Approaches in Ecology: Populations, Communities and Ecosystems.* Chapman & Hall, New York.

DeAngelis, D. L., D. C. Cox and C. C. Coutant. 1979. Cannibalism and size dispersal in young-of-the-year largemouth bass: experiments and model. *Ecological Modelling* 8: 133–148.

DeAngelis, D. L., W. M. Post and C. C. Travis. 1986. *Positive Feedback.* Springer Verlag. Berlin.

de Candolle, A. L. 1855. *Géographie Botanique Raisonée.* Paris.

Delcourt, H. R. and P. A. Delcourt. 1991. *Quaternary Ecology: A Paleoecological Perspective.* Chapman & Hall, London.

Delcourt, H. R., P. A. Delcourt and T. Webb, III. 1983. Dynamic plant ecology: the spectrum of vegetation change in time and space. *Quaternary Science Reviews* 1: 153–175.

Delcourt, P. A. 1980. Goshen Springs: Late-Quaternary vegetation record for southern Alabama. *Ecology* 61: 371–386.

Delhoume, J. P. 1988. *Contribution à l'Etude des Relations Eau sol Végétation en Zone Aride du Nord du Mexique.* Rapport Scientifique, ATP PIREN CNRS, OSTROM, INRA, CIRAD, Paris.

DeLiocourt, F. 1898. De l'amenagement des Sapienieres. *Bulletin of the Society Forest Franch-Conte Belfort.* Besançon, France.

Denslow, J. S. 1980. Gap partitioning among tropical rainforest trees. *Biotropica* 12: 47–55 (suppl.).

Desanker, P. V. and I. C. Prentice. 1994. Miombo – a vegetation dynamics model for the Miombo woodlands of Zambezian Africa. *Forest Ecology and Management* 69: 87–95.

Dethier, M. N. 1984. Disturbance and recovery in intertidal pools: maintenance of mosaic pattern. *Ecological Monographs* 54: 99–118.

Develice, R. L. 1988. Test of a forest dynamics simulator in New Zealand. *New Zealand Journal of Botany* 26: 387–92.

Dexter, E. and A. Chapman. 1995. Climate change and endangered species. *Erinyes* (Newsletter of the Environmental Resources Information Network, Department of the Environment, Sport and Territories, Canberra, Australia) 23: 1 and 6.

Diamond, J. M. 1975. Assembly of species communities (pp. 342–344). In: M. L. Cody and J. M. Diamond (eds.). *Ecology and Evolution of Communities.* Harvard University Press, Cambridge, MA.

Di Castri, F., A. J. Hansen and M. M. Holland (eds.). 1988. A new look at ecotones. *Biology International, Special Issue 17.*

Dickinson, R. E. 1986. How will climate change? (pp 206–270). In: B. Bolin, B. R. Döös, J. Jager and R. A. Warwick (eds.). *The Greenhouse Effect, Climatic Change, and Ecosystems. SCOPE 29.* Wiley, Chichester.

Dickinson, R. E. 1988. Atmospheric Systems and Global Change (pp. 57–80). In: T. Rosswall, R. G. Woodmansee and P. G. Risser (eds.). *Scales and Global Change. SCOPE 35.* John Wiley, Chichester.

Dickinson, R. E. 1992. Land Surface Interaction (pp. 131–150). In: D. Ojima (ed.).

Modeling the Earth System. UCAR/Office for Interdisciplinary Earth Studies, Boulder, CO.

Dickinson, R. E. and A. Henderson-Sellers. 1986. Modeling tropical deforestation: a study of GCM land-surface parameterizations. *Quarterly Journal of Research of the Meteorological Society* 114: 439–462.

Dickinson, R. E and A. Henderson-Sellers. 1988. Modelling tropical deforestation: a study of GCM land-surface parameterizations. *Quarterly Journal of the Royal Meteorological Society* 114: 439–462.

Dickinson, R. E., A. Henderson-Sellers, P. J. Kennedy, M. F. Wilson. 1986. *Biosphere–atmosphere transfer scheme (BATS) for the NCAR community climate model. NCAR Technical Note TN–275+STR.* National Center for Atmospheric Research, Boulder, CO.

Diggle, P. J. 1976. A spatial stochastic model of inter-plant competition. *Journal of Applied Probability* 13: 662–671.

DiStefano, J. J., III, A. R. Stubberud and I. J. Williams. 1967. *Schaum's Outline of Theory and Problems of Feedback and Control Systems.* McGraw-Hill, New York.

Dobson, A. J. 1983. *Introduction to Statistical Modelling.* Chapman & Hall, London.

Dokuchaev, V. V. 1889. *Uchenie o Zonax Prirody* (Teachings about the zones of nature), Vol. 6. Akademie Nauka, Moscow.

Doley, D. 1981. Tropical and subtropical forest and woodlands (pp. 209–323). In: T. T. Kozlowski (ed.). *Water Deficits and Plant Growth.* Academic Press, New York.

Dolman, P. M. and W. J. Sutherland. 1994. The response of bird populations to habitat loss. *Ibis* 137: S38–S46.

Doncaster, C. P. 1981. The spatial distribution of ant's nests on Ramsey Island, New South Wales. *Journal of Animal Ecology* 50: 195–218.

Doyle, T. W. 1981. The role of disturbance in the gap dynamics of a montane rain forest: an application of a tropical forest succession model (pp. 56–73). In: D. C. West, H. H. Shugart and D. B. Botkin (eds.). *Forest Succession: Concepts and Application.* Springer-Verlag, New York.

Doyle, T. W. 1983. Competition and growth relationships in a mixed-aged, mixed-species forest community. PhD dissertation, University of Tennessee, Knoxville.

Drude, O. 1890. *Handbuch der Pflanzengeographie.* Verlag von J. Engelhorn, Stuttgart.

Drury, W. H. and I. C. T. Nesbit. 1973. Succession. *Journal of Arnold Arboretum* 54: 331–368.

Dueser, R. D. and H. H. Shugart. 1978. Microhabitat configurations in forest floor small mammal fauna. *Ecology* 59: 89–98.

Dueser, R. D. and H. H. Shugart. 1979. Niche pattern in a forest-floor small mammal fauna. *Ecology* 60: 108–118.

Dunford, C. 1977. Behavioral limitations of round-tailed ground squirrel density. *Ecology* 58: 1254–1268.

Dunin, F. X. and E. A. Greenwood. 1986. Evaluation of the ventilated chamber for measuring evaporation from a forest. *Hydrological Processes* 1: 47–62.

Dunin, F. X., E. M. O'Loughlin and W. Reyenga. 1988. Interception loss from a eucalypt forest: lysimeter determination of hourly rates for long-term evaluation. *Hydrological Processes* 2: 315–329.

Dyer, M. I., F. di Castri and A. J. Hansen (eds.). 1988. Geosphere–biosphere observatories: their definition and design for studying global change. *Biology International, Special Issue 16.*

Dyrness, C. T. 1982. *Control of Depth to Permafrost and Soil Temperature by the Forest Floor in Black Spruce/Feathermoss Communities. Research Note PNW–396.* United States Forest Service, Washington, DC.

Eck, T. and D. Dye. 1990. Satellite estimation of photosynthetically active radiation at the earth's surface. *Remote Sensing of the Environment* 38: 135–146.

Egler, F. E. 1939. Vegetation zones of Oahu, Hawaii. *Empire Forestry Journal* 18: 1–14.

Egler, F. E. 1954. Vegetation science concepts. I. Initial floristic composition - a factor in old-field vegetation development. *Vegetatio* 4: 412–417.

Eisenbud, M. 1973. *Environmental Radioactivity*, 2nd edn. Academic Press, New York.

Ek, A. R. and R. A. Monserud. 1974a. *FOREST: Computer Model for the Growth and Reproduction Simulation for Mixed Species Forest Stands. Research Report A2635.* College of Agricultural and Life Sciences, University of Wisconsin, Madison, WI.

Ek, A. R. and R. A. Monserud. 1974b. Trials with program FOREST: growth and reproduction simulations for mixed species even- or uneven-aged forest stands (pp. 56–73). In: J. Fries, (ed.). *Growth Models for Tree and Stand Simulation. Department of Forestry Yield Research.* Note 30. Royal College of Forestry, Stockholm, Sweden.

El Bayoumi, M. A., H. H. Shugart and R. W. Wien. 1984. Modeling succession of the eastern Canadian mixed-wood forest. *Ecological Modelling* 21: 175–198.

Ellenberg, H. 1986. *Vegetation Mitteleuropas mit den Alpen in ökologischer Sicht*, 4th edn. Verlag Eugen Ulmer, Stuttgart.

Elliott, C. C. H. 1979. The harvest time method as a means of avoiding quelea damage to irrigated rice in Chad/Cameroun. *Journal of Applied Ecology* 16: 23–35.

Elliott, C. C. H. 1989. The pest status of the quelea. In: R. L. Bruggers and C. C. H. Elliott (eds.). Quelea quelea, *Africa's Bird Pest.* Oxford University Press, New York.

Elton, C. 1927. *Animal Ecology.* MacMillan, New York.

Elton, C. 1958. *The Ecology of Invasions by Animals and Plants.* Methen, London.

Emanuel, W. R., H. H. Shugart and D. C. West. 1978a. Spectral analysis and forest dynamics: the effects of perturbations on long-term dynamics (pp. 195–210). In: H. H. Shugart (ed.). *Time Series and Ecological Processes.* Society for Industrial and Applied Mathematics, Philadelphia, PA.

Emanuel, W. R., D. C. West and H. H. Shugart. 1978b. Spectral analysis of forest model time series. *Ecological Modelling* 4: 323–326.

Emanuel, W. R., G. G. Killough, W. M. Post and H. H. Shugart. 1984. Modeling terrestrial ecosystems and the global carbon cycle with shifts in carbon storage capacity by land use change. *Ecology* 65: 970–983.

Emanuel, W. R., H. H. Shugart and M. P. Stevenson. 1985a. Climate change and the broad scale distribution of terrestrial ecosystem complexes. *Climatic Change* 7: 29–43.

Emanuel, W. R., H. H. Shugart and M. P. Stevenson. 1985b. Response to comment: climatic change and the broad-scale distribution of terrestrial ecosystem complexes. *Climatic Change* 7: 457–460.

Emanuel, W. R., I. Y.-S. Fung, G. G. Killough, B. Moore and P. H. Peng. 1985c. Modeling the global carbon cycle and changes in the atmosphere CO_2 levels (pp. 141–173). In: J. R. Trabalka (ed.). *Atmospheric CO_2 and the Global Carbon Cycle.* Carbon Dioxide Research Division, US Department of the Environment, Washington, DC.

Erwin, D. H., J. W. Valetine and J. J. Sepkoski. 1987. A comparative study of diversification events: the early Paleozoic versus the Mesozoic. *Evolution* 41: 365–389.

Esser, G. 1984. The significance of biospheric carbon pools and fluxes for the atmospheric CO_2: a proposed model structure. *Progress in Biometeorology* 3: 253–294.

Ewald, P. W. and F. L. Carpenter. 1978. Territorial responses to energy manipulations in the Anna's hummingbird. *Oecologia* 31: 277–292.

Farquhar, G. D. and S. von Caemmer. 1982. Modeling of photosynthetic response to environmental conditions (pp. 549–587). In: O. L. Lange, P. S. Noble, C. B. Osmond, and H. Ziegler (eds.). *Physiological Plant Ecology, Encyclopedia of Plant Physiology (NS)*, Vol. 12B. Springer Verlag, Berlin.

Farquhar, G. D. and T. D. Sharkey. 1982. Stomatal conductance and photosynthesis. *Annual Review of Plant Physiology* 33: 317–345.

Farquhar, G. D., S. von Caemmer and J. A. Berry. 1980. A biochemical model of photosynthetic CO_2 fixation in leaves of C_3 species. *Planta* 149: 78–90.

Fenchel, T. 1975. Character displacement and coexistence in mud snails (Hydrobiidae). *Oecologia* 20: 19–32.

Fennessy, M. M., J. L. Kinter, III, B. Kirtman, L. Marx, S. Nigram, E. Schneider, J. Shukla, D. Straus, A. Vernekar, Y. Xue and J. Zhou. 1994. The simulated Indian monsoon: a GCM sensitivity study. *Journal of Climate* 7: 33–41.

Field, C. B. 1991. Ecological scaling of carbon gain to stress and resource availability. In: H. A. Mooney, W. E. Winner and E. J. Pell (eds.). *Responses of Plants to Multiple Stresses*. Academic Press, San Diego, CA.

Finegan, B. 1984. Forest Succession. *Nature* 312: 109–114.

Fischlin, A., H. K. M. Bugmann and D. Gyalistras. 1995. Sensitivity of a forest ecosystem model to climate parametrization schemes. *Environmental Pollution* 87: 267–282.

Forbes, S. A. 1897. The lake as a microcosm. *Bulletin of the Peoria Science Association*. (Reprinted in 1925 in: *Illinois Natural History Survey Bulletin* 15: 537–550).

Forcier, L. K. 1975. Reproductive strategies and the co-occurrence of climax tree species. *Science* 189: 808–810.

Foster, D. H. and G. A. King. 1986. Vegetation pattern and diversity in SE Labrador, Canada: *Betula papyrifera* (birch) forest development in relation to fire history and physiography. *Journal of Ecology* 74: 465–483.

Foster, H. D. and W. W. Ashe. 1908. *Chestnut Oak in the Southern Appalachians. Circular 105*. Department of Agriculture-Forest Service, Washington, DC.

Foster, R. B. 1977. *Tachigalia versicolor* is a suicidal neotropical tree. *Nature* 268: 624–626.

Franklin, J. F., W. H. Moir, G. W. Douglas and C. Wiberg. 1971. Invasion of subalpine meadows by trees in the Cascade Range, Washington and Oregon. *Arctic and Alpine Research* 3: 215–224.

Franklin, J. F., H. H. Shugart and M. E. Harmon. 1987. Tree death as an ecosystem process. *BioScience* 37: 550–556.

Franzblau, M. A. and J. P. Collins. 1980. Test of a hypothesis of territory regulation in an insectivorous bird by experimentally increasing prey abundance. *Oecologia* 46: 164–170.

Friedli, H., H. Lötscher, H. Oeschger, U. Siegenthaler and B. Staaffer. 1986. $^{12}C/^{13}C$ ratio of atmospheric CO_2 in the past two centuries. *Nature* 324: 237–238.

Friend, A. D. 1991. Use of a model of photosynthesis and leaf microenvironment to predict optimal stomatal conductance and leaf nitrogen partitioning. *Plant, Cell and Environment* 14: 895–905.

Friend, A. D., H. H. Shugart and S. W. Running. 1993. A physiology-based model of forest dynamics. *Ecology* 74: 792–797.

Fujii, K. 1969. Numerical taxonomy of ecological characteristics and the niche concept. *Systematic Zoology* 18: 151–153.

Fung, I. Y., C. J. Tucker and K. C. Prentice. 1987. Application of the AVHRR vegetation index to study atmospheric–biospheric exchange of CO_2. *Journal of Geophysical Research* 92: 2999–3015.

Gaffney, P. M. 1975. The roots of the niche concept. *American Naturalist* 109: 490.

Galbraith, H. 1988. The effects of territorial behaviour on Lapwing populations. *Ornis Scandinavica* 19: 134–138.

Ganapathy, R. 1980. A major meteorite impact on the earth 65 million years ago: evidence from the Cretaceous–Tertiary boundary clay. *Science* 209: 921–923.

Garfinkel, D. 1962. Digital computer simulation of ecological systems. *Nature* 194: 856–857.

Gass, C. L., G. Angeher and J. Centra. 1976. Regulation of the food supply by feeding territoriality in the rufous hummingbird. *Canadian Journal of Zoology* 54: 2046–2054.

Gauch, H. G. and R. H. Whittaker. 1972. Coenocline simulation. *Ecology* 53: 446–451.

Gause, G. F. 1932. The ecology of populations. *Quarterly Review of Biology* 7: 27–46.

Gause, G. F. 1934. *The Struggle for Existence.* Williams and Wilkins, Baltimore, MD.

Gause, G. F. 1935. Vérifications expérimentals de la théorie mathematique de la lutte pour la vie. *Actualites Science Indies* 227.

Gear, A. J. and B. Huntley. 1991. Rapid change in the range limits of Scots pine 4000 years ago. *Science* 251: 544–547.

George, M. F., M. J. Burke, H. M. Pellett and A. G. Johnson. 1974. Low temperature exotherms and woody plant distribution. *HortScience* 9: 519–522.

Gerrard, D. J. 1969. Competition quotient: a new measure of the competition affecting individual forest trees. *Michigan State University Agricultural Experiment Station Research Bulletin No. 20.* Michigan State University Press.

Gerrish, G., and D. Mueller-Dombois. 1980. Behavior of native and non-native plants in two tropical rain forests on Oahu, Hawaiian Islands. *Phytocoenologia* 8: 237–295.

Gibb, J. 1956. Food, feeding habit and territoriality of the rock pipit, *Anthus spinoletta*. *Ibis* 98: 506–530.

Giller, P. S. 1984. *Community Structure and the Niche.* Chapman & Hall, London.

Gillon, Y. 1984. *Laboratoire d'Ecologie Tropicale ECOTROP, Rapport général d'activité 1982–1984.* CNRS, Paris.

Gilmanov, T. G. 1992. A new theoretical approach to the ecosystem concept as a differential of the biosphere (pp. 31–33). In: A. Teller, P. Mathy and J. N. R. Jeffers (eds.). *Responses of Forest Ecosystems to Environmental Change.* Elsevier, London.

Gilpin, M. E. and I. Hanski (eds.). 1991. *Metapopulation Dynamics.* Academic Press, London.

Gimingham, C. H., R. J. Hobbs and A. U. Mallik. 1981. Community dynamics in relation to management of heathland vegetation in Scotland. *Vegetatio* 46: 149–155.

Givnish, T. J. (ed.). 1986. *On the Economy of Plant Form and Function.* Cambridge University Press, Cambridge.

Givnish, T. J. and G. J. Vermelj. 1976. Sizes and shapes of liana leaves. *American Naturalist* 110: 743–778.

Gleason, H. A. 1926. The individualistic concept of the plant association. *Bulletin of the Torrey Botany Club* 53: 1–20.

Gleason, H. A. 1939. The individualistic concept of the plant association. *American Midland Naturalist* 21: 92–110.

Glenn-Lewin, D. C., R. K. Peet and T. T. Veblin (eds.). 1992. *Plant Succession: Theory and Prediction.* Chapman & Hall, London.

Goldberg, D. E. 1990. Components of resource competition in plant communities (pp. 27–50). In: J. B. Grace and D. Tilman (eds.). *Perspectives on Plant Competition.* Academic Press, San Diego, CA.

Gollan, T., J. B. Passioura and R. Munns. 1986. Soil water status affects the stomatal conductance of fully turgid wheat and sunflower leaves. *Australian Journal of Plant Physiology* 13: 459–464.

Goodall, D. W. 1963. The continuum concept and the individualistic association. *Vegetatio* 11: 297–316.

Goodall, D. W. 1975. Ecosystem modeling in the desert biome (pp. 73–94). In: B. C. Patten (ed.). *Systems Analysis and Simulation in Ecology, Vol. III.* Academic Press, New York.

Gore, A. 1992. *Earth in the Balance.* Houghton Mifflin, Boston, MA.

Gould, S. J. and R. Lewontin. 1979. The spandrels of San Marco and the panglossian paradigm: a critique of the adaptionist programme. *Proceedings of the Royal Society of London, Series B* 205: 581–598.

Goward, S. N. and K. F. Huemmrich. 1992. Vegetation canopy PAR absorptance and the normalized difference vegetation index: An assessment using the SAIL model. *Remote Sensing of the Environment* 39: 119–140.

Goward, S. N., C. J. Tucker and D. G. Dye. 1985. North American vegetation patterns observed with the NOAA–7 advanced very high resolution radiometer. *Vegetatio* 64: 3–14.

Grace, J. 1987. Climatic tolerance and the distribution of plants. *New Phytologist* 106 (Suppl.): 113–130.

Graham, R. W. 1992. Late Pleistocene faunal changes as a guide to understanding effects of greenhouse warming on the mammalian fauna of North America (pp. 76–90). In: R. L. Peters and T. E. Lovejoy (eds.). *Global Warming and Biological Diversity.* Yale University Press, New Haven, CT.

Grant, P. R. 1972. Convergent and divergent character displacement. *Biological Journal of the Linnean Society* 4: 39–68.

Grayson, D. K. 1984. Nineteenth-century explanations of Pleistocene extinctions: a review and analysis (pp. 5–39). In: P. S. Martin and R. G. Klein (eds.). *Quaternary Extinctions.* University of Arizona Press, Tucson, AZ.

Green, R. H. 1971. A multivariate statistical approach to the Hutchinsonian niche: bivalve molluscs of central Canada. *Ecology* 52: 543–554.

Green, R. H. 1974. Multivariate niche analysis with temporally varying environmental factors. *Ecology* 55: 73–83.

Greenslade, P. J. M. 1983. Adversity selection and the habitat template. *American Naturalist* 122: 352–365.

Greig-Smith, P. 1982. A. S. Watt, FRS: a biographical note (pp. 7–8). In: E. I. Newman (ed.). *The Plant Community as a Working Mechanism. Special Publication No. 1*, British Ecological Society. Blackwell Scientific, Oxford.

Griesbach, A. H. R. 1872. *Die Vegetation der Erde nach ihrer klimatoschen Anordnung* (2 vols.). W. Engelmann, Leipzig.

Griesemer, J. R. 1992. Niche: historical respectives (pp. 231–240). In: E. F. Keller and E. A. Lloyd (eds.). *Keywords in Evolutionary Biology.* Harvard University Press, Cambridge, MA.

Grime, J. P. 1974. Vegetation classification by reference to strategy. *Nature* 250: 26–31.

Grime, J. P. 1977. Evidence for the existence of three primary strategies in plants and its relevance to ecological and evolutionary theory. *American Naturalist* 111: 1169–1194.

Grime, J. P. 1979a. *Plant Strategies and Vegetation Processes.* Wiley, Chichester, UK.

Grime, J. P. 1979b. Competition and the struggle for existence (pp. 123–140). In: R. M. Anderson, B. D. Turner and L. R. Taylor (eds.). *Population Dynamics.* Blackwell Scientific, Oxford.

Grime, J. P. 1993. Vegetation functional classification systems as approaches to predicting and quantifying global vegetation change (pp. 293–305). In: A. M. Solomon and H. H. Shugart (eds.). *Vegetation Dynamics and Global Change.* Chapman & Hall, New York.

Grinnell, J. 1904. The origin and the distribution of the chestnut-backed chickadee. *Auk* 21: 364–382.

Grinnell, J. 1917a. The niche relations of the California thrasher. *Auk* 34: 364–382.

Grinnell, J. 1917b. Field tests and theories concerning distributional control. *American Naturalist* 51: 115–128.

Grinnell, J. 1924. Geography and evolution. *Ecology* 5: 225–229.

Grinnell, J. 1927. The designation of bird's ranges. *Auk* 44: 322–325.

Grinnell, J. 1928. Presence and absence of animals. *University of California Chronicles* 30: 429–450.

Grisebach, A. 1838. Ueber den Einfluss des Climas auf die Begranzung der naturlichen Floren. *Linnaea* 12: 159–200.

Gross, L. J., K. A. Rose, E. J. Rykiel, Jr, W. Van Winkle and E. E. Werner. 1992. Individual-based modeling: summary of a workshop (pp. 511–522). In: D. L. DeAngelis and L. J. Gross (eds.). *Individual-based Models and Approaches in Ecology.* Chapman & Hall, New York.

Grubb, P. J. 1977. The maintenance of species-richness in plant communities: the importance of the regeneration niche. *Biological Review* 52: 107–145.

Gurtin, M. E. and R. C. MacCamy. 1979. Some simple models for nonlinear age-dependent population dynamics. *Mathematical Biosciences* 43: 199–211.

Guthrie, R. D. 1984. Mosaics, allelochemics and nutrients: an ecological theory of late Pleistocene megafaunal extinctions (pp. 259–298). In: P. S. Martin and R. G. Klein (eds.). *Quaternary Extinctions: A Prehistoric Revolution.* University of Arizona Press, Tucson, AZ.

Haffer, J. 1987. Quaternary history of tropical America (pp. 1–18). In: T. C. Whitmore and G. T. Prance (eds.). *Biogeography and Quaternary History in Tropical America. Oxford Monographs on Biogeography No. 3.* Oxford Science Publications, Oxford.

Hallé, F. 1974. Architecture of trees in the rain forest of Morobe District, New Guinea. *Biotropica* 6: 43–50.

Hallé, F. and R. A. A. Oldemann. 1975. *Essay on the Architecture and Dynamics of Growth of Tropical Trees.* Penerbit University, Kuala Lumpur, Malaysia.

Hallé, F., R. A. A. Oldeman and P. B. Tomlinson. 1978. *Tropical Trees and Forests.* Springer-Verlag, Hiedelberg.

Halpin, P. N. and C. M. Secrett. 1994. Potential impacts of climate change on forest production in the humid tropics: a case study in Costa Rica. In: J. Pernetta, R. Leemans, D. Elder and S. Humphrey (eds.). *Impacts of Climate Change on Ecosystems and Species (ICCES): Vol. 2.* The World Conservation Union, Gland, Switzerland.

Hammen, T. van der. 1974. The Pleistocene changes of vegetation and climate in tropical South America. *Journal of Biogeography* 1: 3–26.

Hammen, T. van der. 1985. The Plio-Pleistocene climatic record of the tropical Andes. *Journal of Geological Society of London* 142: 483–489.

Hammen, T. van der. 1988. South America (pp. 307–340). In: B. Huntley and T. Webb, III (eds.). *Vegetation History.* Kluwer, Dordrecht.

Hansen, A. J., P. G. Risser and F. Di Castri. 1992. Epilogue: biodiversity and ecological flows across ecotones (pp. 423–438). In: A. J. Hansen and F. Di Castri (eds.). *Landscape Boundaries. Ecological Studies 92.* Springer-Verlag, New York.

Hansen, J. A. Lacis, D. Rind, G. Russel, P. Stone, I. Fung, R. Reudy and J. Lerner. 1984. Climate sensitivity: analysis of feedback mechanism. In: J. Hansen and R. Thompson (eds.). *Climate Processes and Climate Sensitivity. Geophysical Monogr. 29.* American Geophysical Union, Washington, DC.

Hansen, J., I. Fung, A. Lacis, S. Lebedef, D. Rind, R. Ruedy, G. Russel and P. Stone. 1988. Global climate changes as forecast by the Goddard Institute for Space Studies Three Dimensional Model. *Journal of Geophysical Research* 93: 9341–64.

Hanson, J. S., G. P. Malanson and M. P. Armstrong. 1990. Landscape fragmentation and dispersal in a model of riparian forest dynamics. *Ecological Modelling* 49: 277–296.

Hardin, G. 1960. The competitive exclusion principle. *Science* 131: 1292–1297.

Hardt, R. A. and R. T. T. Forman. 1989. Boundary form effects on woody colonization of reclaimed surface mines. *Ecology* 70: 1252–1260.

Harley, P. C. and T. D. Sharkey. 1991. An improved model of C_3 photosynthesis at high CO_2: reversed O_2 sensitivity explained by lack of glycerate re-entry into the chloroplast. *Photosynthesis Research* 27: 169–178.

Harley, P. C., R. B. Thomas, J. F. Reynolds and B. R. Strain. 1992. Modelling photosynthesis of cotton grown in elevated CO_2. *Plant, Cell and Environment* 15: 271–282.

Harper, J. L. 1977. *Population Biology of Plants.* Academic Press, London.

Harper, J. L. 1982. After description (pp. 11–25). In: E. I. Newman (ed.). *The Plant Community as a Working Mechanism.* Blackwell Scientific, Oxford.

Harrington, J. B. 1987. Climatic change: a review of causes. *Canadian Journal of Forest Research* 17: 1313–1339.

Harris, J. M. and B. A. Bodhaine (eds.). 1983. *Summary Report 1982, Geophysical Monitoring for Climatic Change.* Environmental Research Laboratories/NOAA, US Department of Commerce, Washington, DC.

Harrison, E. A. and H. H. Shugart. 1990. Evaluating performance on an Appalachian oak forest dynamics model. *Vegetatio* 86: 1–13.

Hartshorn, G. S. 1978. Tree falls and tropical forest dynamics (pp. 617–638). In: P. B. Tomlinson and M. H. Zimmermann (eds.). *Tropical Trees as Living Systems.* Cambridge University Press, Cambridge.

Hassell, M. P. and G. C. Varley. 1969. New inductive population models for insect parasites and its bearing on biological control. *Nature* 223: 1133–1136.

Hatch, C. R. 1971. *Simulation of an even-aged red pine stand in Northern Minnesota.* Ph.D. Thesis, University of Minnesota.

Hatton, T. J., L. L. Pierce and J. Walker. 1993. Ecohydrological changes in the Murray–Darling basin. II. Development and tests of a water balance model. *Journal of Applied Ecology* 30: 274–282.

Hatton, T. J., J. Walker, W. R. Dawes, F. X. Dunin. 1992. Simulations of hydroecological responses to elevated CO_2 at the catchment scale. *Australian Journal of Botany* 40: 679–696.

Hayden, B. P. 1994. Global biosphere requirements for general circulation models (pp. 263–276). In: W. K. Michener, J. W. Brunt and S. G. Stafford (eds.). *Environmental Information Management and Analysis: Ecosystem to Global Scales.* Taylor and Francis, London.

Hays, J. D., J. Imbrie and N. J. Shackleton. 1976. Variations in the Earth's orbit: pacemaker of the ice ages. *Science* 194: 1121–1132.

Hearon, K. Z. 1953. The kinetics of linear systems with special reference to periodic reactions. *Bulletin of Mathematical Biophysics* 15: 121–141.

Hegyi, F. 1974. A simulation model for managing jack-pine stands (pp. 74–90). In: J. Fries (ed.). *Growth Models for Tree and Stand Simulation.* Royal College of Forestry, Stockholm, Sweden.

Heinselman, M. L. 1981. Fire and succession in the conifer forests of northern North America. (pp. 374–405). In: D. C. West, H. H. Shugart and D. B. Botkin (eds.). *Forest Succession: Concepts and Application.* Springer Verlag, New York.

Helvey, J. D. and J. H. Patric. 1965. Canopy and litter interception of rainfall by hardwoods of the eastern United States. *Water Resources Research* 1: 193–290.

Hemming, C. F. 1965. Vegetation arcs in Somaliland. *Journal of Ecology* 53: 57–68.

Henderson-Sellers, A. 1990. Predicting generalized ecosystem groups with the NCAR GCM: first steps towards an interactive biosphere. *Journal of Climate* 3: 917–940.

Henderson-Sellers, A. and V. Gornitz. 1984. Possible climatic impacts of land cover transformations, with particular emphasis on tropical deforestation. *Climatic Change* 6: 231–258.

Henderson-Sellers, A., R. E. Dickinson, T. B. Durbidge, P. J. Kennedy, K. McGuffie and A. J. Pitman. 1993. Tropical deforestation: modeling local- to regional-scale climate change. *Journal of Geophysical Research* 98: 7289–7315.

Hepting, G. H. 1968. Diseases of forest and tree crops caused by air pollutants. *Phytopathology* 58: 1098–1101.

Hielkema, J. U., J. Roffey and C. J. Tucker. 1986. Assessment of ecological conditions associated with the 1980/81 desert locust plague upsurge in West Africa using environmental satellite data. *International Journal of Remote Sensing* 7: 1609–1622.

Hilborn, R. 1975. The effect of spatial heterogeneity on the persistence of predator prey interactions. *Theoretical Population Biology* 8: 346–355.

Hilden, O. 1965. Habitat selection in birds. A review. *Annales Zoologici Fennici* 2: 53–74.

Hinde, R. A. 1956. The biological significance of the territories of birds. *Ibis* 98: 340–369.

Hobbs, R. J. and C. J. Legg. 1983. Markov models and initial floristic composition in heathland vegetation dynamics. *Vegetatio* 56: 31–43.

Hobbs, R. J. and H. A. Mooney. 1985. Community and population dynamics of serpentine grassland annuals in relation to gopher disturbance. *Oecologia* 67: 342–351.

Hobbs, R. J. and H. A. Mooney. 1991. Effects of rainfall variability and gopher disturbance on serpentine annual grassland dynamics. *Ecology* 72: 59–68.

Holdridge, L. R. 1967. *Life Zone Ecology*. Tropical Science Center, San José, Costa Rica.

Holdridge, L. R., W. C. Grenke, W. H. Hatheway, T. Liang, J. A. Tosi, Jr. 1971. *Forest Environments in Tropical Life Zones, A Pilot Study*. Pergamon Press, Oxford.

Holling, C. S. 1961. Principles of insect predation. *Annual Review of Entomology* 6: 163–182.

Holling, C. S. 1964. The analysis of complex population processes. *Canadian Entomology* 96: 335–347.

Hool, J. N. 1966. A dynamic programming-Markov chain approach to forest production control. *Forest Science Monograph* 12: 1–26.

Horn, H. S. 1971. *The Adaptive Geometry of Trees*. Princeton University Press, Princeton, NJ.

Horn, H. S. 1975a. Forest succession. *Scientific American* 232: 90–98.

Horn, H. S. 1975b. Markovian properties of forest succession (pp. 196–211). In: M. L. Cody and J. M. Diamond (eds.). *Ecology and Evolution in Communities*. Harvard University Press, Cambridge, MA.

Horn, H. S. 1976. Succession (pp. 187–204). In: R. M. May (ed.). *Theoretical Ecology*. Blackwell Scientific, Oxford.

Horn, H. S. and R. M. May. 1977. Limits to similarity among coexisting competitors. *Nature* 270: 660–661.

Hort, A. 1916. *Enquiry into Plants and Minor Works on Odours and Weather Signs*. By Theophrastus and translated by Sir Albert Hort, vols. I and II. Heinemann, London.

Hosaka, E. Y. 1939. Ecological and floristic studies in Kipapa Gluch, Oahu. *Bishop Museum Occasional Papers*. 13: 175–232.

Houghton, J. T., G. J. Jenkins and J. J. Ephraums (eds.). 1990. *Climate Change – The IPCC Scientific Assessment*. Cambridge University Press, Cambridge.

Houghton, R. A. 1989. The long-term flux of carbon to the atmosphere from changes in land use. Extended abstract from the *Third International Conference on Analysis and Evaluation of Atmospheric CO_2 Data: Present and Past*. Hinterzarten, 16–20 October 1989. Environmental Pollution Monitoring and Research Programme No. 59. World Meteorological Organization.

Houghton, R. A. 1992. Effects of land-use change, surface temperature and CO_2 concentration on terrestrial stores of carbon. Paper presented at the *IPCC-sponsored International Workshop on Biotic Feedbacks in the Global Climatic System*. Woods Hole, MA. October 26–29, 1992.

Houghton, R. A., J. E. Hobbie, J. M. Melillo, B. Moore, B. J. Peterson, G. R. Shaver and G. M. Woodwell. 1983. Changes in the carbon content of the terrestrial biota and the soils between 1860 and 1980: net release of CO_2 to the atmosphere. *Ecological Monographs* 53: 235–262.

Howell, T. R. 1952. Natural history and differentiation in the Yellow-bellied Sapsucker. *Condor* 54: 237–282.

Huang, S., S. J. Titus and D. P. Weins. 1992. Comparison of nonlinear height-diameter functions for major Alberta tree species. *Canadian Journal of Forestry Research* 22: 1297–1304.

Huffaker, C. B. 1958. Experimental studies on predation: dispersion factors and predator–prey oscillations. *Hildigardia* 27: 343–383.

Humboldt, A. von. 1807. *Ideen zu einer Geographie der Pflanzen*. F. G. Cotta, Tübingen (reprinted in 1963 by Wissenschaftl. Buchges, Darmstadt).

Hunt, E. R. and S. W. Running. 1992. Simulated dry matter yields for aspen and spruce stands in the North American boreal forest. *Canadian Journal of Remote Sensing* 18: 126–133.

Hunter, M. L., Jr. 1990. *Wildlife, Forests and Forestry: Principles of Managing Forests for Biological Diversity*. Prentice Hall, Englewood Cliffs, NJ.

Huntley, B. 1988. Europe (pp. 341–383). In: B. Huntley and T. Webb, III (eds.). *Vegetation History*. Kluwer, Dordrecht.

Huntley, B. 1990. Studying global change - the contribution of Quaternary palynology. *Global and Planetary Change* 82: 53–61.

Huntley, B. 1994. Plant species' response to climate change: implications for the conservation of European birds. *Ibis* 137: S127-S138.

Huntley, B. and H. J. B. Birks. 1983. *An Atlas of Past and Present Pollen Maps for Europe: 0–13 000 Years Ago*. Cambridge University Press, Cambridge.

Huntley, B. and T. Webb, III. 1988. *Vegetation History*. Kluwer, Dordrecht.

Huntley, B., P. J. Bartlein and I. C. Prentice. 1989. Climatic control of the distribution of beech (*Fagus* L.) in Europe and North America. *Journal of Biogeography* 16: 551–560.

Huntley, B., D. R. Ascroft, P. M. Berry, W. P. Cramer and A. P. MacDonald. 1995. Modeling present and future ranges of some European higher plants using climate response surfaces. *Journal of Biogeography* 22: 967–1002.

Huston, M. 1979. A general hypothesis of species diversity. *American Naturalist* 113: 81–101.

Huston, M. A. and D. L. DeAngelis. 1987. Size bimodality in monospecific plant populations: a critical review of potential mechanisms. *American Naturalist* 129: 678–707.

Huston, M., D. L. DeAngelis and W. M. Post. 1988. New computer models unify ecological theory. *BioScience* 38: 682–691.

Huston, M. A. and T. M. Smith. 1987. Plant succession: life history and competition. *American Naturalist* 130: 168–198.

Hutchinson, G. E. 1957. Concluding remarks. *Cold Spring Harbor Symposium on Quantitative Biology* 22: 415–427.

Hutchinson, G. E. 1959. Homage to Santa Rosalia, or why are there so many kinds of animals? *American Naturalist* 93: 145–159.

Hutchinson, G. E. 1965. *The Ecological Theater and the Evolutionary Play*. Yale University Press, New Haven, CT.

Imbrie, J. and J. Z. Imbrie. 1980. Modelling the climatic response to orbital variations. *Science* 207: 943–953.

Imbrie, J., J. D. Hays, D. G. Martinson, A. McIntyre, A. C. Mix, J. J. Morley, N. G. Pisias, W. L. Prell and N. J. Shackleton. 1984. The orbital theory of Pleistocene climate: support from revised chronology of marine $\delta^{18}O$ record (pp. 269–305). In: A. Berger, J. Imbrie, J. Hays, G. Kukla and B. Saltzman (eds.). *Milankovitch and Climate,*

Part 1. NATO ASI Series, Series C, Mathematics and Physical Science Vol. 126. D. Reidel, Dordrecht.

Innis, G. S. 1975. Role of total systems models in the Grassland Biome study (pp. 14–48). In: B. C. Patten (ed.). *Systems Analysis and Simulation in Ecology,* Vol. III. Academic Press, New York.

Isagi, Y. and N. Nakagoshi. 1990. A Markov approach for describing post-fire succession of vegetation. *Ecological Research* 5: 163–171.

Jablonski, D. 1986. Background and mass extinctions: the alternation of macroevolutionary regimes. *Science* 231: 129–133.

Jacobson, G. L. and E. C. Grimm. 1986. A numerical analysis of Holocene forest and prairie vegetation in central Minnesota. *Ecology* 67: 958–966.

Jaeger, M. M., R. L. Bruggers, B. E. Johns and W. A. Erikson. 1986. Evidence of itinerant breeding in the red-billed quelea, *Quelea quelea,* in the Ethiopian Rift Valley. *Ibis* 128: 469–482.

Jalas, J. and J. Suominen (eds.). 1973. *Atlas Florae Europaeae,* Vol. 2, *Gymnospermae (Pinaceae to Ephedraceae).* Societas Biologica Fennica Vanamo, Helsinki.

Jalas, J. and J. Suominen (eds.). 1976. *Atlas Florae Europaeae,* Vol. 3, *Salicaceae to Balanophoraceae.* Societas Biologica Fennica Vanamo, Helsinki.

Jalas, J. and J. Suominen (eds.). 1979. *Atlas Florae Europaeae,* Vol. 4, *Polygonaceae.* Societas Biologica Fennica Vanamo, Helsinki.

James, F. C. and H. H. Shugart. 1971. A quantitative method of habitat description. *Audubon Field Notes* 24: 727–736.

James, F. C., R. F. Johnston, N. O. Warmer, G. J. Niemi and W. J. Broecklin. 1984. The Grinnellian niche of the wood thrush. *American Naturalist* 124: 17–47.

Jarvis, P. G. and J. W. Leverenz. 1983. Productivity of temperate, deciduous and evergreen forests (pp. 234–280). In: O. L. Lange, P. S. Nobel, C. B. Osmund and H. Ziegler (eds.). *Ecosystems Processes: Mineral Cycling and Man's Influence.* Springer Verlag, Berlin.

Jarvis, P. G. and K. G. McNaughton. 1986. Stomatal control of transpiration: scaling up from leaf to region. *Advances in Ecological Research* 15: 1–49.

Jefferson, T. 1799. A memoir on the discovery of certain bones of a quadruped of the clawed kind in the western part of Virginia. *Transactions of the American Philosophical Society* 4: 246–260.

Jenkins, M. 1992. Species extinctions (pp. 192–205). In: B. Groombridge (ed.). *Global Biodiversity: Status of the Earth's Living Resources.* Chapman & Hall, London.

Jenne, R. L. 1992. Climate model description and impact on terrestrial climate. In: S. K. Majumdar, L. S. Kalkstein, B. Yarnal, E. W. Miller and L. M. Rosenfeld. *Climate Change: Implications, Challenges and Mitigation Measures.* Pennsylvania Academy of Sciences, PA.

Johnson, E. A. 1977a. A multivariate analysis of the niches of plant populations in raised bogs. I. Niche dimensions. *Canadian Journal of Botany* 55: 1201–1210.

Johnson, E. A. 1977b. A multivariate analysis of the niches of plant populations in raised bogs. II. Niche width and overlap. *Canadian Journal of Botany* 55: 1210–1220.

Johnson, W. C. and D. M. Sharpe. 1976. Forest dynamics in the northern Georgia piedmont. *Forest Science* 22: 307–322.

Jones, E. W. 1945. The structure and reproduction of the virgin forests of the north temperate zone. *New Phytologist* 44: 130–148.

Jones, E. W. 1955–1956. Ecological studies on the rain forest of southern Nigeria. IV.

The plateau forest of the Okumu forest reserve. *Journal of Ecology* 43: 564–594; 44: 83–117.

Jordan, C. F. 1986. Ecological effects of nuclear radiation (pp. 331–344). In: G. H. Orians (ed.). *Ecological Knowledge and Environmental Problem Solving: Concepts and Case Studies.* National Academy Press, Washington, DC.

Jordan, D. B. and W. L. Ogren. 1984. The CO_2/O_2 specificity of ribulose 1,5-biphosphate carboxylase/oxygenase: dependence on ribulose-biphosphate concentration, pH and temperature. *Planta* 161: 308–313.

Jørgensen, S. E. 1986. *Fundamentals of Ecological Modeling.* Elsevier, Amsterdam.

Justice, C. O., J. R. G. Townshend, B. N. Holben and C. J. Tucker. 1985. Analysis of the phenology of global vegetation using meteorological satellite data. *International Journal of Remote Sensing* 6: 1271–1318.

Justice, C. O., B. N. Holben and M. D. Gwynne. 1986. Monitoring East African vegetation using AVHRR data. *International Journal of Remote Sensing* 7: 1453–1474.

Justice, C. O., T. Eck, B. N. Holben and T. Tancre. 1991. The effect of water vapour on the NDVI derived for the Sahelian region from NOAA-AVHRR data. *International Journal of Remote Sensing* 12: 1165–1185.

Karieva, P. and M. Anderson. 1988. Spatial aspects of species interactions: the wedding of models and experiments (pp. 35–50). In A. Hastings (ed.). *Community Ecology.* Springer-Verlag, New York.

Karlson, R. H. 1978. Predation and space utilization patterns in a periodically disturbed habitat. *Bulletin of Marine Science* 30: 894–900.

Kaye, S. V. and S. J. Ball. 1969. Systems analysis of a coupled compartment model for radionuclide transfer in a tropical environment (pp. 731–739). In: D. J. Nelson and F. C. Evans (eds.). *Symposium on Radioecology. Proceedings of the Second National Symposium.* US Atomic Energy Commission, Washington, DC (now available as CONF 670–503, US Department of Commerce, Springfield, VA).

Keeling, C. D. 1983. The global carbon cycle: what we know and could know from atmospheric, biospheric and oceanic observations (pp. II. 3–II. 62). In: *Proceedings of the CO_2 Research Conference: Carbon Dioxide, Science, and Concensus,* DOE CONF–820970. NTIS, Springfield, VA.

Keeling, C. D., R. B. Bascastow, A. F. Carter, S. C. Piper, T. P. Whorf, M. Heimann, W. G. Mook and H. Roeloffzen. 1989. A three-dimensional model of atmospheric CO_2 transport based on observed winds: 4. Mean annual gradients and interannual variations (pp. 305–363). In: D. H. Peterson (ed.). *Aspects of Climate Variability in the Pacific and Western Americas. AGU Monograph No. 5.* AGU, Washington, DC

Keeling, C. D., T. P. Whorf, M. Wahlen and J. van der Pilcht. 1995. Interannual extremes in the rate of rise of atmospheric carbon dioxide since 1980. *Nature* 375: 666–670.

Keeling, R. F. and S. R. Shertz. 1992. Seasonal and interannual variations in atmospheric oxygen and implications for the global carbon cycle. *Nature* 358: 723–727.

Keen, B. 1959. *The Life of Admiral Christopher Columbus.* Rutgers University Press, Brunswick, NJ.

Ker, J. W. and J. H. G. Smith. 1955. Advantages of the parabolic expression of height-diameter relationships. *Forestry Chronicles* 31: 235–246.

Kercher, J. R. and M. C. Axelrod. 1984. A process model of fire ecology and succession in a mixed-conifer forest. *Ecology* 65: 1725–1742.

Kercher, J. R., M. C. Axelrod, and G. E. Bingham. 1980. Forecasting effects of sulfur dioxide pollution on growth and succession (pp. 200–202). In: P. R. Miller (ed.). *Proceedings of the Symposium on Effects of Air Pollutants on Mediterranean and Temperate Forest Ecosystems. General Technical Report, PSW–43.* US Department of Agriculture-Forest Service, Washington, DC.

Kessell, S. R. 1976. Gradient modeling: a new approach to fire modeling and wilderness resource management. *Environmental Management* 1: 39–48.

Kessell, S. R. 1979a. *Gradient Modeling: Resource and Fire Management.* Springer-Verlag, New York.

Kessell, S. R. 1979b. Phytosociological inference and resource management. *Environmental Management* 3: 29–40.

Kessell, S. R. and M. W. Potter. 1980. A quantitative succession model for nine Montana forest communities. *Environmental Management* 4: 227–240.

Khalil, A. A. M. and J. Grace. 1993. Does xylem sap ABA control the stomatal behaviour of water-stressed sycamore (*Acer pseudoplatanus* L.) seedlings? *Journal of Experimental Botany* 44: 1127–1134.

Kienast, F. and N. Kuhn. 1989. Simulating forest succession along ecological gradients in southern central Europe. *Vegetatio* 79: 7–20.

King, J. R. and L. R. Mewaldt. 1987. The summer biology of an unstable insular population of White-crowned Sparrows in Oregon. *Condor* 89: 549–565.

Kittredge, J. 1948. *Forest Influences.* McGraw-Hill, New York.

Kliejunas, J. T. and W. H. Ko. 1973. Root rot of 'ohi'a (*Metrosideros collina* subsp. *polymorpha*) caused by *Phytophythora cinnamomi. Plant Disease Reports* 57: 383–384.

Kliejunas, J. T. and W. H. Ko. 1974. Deficiency of inorganic nutrients as contributing factor to 'ohi'a decline. *Phytopathology* 64: 891–896.

Klir, G. J. 1972. *Trends in General Systems Theory.* Wiley, New York.

Knapp, A. H. 1961. *The Effect of Deposition Rate and Cumulative Soil Level on the Concentration of Strontium–90 in US Milk and Food Supplies. Rep. TID–13945.* USAEC, Washington, DC.

Knight, D. H. 1975. A phytosociological analysis of species-rich tropical forest on Barro Colorado Island, Panama. *Ecological Monographs* 45: 259–284.

Knuchel, H. 1953. *Planning and Control in the Managed Forest.* Oliver and Boyd, Edinburgh.

Koop, H. and P. Hilgren. 1987. Forest dynamics and regeneration mosaic shifts in unexploited beech (*Fagus sylvatica*) stands at Fontainebleau (France). *Forest Ecology and Management* 20: 135–150.

Köppen, W. and R. Geiger. 1930. *Handbuch der Climatologie.* Teil I D, Borntraeger, Berlin.

Korzukhin, M. D. and M. Ya. Antonovski. 1992. Population-level models of forest dynamics (pp. 334–372). In: H. H. Shugart, R. Leemans and G. B. Bonan (eds.). *A Systems Analysis of the Global Boreal Forest.* Cambridge University Press, Cambridge.

Kozlowski, T. T. 1971a. *Growth and Development of Trees: Vol. I. Seed Germination, Ontogeny and Shoot Growth.* Academic Press, New York

Kozlowski, T. T. 1971b. *Growth and Development of Trees: Vol. II. Cambial Growth, Root Growth and Reproductive Growth.* Academic Press, New York.

Kozlowski, T. T. 1981. Impacts of air pollution of forest ecosystems. *BioScience* 30: 88–93.

Kozlowski, T. T., P. J. Kramer and S. G. Pallardy. 1991. *The Physiological Ecology of Woody Plants*. Academic Press, San Diego, CA.

Kramer, P. J. 1981. Carbon dioxide concentration, photosynthesis, and dry matter production. *BioScience* 31: 29–33.

Krebs, J. R. 1971. Territory and breeding density of the Great Tit, *Parus major* L. *Ecology* 52: 2–22.

Küchler, A. W. 1967. *Vegetation Mapping*. Ronald Press, New York.

Kullman, L. 1986. Recent tree-limit history of *Picea abies* in the southern Swedish Scandes. *Canadian Journal of Forestry Research* 16: 761–771.

Lack, D. 1953. Darwin's finches. *Scientific American* 188: 66–72.

Lack, D. 1966. *Population Studies of Birds*. Clarendon Press, Oxford.

Ladd, J. H., J. M. Oades and M. Amato. 1981. Microbial biomass formed from ^{14}C and ^{15}N-labeled plant material decomposition in soils in the field. *Soil Biology and Biochemistry* 13: 119–126.

Laemmlen, F. F. and R. V. Bega. 1972. Decline of 'ohi'a and koa forests in Hawaii. *Phytopathology* 62: 770.

Lamb, H. F. and M. E. Edwards. 1988. The Arctic (pp. 519–555). In: B. Huntley and T. Webb, III (eds.). *Vegetation History*. Kluwer, Dordrecht.

Landsberg, J. J. 1986. *Physiological Ecology of Forest Production*. Academic Press, New York.

Larcher, W., U. Heber and K. A. Santarius. 1973. Limiting temperatures for life functions (pp. 195–292). In: H. Precht, J. Christopherson, H. Hensel and W. Larcher (eds.). *Temperature and Life*. Springer-Verlag, Berlin.

Laws, R. M. 1970. Elephants as agents of habitat and landscape change in East Africa. *Oikos* 21: 1–15.

Lean, J. and D. A. Warrilow. 1989. Simulation of the regional climatic impact of tropical deforestation. *Nature* 342: 411–413.

Lee, Y. 1967. Stand models for lodgepole pine and limits to their application. *Forest Chronicles* 43: 387–388.

Leemans, R. and W. P. Cramer. 1991. *The IIASA Climate Database for Mean Monthly Values of Temperature, Precipitation and Cloudiness on a Terrestrial Grid*. RR–91–18. International Institute for Applied Systems Analysis, Laxenburg, Austria.

Leemans, R. and P. N. Halpin. 1992. Biodiversity and global change (pp. 254–255). In: B. Groombridge (ed.). *Global Biodiversity: Status of the Earth's Living Resources*. Chapman & Hall, London.

Leemans, R. and I. C. Prentice. 1987. Description and simulation of tree-layer composition and size distributions in a primaeval *Picea–Pinus* forest. *Vegetatio* 69: 147–156.

Leibundgut, H. 1959. Uber Zweck und Methoden der Struktur-und-Zuwachsanalyse von Urwalden. *Schweizer Zeitschrift Forstwesen* 122: 45–66.

Leibundgut, H. 1978. Uber der Dynamik Europaischer Urwalder. *Allgraft Forstzeitschrift* 24: 686–690.

Leith, H. 1975. Modeling the primary productivity of the world (pp. 237–263). In: H. Leith and R. H. Whittaker (ed.). *Primary Productivity of the Biosphere*. Springer Verlag, New York.

Leopold, A. 1933. *Game Management*. Charles Scribner's Sons, New York.

Leprune, J. C. 1992. Etude de quelques brousse tigrées sahéliennew: structure, dynamique, écologie (pp. 221–244). In: E. Le Floc'h, M. Grouzis, A. Cornet and J.

C. Bille (eds.). *L'aridité, une Contrainte au Développement.* OSTROM, Coll Didactiques, Paris.

Levandowsky, M. 1977. The white queen speculation. *Quarterly Review of Biology* 52: 383–386.

Levin, S. I. 1992. The problem of scale in ecology. *Ecology* 73: 1943–1967.

Levins, R. 1968. *Evolution in Changing Environments.* Princeton University Press, Princeton, NJ.

Levins, R. 1970. Extinction (pp. 77–107). In: M. Gerstenhaber (ed.). *Some Mathematical Problems in Biology.* American Mathematical Society, Providence, Rhode Island.

Levins, R. 1974. The quantitative analysis of partially specified systems. *Annals of the New York Academy of Science* 231: 123–138.

Levins, R. 1975. Evolution in communities near equilibrium (pp. 16–50). In: M. L. Cody and J. Diamond (eds.). *Ecology and the Evolution of Communities.* Harvard University Press, Cambridge, MA.

Levitt, J. 1980. *Responses of Plants to Environmental Stress*, Vol. I. *Chilling, Freezing and High Temperature Stresses*, 2nd edn. Academic Press, New York.

Lewin, R. 1986. Mass extinctions select different victims. *Science* 231: 219–220.

Lewton-Brain, L. 1909. The Maui forest trouble. *Hawaiian Planter's Record* 1: 92–95.

Lidicker, W. Z., Jr. 1975. The role of dispersal in the demography of small mammals (pp. 103–128). In: F. B. Golley, K. Petrusewicz and L. Ryszkowski (eds.). *Small Mammals: Their Productivity and Population Dynamics.* Cambridge University Press, New York.

Lieberman, M. and D. Lieberman. 1989. Forests are not just Swiss cheese: canopy stereogeometry of non-gaps in tropical forests. *Ecology* 70: 550–552.

Likens, G. E., and T. J. Butler. 1981. Recent acidification of precipitation in North America. *Atmospheric Environment* 15: 1103–1109.

Lin, J. Y. 1970. *Growing space index and stand simulation of young western hemlock in Oregon.* PhD Thesis, Duke University, Durham, NC.

Lindeman, R. L. 1942. The trophic–dynamic aspect of ecology. *Ecology* 23: 399–418.

Linder, S. 1985. Potential and actual management in Australian forest stands (pp. 11–35). In: *Research for Forest Management.* CSIRO, Melbourne, Australia.

Lindsay, A. 1939. Unpublished data tabularized in: Borough, C. J., W. D. Incoll, J. R. May and T. Bird. 1978. Yield Statistics (pp. 201–205). In: W. Hillis and A. G. Brown (eds.). *Eucalypts for Wood Production.* CSIRO, Melbourne, Australia.

Linthicum, K. J., C. L Bailey, F. G. Davies and C. J. Tucker. 1987. Detection of Rift Valley fever viral activity in Kenya by satellite remote sensing imagery. *Science* 235: 1651–1659.

Lippe, E., J. T. de Smidt and D. C. Glenn-Lewin. 1985. Markov models and succession: a test from a heathland in the Netherlands. *Journal of Ecology* 73: 775–791.

Litvak, M. K and R. I. C. Hansell. 1990. A community perspective on the multi-dimensional niche. *Journal of Animal Ecology* 59: 931–940.

Lloyd, G. H. 1962. The distribution of squirrels in England and Wales. *Journal of Animal Ecology* 31: 157–161.

Lomnicki, A. 1988. *Population Ecology of Individuals.* Princeton University Press, Princeton, NJ.

Lotka, A. J. 1925. *Elements of Physical Biology.* Williams and Wilkins, Baltimore, MD.

Lovelock, J. E. and L. Margulis. 1974. Atmospheric homeostasis by and for the biosphere: the Gaia hypothesis. *Tellus* 26: 1–10.

Low, W. A., W. J. Müller and M. L. Dudzinski. 1981. Grazing intensity of cattle on a complex of rangeland communities in central Australia. *Australian Rangelands Journal* 2: 76–82.

Lyon, H. L. 1909. The forest disease on Maui. *Hawaiian Planter's Record* 1: 151–159.

Lyon, H. L. 1918. The forests of Hawaii. *Hawaiian Planter's Record* 20: 267–281.

Lyon, H. L. 1919. Some observations on the forest problems of Hawaii. *Hawaiian Planter's Record* 21: 289–300.

Lyons, J. M. 1973. Chilling injury in plants. *Annual Review of Plant Physiology* 24: 445–466.

MacArthur, R. H. 1968. The theory of the niche (pp. 159–176). In: R. C. Lewontin (ed.). *Theoretical and Mathematical Biology.* Blaisdell, New York.

MacArthur, R. H. 1970. Species packing and competitive equilibrium for many species. *Theoretical Population Biology* 1: 1–11.

MacArthur, R. H. 1972. *Geographical Ecology.* Harper and Row, New York.

MacArthur, R. H. and R. Levins. 1967. The limiting similarity, convergence and divergence of coexisting species. *American Naturalist* 101: 377–385.

MacArthur, R. H. & E. O. Wilson. 1967. *The Theory of Island Biogeography.* Princeton University Press, Princeton, NJ.

MacCracken, M. C. and F. M. Luther. 1985. *Projecting Climate Effects of Increasing Carbon Dioxide (DOE/ER–0237).* US Department of Energy, Washington, DC.

Maglio, V. J. 1973. Origin and evolution of the Elephantidae. *Transactions of the American Philosophical Society (New Series)* 63: 1–149.

Malanson, G. P. 1996. Effects of dispersal and mortality on diversity in a forest stand model. *Ecological Modelling* 87: 103–110.

Malanson, G. P. and M. P. Armstrong. 1996. Dispersal probability and forest diversity in a fragmented landscape. *Ecological Modelling* 87: 91–102.

Malanson, G. P. and D. M. Cairns. 1995. Effects of increased cloud-cover on a montane forest landscape. *Ecoscience* 2: 75–82.

Malanson, G. P., M. P. Armstrong and D. A. Bennett. 1994. Fragmented forest response to climatic warming and disturbance (pp. 243–247). In: M. F. Goodchild, L. T. Steyaert, B. O. Parks, M. P. Crane, C. A. Johnston, D. R. Maidment and S. J. Glendinning (eds.). *GIS and Environmental Modeling.* GIS World, Fort Collins, CO.

Malingreau, J.-P. 1986. Global vegetation dynamics: satellite observations over Asia. *International Journal of Remote Sensing* 7: 1121–1146.

Malingreau, J.-P. and C. J. Tucker. 1988. Large-scale deforestation in the southern Amazon Basin of Brazil. *Ambio* 17: 49–55.

Manabe, S. 1969. The atmospheric circulation and the hydrology of the earth's surface. *Monthly Weather Review* 97: 739–774.

Manabe, S. and R. J. Stouffer. 1980. Sensitivity of a global climate to an increase in CO_2 concentration in the atmosphere. *Journal of Geophysical Research* 85: 5529–5554.

Manabe, S. and R. T. Wetherald. 1987. Large scale changes in soil wetness induced by an increase in carbon dioxide. *Journal of Atmospheric Science* 44: 1211–1235.

Mankin, J B., R. V. O'Neill, H. H. Shugart and B. W. Rust. 1977. The importance of validation in ecosystem analysis (pp. 63–71). In: G. S. Innis (ed.). *New Directions in the Analysis of Ecological Systems. Part I.* Simulation Councils of America, La Jolla, CA.

Margulis, C. R. and M. P. Austin (eds.). 1991. *Nature Conservation: Cost Effective Biological Surveys and Data Analysis.* CSIRO, Canberra, Australia.

Margulis, C. R. and J. L. Stein. 1989. Patterns in the distributions of species and the selection of nature reserves: an example from the *Eucalyptus* forests in south-eastern New South Wales, Australia. *Biology and Conservation* 50: 219–238.

Margulis, C. R., A. O. Nicholls and M. P. Austin. 1987. Diversity of *Eucalyptus* species predicted by multivariable environmental gradient. *Oecologia* 71: 229–238.

Marks, P. L. 1974. The role of pin cherry (*Prunus pennsylvanica*) in the maintenance of stability in northern hardwood ecosystems. *Ecological Monographs* 44: 73–88.

Marland, G., T. A. Boden, R. C. Griffin, S. F. Huang, P. Kanciruk and T. R. Nelson. 1989. *Estimates of CO$_2$ Emissions from Fossil Fuel Burning and Cement Manufacturing, Based on the United Nations Energy Statistics and the US Bureau of Mines Cement Manufacturing Data.* ORNL/CDIAC–25. Carbon Dioxide Information Analysis Center, Oak Ridge National Laboratory, Oak Ridge, TN.

Marr, J. W. 1977. The development and movement of tree islands near the upper limits of tree islands near the upper limit of tree growth in the Southern Rocky Mountains. *Ecology* 58: 1159–1164.

Martel, Y. A. and E. A. Paul. 1974. Effects of cultivation on the organic matter of grassland soils determined by fractionation and radio-carbon dating. *Canadian Journal of the Soil Society* 54: 419–426.

Martin, P. 1992. EXE: a climatically sensitive model to study climate change and CO$_2$ enrichment effects on forests. *Australian Journal of Botany* 40: 717–735.

Martin, P. H. 1996. Climate change, water stress and fast forest response: a sensitivity study. *Climatic Change* 34: 289–313.

Martin, P. S. and R. G. Klein (eds.). 1984. *Quaternary Extinctions: A Prehistoric Revolution.* University of Arizona Press, Tucson, AZ.

Martin, P. S. and H. E. Wright, Jr. 1967. *Pleistocene Extinctions: The Search for a Cause.* Yale University Press, New Haven, CT.

Martin, W. E. 1969. Radioecology and the feasibility of nuclear canal excavation. (pp. 9–22). In: D. J. Nelson and F. C. Evans (eds.). *Symposium on Radioecology. Proceedings of the Second National Symposium.* US Atomic Energy Commission, Washington, DC (now available as CONF 670–503, US Department of Commerce, Springfield, VA).

May, R. M. 1973. *Stability and Complexity in Model Ecosystems.* Princeton University Press, Princeton, NJ.

Maynard Smith, J. and M. Slatkin. 1973. The stability of predator–prey systems. *Ecology* 54: 384–391.

McCullagh, P. and J. A. Nelder. 1983. *Generalized Linear Modelling.* Chapman & Hall, London.

McDonald, K. A. and J. H. Brown. 1992. Using montane mammals to model extinctions due to global change. *Conservation Biology* 6: 409–415.

McGuire, A. D., J. M. Melillo, L. A. Joyce, D. W. Kicklighter, A. L. Grace, B. Moore, III and C. Vörösmarty. 1992. Interactions between carbon and nitrogen dynamics in estimating net primary productivity for potential vegetation in North America. *Global Biogeochemical Cycles* 6: 101–124.

McIntosh, R. P. 1957. The York woods, a case study of forest succession. *Ecology* 38: 29–37.

McIntosh, R. P. 1967. The continuum concept of vegetation. *Botany Review* 33: 130–187.

McIntosh, R. P. 1981. Succession and ecological theory (pp. 10–23). In: D. C. West, H. H. Shugart and D. B. Botkin (eds.). *Forest Succession: Concepts and Application*. Springer-Verlag, New York.

McIntosh, R. P. 1985. *The Background of Ecology: Concept and Theory*. Cambridge University Press, Cambridge.

McLaughlin, S. B., D. C. West, H. H. Shugart and D. S. Shriner. 1978. Air pollution effects on forest succession: application of a mathematical model. *71st Annual Meeting of the Air Pollution Control Association* 78–24. 5.

M'Closky, R. T. 1976. Community structure of sympatric rodents. *Ecology* 57: 728–739.

McMurtrie, R. E. and Y. P. Wang. 1993. Mathematical models of the photosynthetic response of tree stands to rising CO_2 concentrations and temperatures. *Plant, Cell and Environment* 16: 1–13.

McMurtrie, R. E, D. A. Rook and F. M. Kelliher. 1990. Modeling the yield of *Pinus radiata* on a site limited by water and nitrogen. *Forest Ecology and Management* 30: 38–413.

McNaughton, S. J., R. W. Ruess and S. W. Seagle. 1988. Large mammals and process dynamics in African ecosystems. *BioScience* 88: 794–800.

Meentenmeyer, V. 1978. Macroclimate and lignin control of litter decomposition rates. *Ecology* 59: 465–472.

Melillo, J. M., A. D. McGuire, D. W. Kicklighter, B. Moore, III, C. J. Vörösmarty and A. L. Schloss. 1993. Global climate change and terrestrial net primary productivity. *Nature* 363: 234–240.

Menaut, J. C., J. Gignoux, C. Prado and J. Clobert. 1990. Tree community dynamics in a humid savanna of the Côte- d'Ivoire: modelling the effects of fire and competition with grass and neighbors. *Journal of Biogeography* 17: 471–481.

Meyer, H. A. 1952. Structure, growth and drain in balanced unevenaged forests. *Journal of Forestry* 50: 85–92.

Mielke, D. L., H. H. Shugart and D. C. West. 1978. *A Stand Model for Upland Forests of Southern Arkansas*. ORNL/TM–6225. Oak Ridge National Laboratory. Oak Ridge, TN.

Milankovitch, M. M. 1941. *Canon of Insolation and the Ice-age Problem*. Köningliche Serbische Adademie, Beograd. (English translation by the Isreal Program for Scientific Translation, published by the US National Science Foundation, 1969, Washington, DC).

Miles, J., D. D. French and Z. Xu. 1983. A preliminary study of the successional status of a stand of mixed broadleaved/ *Pinus koraiensis* forest in Changbaishan, Northeast China. *Forestry and Ecosystem Research* 3: 54–72. (in Chinese).

Miller, P. C., B. D. Collier and F. L. Bunnell. 1975. Development of ecosystem modeling in the Tundra Biome (pp. 95–116). In: B. C. Patten (ed.). *Systems Analysis and Simulation in Ecology*, Vol. III. Academic Press, New York.

Miller, P. R. and J. R. McBride. 1975. Effects of air pollutants of forests (pp. 196–236). In: J. B. Mudd and T. T. Kozlowski (eds.). *Responses of Plants to Air Pollution*. Academic Press, New York.

Mintz, Y. 1984. The sensitivity of numerically simulated climates to land-surface boundary conditions (pp. 79–105). In: J. T. Houghton (ed.). *The Global Climate*. Cambridge University Press, Cambridge.

Miracle, M. R. 1974. Niche structure in zooplankton: a principal components approach. *Ecology* 55: 1306–1316.

Mitchell, J. F. B. 1983. The seasonal response of a general circulation model to changes in CO_2 and sea temperatures. *Quarterly Journal of the Royal Meterorological Society* 109: 113–152.

Mitchell, K. J. 1969. Simulation of growth of even-aged stands of white spruce. *Yale University School Forestry Bulletin* 75: 1–48.

Mitchell, K. J. 1975. Dynamics and simulated yield of Douglas-fir. *Forest Science Monograph* 17: 1–39.

Möbius, K. 1877. *Die Auster und die Austernwirtschaft*. Wiegundt Hempel Parry, Berlin.

Montaña, C. 1988. Las foraciones vegetales (pp. 167–198). In: C. Montaña (ed.). *Estudio Integrado de los Recursos Vegetación, Suelo Y Aqua en la Reserva de la Biosfera de Mapimí*. Instituto de Ecologia, Xalapa.

Montaña, C. 1992. The colonization of bare areas in two-phase mosaics of an arid ecosystem. *Journal of Ecology* 80: 315–327.

Montaña, C., J. Lopez-Portillo and A. Mauchamp. 1990. The response of two woody species to the conditions created by a shifting ecotone in an arid ecosystem. *Journal of Ecology* 78: 789–798.

Monteith, J. L. 1972. Solar radiation and productivity in tropical ecosystems. *Journal of Applied Ecology* 9: 747–766.

Monteith, J. L. 1973. *Principles of Environmental Physics*. Elsevier, New York.

Monteith, J. L. 1977. Climate and the efficiency of crop production in Britain. *Philosophical Transactions of the Royal Society of London, Series B* 281: 277–294.

Monteith, J. L. 1981a. Does light limit crop production? (pp. 23–38). In: C. B. Johnson (ed.). *Physiological Processes Limiting Plant Productivity*. Butterworth, London.

Monteith, J. L. 1981b. Evaporation and the environment (pp. 205–234). In: C. E. Fogg (ed.). The *State and Movement of Water in Living Organisms*. Cambridge University Press, Cambridge.

Moore, A. D. 1989. On the maximum growth equation used in gap simulation models. *Ecological Modelling* 45: 63–67.

Moore, A. D. and I. R. Noble. 1990. An individualistic model of vegetation dynamics. *Journal of Environmental Management* 30: 61–81.

Moore, A. D. and I. R. Noble. 1993. Automatic model simplification: the generation of replacement sequences and their use in vegetation modeling. *Ecological Modelling* 70: 137–157.

Moore, J. A., C. A. Budelsky and R. C. Schlesinger. 1973. A new index representing individual tree competitive status. *Canadian Journal of Forestry Research* 3: 495–500.

Morrison, D. F. 1967. *Multivariate Statistical Methods*. McGraw-Hill, New York.

Morse, D. H. 1976. Variables affecting the density and territory size of breeding spruce-woods warblers. *Ecology* 57: 290–301.

Morton, A. G. 1981. *History of Botanical Science*. Academic Press, London.

Mueller-Dombois, D. 1980. The 'ohi'a dieback phenomenon in the Hawaiian rain forest (pp. 153–161). In: J. Cairns (ed.). *The Recovery Process in Damaged Ecosystems*. Ann Arbor Science, Ann Arbor, MI.

Mueller-Dombois, D. 1982. Canopy dieback in indigenous forests of Pacific Islands: Hawaii, Papua New Guinea and New Zealand. *Hawaiian Botanical Society Newsletter* 21: 2–6.

Mueller-Dombois, D. 1991. The mosaic theory and spatial dynamics of natural dieback and regeneration in Pacific forests (pp. 46–60). In: H. Remmert (ed.). *The Mosaic-cycle Concept of Ecosystems.* Springer Verlag, Berlin.

Munro, D. D. 1974. Forest growth models: a prognosis (pp. 7–21). In: J. Fries (ed.). *Growth Models for Tree and Stand Simulation. Research Note 30.* Royal College of Forestry, Stockholm, Sweden.

Myers, J. H. 1976. Distribution and dispersal in populations capable of resource depletion: a simulation model. *Oecologia* 23: 255–269.

Myers, J. H. and B. J. Campbell. 1976. Distribution and dispersal in populations capable of resource depletion. *Oecologia* 24: 7–20.

Myers, J. P., P. G. Conners and F. A. Pitelka. 1979. Territory size in wintering sanderlings: the effect of prey abundance and intruder density. *Auk* 96: 551–561.

Neftel, A., E. Moor, H. Oescheger and B. Stauffer. 1985. Evidence from polar ice cores for the increase in atmospheric CO_2 for the past 2 centuries. *Nature* 315: 45–47.

Neilson, R. P., G. A. King and G. Koerper. 1992. Toward a rule-based biome model. *Landscape Ecology* 7: 27–43.

Nelder, A. and R. W. M. Wedderburn. 1972. Generalized linear models. *Journal of the Royal Statistical Society A* 135: 370–384.

Nemani, R. R. and S. W. Running. 1989. Estimating regional surface resistance to evapotranspiration from NDVI and Thermal-IR AVHRR data. *Journal of Climate and Applied Meteorology* 28: 276–294.

Newman, E. I. (ed.). 1982. *The Plant Community as a Working Mechanism.* Special Publication No. 1, British Ecological Society. Blackwell Scientific, Oxford.

Newnham, R. M. 1964. The Development of a Stand Model for Douglas-Fir. PhD Thesis, University of British Columbia, Vancouver.

Newton, I. 1972. *Finches.* Collins, London.

Newton, I. 1986. *The Sparrowhawk.* Pyser, Calton.

Nicholson, P. H. 1981. Fire and the Australian aborigine – An enigma (pp. 55–76). In: A. M. Gill, R. H. Groves and I. R. Noble (eds.). *Fire and the Australian Biota.* Australian Academy of Sciences, Canberra.

Nilsson, T. 1983. *The Pleistocene.* Ferdinand Enke Verlag, Stuttgart.

Nix, H. A. 1986. A biogeographical analysis of Australian elaphid snakes (pp. 4–15). In: R. Longmore (ed.). *Atlas of Australian Elaphid Snakes.* Bureau of Flora and Fauna, Canberra, Australia.

Noble, I. R. 1975. *Computer Simulations of Sheep Grazing in the Arid Zone.* PhD Thesis, University of Adelaide.

Noble, I. R. 1993. A model of responses of ecotones to climate change. *Ecological Applications* 3: 396–403.

Noble, I. R. and R. O. Slatyer. 1978. The effect of disturbances on plant succession. *Proceedings of the Ecological Society of Australia* 10: 135–145.

Noble, I. R. and R. O. Slatyer. 1980. The use of vital attributes to predict successional changes in plant communities subject to recurrent disturbances. *Vegetatio* 43: 5–21.

Noble, I. R., A. D. Moore and M. J. Strasser. 1988. Predicting vegetation dynamics based on structural and functional attributes. *Proceedings of the International Symposium on Vegetation Structure*, Utrecht.

Nobre, C. A., P. J. Sellers and J. Shukla. 1991. Amazonian deforestation and climate change. *Journal of Climate* 4: 957–988.

Norton-Griffiths, M. 1979. The influence of grazing, browsing, and fire on the vegetation dynamics of the Serengeti (pp. 310–352). In: A. R. E. Sinclair and M. Norton-Griffiths (eds.). *Serengeti: Dynamics of an Ecosystem*. University of Chicago Press, Chicago, IL.

O'Brien, S. T., B. P. Hayden and H. H. Shugart. 1992. Global change, hurricanes, and a tropical forest. *Climatic Change* 22: 175–190.

Odum, E. P. 1953. *Fundamentals of Ecology*. W. B. Saunders, Philadelphia, PA.

Odum, E. P. 1968. Energy flow in ecosystems: a historical review. *American Zoologist* 8: 11–18.

Odum, E. P. 1971. *Fundamentals of Ecology*, 3rd edn. W. B. Saunders, Philadelphia, PA.

Odum, H. T. 1955. Trophic structure and productivity of Silver Springs, Florida. *Ecological Monographs* 27: 55–112.

Odum, H. T. 1983. *Systems Ecology*. Wiley, New York.

Oja, T. 1983. Metsa suktsessiooni ja tasandilise struktuuri imiteerimisest. *Yearbook of the Estonian Naturalist Society* 69: 110–117.

Ojima, D. (ed.). 1992. *Modeling the Earth System*. UCAR/Office for Interdisciplinary Earth Studies, Boulder, CO.

Old, K. M., G. A. Kile and C. P. Ohmart (eds.). 1981. *Eucalypt Dieback in Forests and Woodlands*. CSIRO, Melbourne, Australia.

Oldeman, R. A. A. 1978. Architecture and energy exchange of dicotyledonous trees in the forest (pp. 525–560). In: P. B. Tomlinson and M. H. Zimmerman (eds.). *Tropical Trees as Living Systems*. Cambridge University Press, Cambridge.

Oldeman, R. A. A. 1991. *Forests: Elements of Silvology*. Springer Verlag, Berlin.

Oldeman, R. A. A. and J. van Dijk. 1991. Diagnosis of the temperament of tropical rain forest trees (pp. 21–66). In: A. Gómez-Pompa, T. C. Whitmore and M. Hadley (eds.). *Rain Forest Regeneration and Management. Man and the Biosphere Series*. Vol. 6. UNESCO, Paris.

Oliver, C. D. 1981. Forest development in North America. *Forest Ecology and Management* 3: 153–168.

O'Loughlin, E. M. 1990. Modelling soil water status in complex terrain. *Agricultural and Forest Meteorology* 50: 23–38.

O'Loughlin, E. M., D. L. Short and W. R. Dawes. 1989. Modelling the hydrological response of catchments to landuse change (pp. 335–340). In: *Hydrology and Water Resources Symposium*, Institution of Engineers, Australia, Publication Number 89/19. (Christchurch, New Zealand).

Olson, J. S. 1963. Analog computer models for movement of nuclides through ecosystems (pp. 121–125). In: V. Shultz and A. W. Klement, Jr (eds.). *Proceedings of the First National Symposium on Radioecology*. Reinhold, New York.

Olson, J. S., J. A. Watts and L. J. Allison, 1983. *Carbon in Live Vegetation of Major World Ecosystems. ESD Publication No. 1997*. Oak Ridge National Laboratory, Oak Ridge, TN.

O'Neill, R. V. 1975. Modeling in the eastern deciduous forest biome (pp. 49–72). In: B. C. Patten (ed.). *Systems Analysis and Simulation in Ecology*, Vol. III. Academic Press, New York.

O'Neill, R. V. 1979. Transmutations across hierarchical levels (pp. 59–78). In: G. S. Innis and R. V. O'Neill (eds.). *Systems Analysis of Ecosystems*. International Co-operative, Fairland, MD.

O'Neill, R. V. 1988. Hierarchy theory and global change (pp. 29–45). In: T. Rosswall, R. G. Woodmansee and P. G. Risser (eds.). *Scales and Global Change, SCOPE 35*. Wiley, Chichester.

O'Neill, R. V., D. L. DeAngelis, J. B. Waide and T. F. H. Allen. 1986. *A Hierarchical Concept of Ecosystems*. Princeton University Press, Princeton.

Orians, G. H. 1986. Site characteristics favoring invasions (pp. 133–148). In: H. A. Mooney and J. A. Drake (eds.). *Ecology of Biological Invasions of North America and Hawaii*. Springer Verlag, New York.

Orians, G. H. and O. T. Solbrig. 1977. A cost-income model of leaves and roots with special reference to arid and semiarid areas. *American Naturalist* 111: 677–690.

Oshima, Y., M. Kimura, H. Iwake and S. Kuroiwa. 1958. Ecological and physiological studies on the vegetation of Mt. Shimagaree. I. Preliminary survey of the vegetation of Mt. Shimagaree. *Botanical Magazine of Tokyo* 71: 289–300.

Overpeck, J. T. and P. J. Bartlein. 1989. Assessing the response of vegetation to future climate change: response surfaces and paleoecological model validation. In: J. B. Smith and D. A. Tirpak (eds.). *The Potential Effects of Global Climate Change on the United States* (EPA–230–05–89–054). US Environmental Protection Agency, Washington, DC.

Overpeck, J. T., D. Rind and R. Goldberg. 1990. Climate-induced changes in forest disturbance and vegetation. *Nature* 343: 51–53.

Overton, W. S. 1975. The ecosystem modeling approach in the coniferous forest biome (pp. 117–138). In: B. C. Patten (ed.). *Systems Analysis and Simulation in Ecology*, Vol. III. Academic Press, New York.

Owen-Smith, R. N. 1988. *Megaherbivores: The Influence of Very Large Body Size on Ecology*. Cambridge University Press, Cambridge.

Pacala, S. W. and J. A. Silander. 1985. Neighborhood models of plant populations dynamics. I. Single-species models of annuals. *American Naturalist* 125: 385–411.

Pacala, S. W., C. D. Canham and J. A. Silander, Jr. 1993. Forest models defined by field measurements: I. The design of a northeastern forest simulator. *Canadian Journal of Forestry Research* 23: 1980–1988.

Paijmans, K. (ed.). 1976. *New Guinea Vegetation*. Australian National University Press, Canberra, Australia.

Paine, R. T. and S. A. Levin. 1981. Intertidal landscapes: disturbance and the dynamics of pattern. *Ecological Monographs* 51: 145–178.

Paleg, L. G. and D. Aspinall (eds.). 1981. *The Physiology and Biochemistry of Drought Resistance in Plants*. Academic Press, Sydney, Australia.

Palmer, M. W. 1988. Fractal geometry: a tool for describing spatial pattern of plant communities. *Vegetatio* 75: 91–102.

Papp, R. P., J. T. Kliejunas, R. S. Smith, Jr and R. F. Scharpf. 1979. Association of *Plagithmysus bilineatus* (Coleoptera: Cerambycidae) and *Phytophthora cinnamomi* with the decline of 'ohi'a–lehua forests on the island of Hawaii. *Forestry Science* 25: 187–196.

Parker, G. A. and W. J. Sutherland. 1986. Ideal free distributions when individuals differ in competitive ability: phenotype limited ideal free models. *Animal Behaviour* 34: 1222–1242.

Parkhurst, D. G. and O. L. Loucks. 1972. Optimal leaf size in relation to environment. *Journal of Ecology* 60: 505–537.

Parry, M. 1992. The potential effect of climate changes on agriculture and land use. *Advances in Ecological Research* 22: 63–92.

Parsons, R. F. 1968a. The significance of growth rate comparisons for plant ecology. *American Naturalist* 102: 295–297.

Parsons, R. F. 1968b. Ecological aspects of growth and mineral nutrition of three mallee species of *Eucalyptus*. *Oecologia Planta* 3: 121–136.

Parton, W. J. 1978. Abiotic section of ELM (pp. 31–53) In: G. S. Innis (ed.). *Grassland Simulation Model. Ecological Studies 26*. Springer-Verlag, New York.

Parton, W. J., D. S. Schimel, C. V. Cole and D. Ojima. 1987. Analysis of factors controlling soil organic levels of grasslands of the Great Plains. *Soil Science Society of America Journal* 51: 1173–1179.

Parton, W. J., J. W. B. Stewart and C. V. Cole. 1988. Dynamics of C, N, P and S in grassland soils: a model. *Biogeochemistry* 5: 109–131.

Parton, W. J., J. M. O. Scurlock, D. S. Ojima, T. G. Gilmanov, R. J. Scholes, D. S. Schimel, T. Kirchner, J.-C. Menaut, T. Seastedt, E. Garcia Moya, A. Kamnalrut and J. I. Kinyamario. 1993. Observations and modeling of biomass and soil organic matter dynamics for the grassland biome worldwide. *Global Biogeochemical Cycles* 7: 785–809.

Parton, W. J., J. M. O. Scurlock, D. S. Ojima, D. S. Schimel, D. O. Hall and SCOPE-GRAM Group members (M. B. Coughenour, E. G. Moya, T. G. Gilmanov, A. Kamnalrut, J. I. Kinyamario, T. G. F. Kittel, J.-C. Menaut, O. E. Sala and J. A. van Veen). 1995. Impact of climate change on grassland production and soil carbon worldwide. *Global Change Biology* 1: 13–22.

Pastor, J. and W. M. Post. 1986. Influences of climate, soil moisture, and succession on forest carbon and nitrogen cycles. *Biogeochemistry* 2: 3–27.

Pastor, J. and W. M. Post. 1988. Response of northern forests to CO_2 induced climate change. *Nature* 334: 55–58.

Patten, B. C. 1971. A primer for ecological modeling and simulation with analog and digital computers (pp. 4–121). In: B. C. Patten (ed.). *Systems Analysis and Simulation in Ecology*, Vol. I. Academic Press, New York

Pearlstine, L., H. McKellar and W. Kitchens. 1985. Modeling the impacts of river diversion on bottomland forest communities in the Santee River Floodplain, South Carolina. *Ecological Modelling* 29: 283–302.

Pearson, R. G. 1981. Recovery and recolonization of coral reefs. *Marine Ecology* 4: 105–122.

Pelz, D. R. 1978. Estimating tree growth with tree polygons (pp. 172–178). In: J. Fries, H. E. Burkhart and T. A. Max (eds.). *Growth Models for Forecasting of Timber Yields*. School of Forestry and Wildlife Resources, Virginia Polytechnical Institute and State University FWS–1–78, Blacksburg, VA.

Penman, H. L. 1948. Natural evaporation from open water, soil and grass. *Proceedings of the Royal Society of London, Series A* 193: 120–145.

Peterken, G. F. and E. W. Jones. 1987. Forty years of change in Lady Park Wood: the old-growth stands. *Journal of Ecology* 75: 477–512.

Peters, R. H. 1976. Tautology in evolution and ecology. *American Naturalist* 110: 1–12.

Peters, R. H. 1991. *A Critique for Ecology*. Cambridge University Press, Cambridge.

Peters, R. L. 1992. Conservation of biotic diversity in the face of climate change (pp. 15–30). In: R. L. Peters and T. E. Lovejoy (eds.). *Global Warming and Biological Diversity.* Yale University Press, New Haven, CT.

Petersen, C. G. J. 1918. The sea bottom and its production of fish food. *Reports of the Danish Biological Station* No. 25, pp. 1–62.

Phillips, J. 1934. Succession, development, the climax and the complex organism: an analysis of concepts. I. *Journal of Ecology* 22: 554–571.

Phillips, J. 1935a. Succession, development, the climax and the complex organism: an analysis of concepts. II. *Journal of Ecology* 23: 210–246.

Phillips, J. 1935b. Succession, development, the climax and the complex organism: an analysis of concepts. III. *Journal of Ecology* 23: 488–508.

Phipps, R. L. 1979. Simulation of wetlands forest dynamics. *Ecological Modelling* 7: 257–288.

Pianka, E. R. 1976. Competition and niche theory (pp. 114–141). In: R. M. May (ed.). *Theoretical Ecology: Principles and Applications.* Blackwell, Oxford.

Pianka, E. R. 1981. Competition and niche theory (pp. 167–196). In: *Theoretical Ecology: Principles and Applications,* 2nd edn. Blackwell, Oxford.

Pianka, E. R. 1994. *Evolutionary Ecology,* 5th edn. HarperCollins, New York.

Pickett, S. T. A., S. L. Collins and J. J. Armesto. 1987. Models, mechanisms and pathways of succession. *Botany Review* 53: 335–371.

Pickup, G. 1985. The erosion cell – a geomorphic approach to landscape classification in range assessment. *Australian Rangeland Journal* 7: 114–121.

Pickup, G. and V. H. Chewings. 1988. Estimating the distribution of grazing and patterns of cattle movement in a large arid zone paddock. *Internation Journal of Remote Sensing* 9: 1469–1490.

Pielou, E. C. 1977. *Mathematical Ecology,* 2nd edn. Wiley, New York.

Podger, F. D. 1972. *Phytophthora cinnamomi,* a cause of lethal disease in indigenous plant communities in Western Australia. *Phytopathology* 62: 972–981

Pollack, J. B., O. B. Toon, T. P. Ackerman, C. P. McKay and R. P. Turco. 1983. Environmental effects of an impact-generated dust cloud: implications for the Cretaceous–Tertiary extinctions. *Science* 219: 287–289.

Porter, J. H. and R. D. Dueser. 1982. Niche overlap and competition in an insular small mammal fauna: A test of the niche overlap hypothesis. *Oikos* 39: 228–236.

Post, W. M. and J. Pastor. 1990. An individual-based forest ecosystem model for projecting forest response to nutrient cycling and climate changes (pp. 61–74). In: L. Wensel and G. Biging (eds.). *Forest Simulation Systems: Proceedings of the IUFRO Conference,* Berkeley, California, 2–5 Nov. 1988. University of California, Division of Agriculture and Natural Resources, Bulletin 1927, Berkeley, CA.

Post, W. R., W. R. Emanuel, P. J. Zinke, A. G. Stangenberger. 1982. Soil carbon pools and world life zones. *Nature* 298: 156–159.

Post, W. M., J. Pastor, P. J. Zinke and A. G. Stangenberger. 1985. Global patterns of soil nitrogen. *Nature* 317: 613–616.

Post, W. M., T. -H. Peng, W. R. Emanuel, A. W. King, V. H. Dale and D. L. DeAngelis. 1990. The global carbon cycle. *American Scientist* 78: 310–326.

Potter, C. S., T. Randerson, C. B. Field, P. A. Matson, P. M. Vitousek, H. A. Mooney and S. A. Klooster. 1993. Terrestrial ecosystem production: a process model based on global satellite and surface data. *Global Biogeochemical Cycles* 7: 811–841.

Potter, M. W., S. R. Kessell and P. J. Cattelino. 1979. FORPLAN: A FORest Planning LANguage and simulator. *Environmental Management* 3: 59–72.

Prentice, I. C. 1992. Climatic change and long-term vegetation dynamics (pp. 293–339). In: D. C. Glenn-Lewin, R. A. Peet and T. T. Veblen (eds.). *Vegetation Dynamics Theory.* Chapman & Hall, New York.

Prentice, I. C., W. P. Cramer, S. P. Harrison, R. Leemans, R. A. Monserud and A. M. Solomon. 1993. A global biome model based on plant physiology and dominance, soil properties and climate. *Journal of Biogeography* 19: 117–134.

Prentice, K. C. and I. Y. Fung. 1990. Bioclimatic simulations test the sensitivity of terrestrial carbon storage to perturbed climates. *Nature* 346: 48–51.

Preston, F. W. 1962. The canonical distribution of commonness and rarity: Part I. *Ecology* 43: 185–215.

Prince, S. D. 1991. Satellite remote sensing of primary production: comparison of results for Sahelian grasslands. *International Journal of Remote Sensing* 12: 1301–1311.

Prince, S. D., S. Goward, D. Dye and C. Daughtry. 1991. Models for remote sensing of primary production at regional and global scales. *International Geosciences and Remote Sensing Symposium '90.* Vol. 1.

Prince, S. D., C. O Justice and B. Moore, III. 1994. *Monitoring and Modeling of Terrestrial Net and Gross Production. Joint IGBP-DIS-GAIM Working Papers.* WP 1. Institute for the Study of Earth, Oceans and Space, University of New Hampshire, Durham, NH.

Pulliam, H. R. 1988. Sources, sinks and population regulation. *American Naturalist* 132: 652–661.

Putz, F. E. 1983. Treefall pits and mounds, buried seeds, and the importance of soil disturbance to pioneer trees on Barro Colorado Island, Panama. *Ecology* 64: 1069–1074.

Putz, F. E. and S. Appanah. 1987. Buried seeds, newly dispersed seeds and the dynamics of lowland forest in Malaysia. *Biotropica* 19: 326–333.

Quinn, P. J. and W. P. Williams. 1978. Plant lipids and their role in membrane function. *Progress in Biophysics and Molecular Biology.* 34: 109–173.

Rackham, O. 1992. Mixtures, mosaics and clones: the distribution of trees within European woods and forests (pp. 1–20). In: M. G. R. Cannell, D. C. Malcolm and P. A. Robertson (eds.). *The Ecology of Mixed-species Stands of Trees.* Blackwell Scientific, Oxford.

Raich, J. W. and W. H. Schlesinger. 1992. The global carbon dioxide flux in soil respiration in relation to vegetation and climate. *Tellus* B 44: 81–99.

Raich, J. W., E. B. Rastetter, J. M. Melillo, D. W. Kicklighter, P. A. Steudler, B. J. Peterson, A. L. Grace, B. Moore, III and C. Vörösmarty. 1991. Potential net productivity in South America: application of a global model. *Ecological Applications* 1: 399–429.

Ramanathan, V., H. B. Singh, R. J. Cicerone and J. T. Kiehl. 1985. Trace gas trends and their potential role in climate change. *Journal of Geophysical Research* 90: 5547–5566.

Ramensky, L. G. 1924. Vestnik opytnogo dela Sredne-Chernoz [Basic lawfulness in the structure of vegetation cover]. Excerpted in: E. J. Kormandy (ed.). *Readings in Ecology* (pp. 151–152). Prentice-Hall, Englewood Cliffs, NJ.

Ranney, J. W. 1978. Edges of forest islands: structure, composition, and importance to regional forest dynamics. PhD Dissertation, University of Tennessee, Knoxville, TN.

Rastetter, E. B. 1991. A spatially explicit model of vegetation–habitat interactions on barrier islands (pp. 353–378). In: M. G. Turner and R. H. Gardner (eds.). *Quantitative Methods in Landscape Ecology. Ecological Studies 82.* Springer-Verlag, New York.

Rastetter, E. B., M. G. Ryan, G. R. Shaver, J. M. Melillo, K. J. Nadelhoffer, J. E. Hobbie and J. D. Aber. 1991. A general biogeochemical model describing the responses of the C and N cycles in terrestrial ecosystems to changes in CO_2, climate and N deposition. *Tree Physiology* 9: 101–126.

Raunkiaer, O. 1934. *The Life Forms of Plants and Statistical Plant Geography.* Clarendon Press, Oxford.

Raup, D. M. and J. J. Sepkoski. 1984. Periodicity of extinctions in the geologic past. *Proceedings of the National Academy of Sciences of the USA, Physical Sciences* 81: 801–805.

Raup, H. M. 1964. Some problems in ecological theory and their relation to conservation. *Journal of Ecology* 52(suppl.): 19–28.

Reed, K. L. 1980. An ecological approach to modeling growth of forest trees. *Forest Science* 26: 33–50.

Remmert, H. 1985. Was geschieht im Klima Stadium? *Naturwissenschaften* 72: 505–512.

Remmert, H. 1991. *The Mosaic-cycle Concept of Ecosystems.* Springer Verlag, Berlin.

Reynolds, J. C. 1985. Details of the geographical replacement of the red squirrel (*Sciurus vulgaris*) by the grey squirrel (*Sciurus carolinensis*) in Eastern England. *Journal of Animal Ecology* 54: 149–162.

Reynolds, J. F., D. Bachelet, P. W. Leadley and D. Moorhead. 1986. Response of vegetation to carbon dioxide. Assessing the effects of elevated carbon dioxide on plants: toward the development of a generic plant growth model. *Progress Report 023.* US Department of Energy, Washington, DC.

Reynolds, J. F., D. W. Hilbert, J. L. Chen, P. C. Harley, P. R. Kemp and P. W. Leadley. 1992. *Modeling the Response of Plants to Elevated CO_2 and Climate Change.* DOE/ER–60490Y-H1, Carbon Dioxide Research Division, US Department of Energy, Washington, DC.

Rice, J., R. D. Ohmart and B. W. Anderson. 1981. Bird community use of riparian habitats: the importance of temporal scale in interpreting discriminant function analysis (pp. 186–196). In: D. E. Capen (ed.). *The Use of Multivariate Statistics in Studies of Wildlife Habitat.* Gen. Tech. Rept. RM–87. Rocky Mountain Forest and Range Experimental Station. US Department of Agriculture-Forest Service, Ft. Collins, CO.

Rice, J., R. D. Ohmart and B. W. Anderson. 1983a. Turnovers in species composition of avian communities in contiguous riparian habitats. *Ecology* 64: 1444–1455.

Rice, J., R. D. Ohmart and B. W. Anderson. 1983b. Habitat selection attributes of an avian community: a discriminant function analysis. *Ecological Monographs* 53: 263–290.

Rice, J., B. W. Anderson and R. D. Ohmart. 1984. Comparison of the importance of different habitat attributes to avian community organization. *Journal of Wildlife Management* 48: 895–911.

Rice, J., R. D. Ohmart and B. W. Anderson. 1986. Limits in a data rich model: modeling experience with habitat management on the Colorado River (pp. 79–86). In: J. Verner, M. L. Morrison and C. J. Ralph (eds.). *Wildlife 2000: Modeling Habitat Relationships of Terrestrial Vertebrates.* University of Wisconsin Press, Madison, WI.

Rich, P. H. 1988. The origin of ecosystems by means of subjective selection (pp. 19–28). In: L. R. Pomeroy and J. J. Alberts (eds.). *Concepts of Ecosystem Ecology.* Springer-Verlag, New York.

Robbins, R. G. and R. Pullen. 1965. Vegetation of the Wabag-Tari area. *CSIRO. Australian Land Research Series* 15: 100–115.

Robertson, G. P. 1987. Geostatistics in ecology: interpolating with known variance. *Ecology* 78: 340–355.

Robock, A. 1978. Internally and externally caused climatic change. *Atmospheric Science* 35: 1111–1122.

Rohlf, F. J. and D. Davenport. 1969. Simulation of simple models of animal behavior with a digital computer. *Journal of Theoretical Biology* 23: 400–424.

Root, R. B. 1967. The niche exploitation pattern of the blue-gray gnatcatcher. *Ecological Monographs* 37: 317–350.

Rosenburg, N. J., B. A. Kimball, B. Martin and C. F. Cooper. 1990. From climate and CO_2 enrichment to evapotranspiration (pp. 151–175). In: P. E. Waggoner (ed.). *Climate Change and US Water Resources.* Wiley, New York.

Rosenzweig, M. L. 1968. Net primary productivity of terrestrial communities: prediction from climatological data. *American Naturalist* 102: 67–74.

Rosenzweig, M. L. 1969. Why the prey curve has a hump. *American Naturalist* 103: 81–87.

Rosenzweig, M. L. and R. H. MacArthur. 1963. Graphical representation and stability conditions of predator–prey interactions. *American Naturalist* 97: 209–223.

Rottenberry, J. T. and J. A. Wiens. 1980. Habitat structure, patchiness and avian communities in North America. *Ecology* 61: 1228–1250.

Roughgarden, J. 1976. resource partitioning among competing species (a coevolutionary approach. *Theoretical Population Biology* 9: 388–424.

Rübel, E. F. 1930. *Pflanzengesellschaften der Erde.* Verlag Has Huber, Berlin.

Runkle, J. R. 1989. Synchrony of regeneration, gaps and latitudinal differences in tree species diversity. *Ecology* 70: 546–547.

Running, S. W. and J. C. Coughlan. 1988. A general model of forest ecosystem processes for regional applications I. Hydrological balance, canopy gas exchange and primary production processes. *Ecological Modelling* 42: 125–154.

Running, S. W. and S. T. Gower. 1991. FOREST-BCG, a general carbon model of forest ecosystem processes for regional applications. II. Dynamic carbon allocation and nitrogen budgets. *Tree Physiology* 9: 147–160.

Running, S. W. and R. R. Nemani. 1988. Relating seasonal patterns of the AVHRR vegetation index to simulated photosynthesis and transpiration of forests in different climates. *Remote Sensing of the Environment* 24: 347–367.

Running, S. W., R. R. Nemani, D. L. Peterson, L. E. Band, D. F. Potts, L. L. Peirce and M. A. Spanner. 1989. Mapping regional forest evapotranspiration and photosynthesis by coupling satellite data with ecosystem simulation. *Ecology* 70: 1090–1101.

Russell, G., P. G. Jarvis and J. L. Monteith. 1989. Absorption of radiation by canopies and stand growth. In: G. Russell, P. G. Jarvis and B. Marshall (eds.). *Plant Canopies: Their Growth, Form and Function.* Cambridge University Press, Cambridge.

Ryan, M. G., R. E. McMurtrie, G. Ågren, E. R. Hunt, Jr, J. D. Aber, A. D. Friend, E. B. Rastetter and W. M. Pulliam. 1997a. Comparing models of ecosystem function for temperate conifer forests: I. Model description and validation. In: *Effects of Climate Change on Forests and Grasslands*. Wiley, New York.

Ryan, M. G., R. E. McMurtrie, G. Ågren, E. R. Hunt, Jr, J. D. Aber, A. D. Friend, E. B. Rastetter, W. M. Pulliam. 1997b. Comparing models of ecosystem function for temperate conifer forests: II. Simulations of the effect of climate change. In: *Effects of Climate Change on Forests and Grasslands*. Wiley, New York.

Sage, R. F. 1995. Was low atmospheric CO_2 during the Pleistocene a limiting factor for the origin of agriculture? *Global Change Biology* 1: 93–106.

Sakai, A. and C. J. Weiser. 1973. Freezing resistance of trees in North America with reference to the tree regions. *Ecology* 54: 118–126.

Salmonson, M. G. and R. P. Balda. 1977. Winter territoriality of Townsend's Solitaires (*Myadestes townsendi*) in a Pinon–Juniper–Ponderosa ecosystem. *Condor* 79: 148–161.

Saltzman, B. 1983. Climatic systems analysis. *Advances in Geophysics* 25: 173–233.

Sammi, J. C. 1969. Graphic stand tables. *Journal of Forestry* 67: 498–500.

Sanford, R. L., Jr, W. J. Parton and D. J. Lodge. 1991. Hurricane effects on soil organic matter dynamics and forest production in the Luquillo Experimental Forest, Puerto Rico: results of simulation modeling. *Biotropica* 23: 364–372.

Sarmiento, J. L. and M. Bender. 1994. Carbon biogeochemistry and climate change. *Photosynthesis Research* 39: 209–234.

Sarmiento, J. L. and U. Siegenthaler. 1992. New production and the global carbon cycle (pp. 317–332). In: P. G. Falkowski and A. D. Woodhead (eds.). *Primary Productivity and the Biogeochemical Cycles of the Sea*. Plenum Press, New York.

Sarmiento, J. L., J. C. Orr and U. Siegenthaler. 1992. A perturbation simulation of CO_2 uptake in an ocean general circulation model. *Journal of Geophysical Research* 97: 3621–3646.

Saxon, E. C. 1983. Mapping the habitats of rare animals in the Tanami Wildlife Sanctuary (central Australia): an application of satellite imagery. *Biological Conservation* 27: 243–257.

Schimper, A. F. W. 1898. *Pflanzengeographie auf physiologischer Grundlage*. Fischer-Verlag, Jena.

Schlesinger, M. E. and Z.-C. Zhao. 1988. Seasonal Climatic Changes Induced by Doubled CO_2 as Simulated by the OSU Atmospheric GCM/Mixed Layer Ocean Model. Climate Research Institute, Oregon State University, Corvallis, OR.

Schlesinger, M. E. and Z.-C. Zhao. 1989. Seasonal climate changes induced by a doubled CO_2 as simulated by the OSU atmospheric GCM-mixed layer ocean model. *Journal of Climatology* 2: 459–495.

Schlesinger, W. H. 1984. Soil organic matter: a source of atmospheric CO_2 (pp. 111–127). In: G. M. Woodwell (ed.). *The Role of Terrestrial Vegetation in the Global Carbon Cycle*. Wiley, New York.

Schlesinger, W. H., J. F. Reynolds, G. L. Cunningham, L. F. Huenneke, W. M. Jarrell, R. A. Virginia and W. G. Whitford. 1990. Biological feedbacks in global desertification. *Science* 147: 1043–1048.

Schoener, T. W. 1971. Theory of feeding strategies. *Annual Review of Ecology and Systematics* 2: 369–404.

Schoener, T. W. 1983. Field experiments on interspecific competition. *American Naturalist* 122:240–285.

Schoener, T. W. 1985. Some comments on Connell's and my reviews of field experiments on interspecific competition. *American Naturalist* 125:730–740.

Schoener, T. W. 1988. Sizes of feeding territories among birds. *Ecology* 49:123–131.

Schouw, J. F. 1823. *Grundzüge einer allgemeinen Pflanzengeographie*, Reimer. Berlin

Schulze, E.-D. 1982. Plant life forms and their carbon, water and nutrient relations (pp. 616–676). In: O. L. Lange, P. S. Nobel, C. B. Osmond and H. Ziegler (eds.). *Encyclopedia of Plant Physiology*, vol. 12B. Springer-Verlag, Berlin.

Schwartz, M. D. and T. R. Karl. 1990. Spring phenology: Nature's experiment to detect the effect of 'green-up' on the surface maximum temperatures. *Monthly Weather Review* 118:883–890.

Scott, P. A., R. I. C. Hansell and D. C. F. Fayle. 1987. Establishment of white spruce populations and responses to climatic change at the treeline, Churchill, Manitoba, Canada. *Arctic and Alpine Research* 19:45–51.

Scurfield, G. 1960. Air pollution and tree growth. *Forestry Abstracts* 21:339–349.

Seagle, S. W. and H. H. Shugart. 1985. Landscape dynamics and the species-area curve. *Journal of Biogeography* 12:499–508.

Seidel, S. and D. Keyes. 1983. *Can We Delay a Greenhouse Warming?* Environmental Protection Agency, Washington, DC.

Sellers, P. J. 1985. Canopy reflectance, photosynthesis and transpiration. *International Journal of Remote Sensing* 6:135–1372.

Sellers, P. J. 1987. Canopy reflectance, photosynthesis and transpiration II. *Remote Sensing of the Environment* 21:143–183.

Sellers, P. J., Y. Mintz, Y. C. Sud and A. Dalcher. 1986. The design of a simple biosphere model (SiB) for use within general circulation models. *Journal of Atmospheric Science* 43:505–531.

Sellers, P. J., S. O. Los, C. J. Tucker, C. O. Justice, D. A. Dazlich, G. J. Collatz and D. A. Randall. 1996. A revised land surface parameterization (SiB–2) for atmospheric general circulation models. Part 2. The generation of global fields of terrestrial biophysical parameters from satellite data. *Journal of Climate* 9:706–737.

Semper, K. 1881. *Animal Life as Affected by the Natural Conditions of Existence.* Appleton, New York.

Senser, M. and E. Beck. 1977. On the mechanisms of frost injury and frost hardening of spruce chloroplasts. *Planta* 137:195–201.

Shao, G. 1991. Moisture-therm indices and optimum-growth modeling for the main species in Korean pine/deciduous mixed forests. *Scientia Silvae Sinicae* 21:21–27.

Shao, G., H. H. Shugart and T. M. Smith. 1996. A role-type model (ROPE) and its application in assessing climate change impacts on forest landscapes. *Vegetatio* 121:135–146.

Shao, G., S. Zhao and G. Zhao. 1991. Application of GIS in simulation of forested landscape communities: a case study in Changbaishan Biosphere Reserve. *Chinese Journal of Applied Ecology* 2:103–107.

Sharkey, T. D. 1985. Photosynthesis in intact leaves of C_3 plants: physics, physiology and rate limitations. *Botanical Review* 51:53–105.

Sharpe, P. J. H., J. Walker, L. K. Penridge and H. Wu. 1985. A physiologically-based continuous-time Markov approach to plant growth modeling in semi-arid woodlands. *Ecological Modelling* 29:189–213.

Sharpe, P. J. H., J. Walker, L. K. Penridge H. Wu and E. J. Rykiel. 1986. Spatial considerations in physiological models of tree growth. *Tree Physiology* 2: 403–421.

Shear, G. M. and W. D. Stewart. 1934. Moisture and pH studies of the soil under forest trees. *Ecology* 15: 350–358.

Shelford, V. E. 1913. *Animal Communities in Temperate America.* University of Chicago Press, Chicago, IL.

Sheppard, C. W. and A. S. Householder. 1951. The mathematical basis of the interpretation of tracer experiments in closed, steady-state systems. *Journal of Applied Physics* 22: 510–520.

Shinozaki, K., K. Yoda, K. Hozumi and T. Kira. 1964. A quantitative analysis of plant form – the pipe model theory. I. Basic analysis. *Japanese Journal of Ecology* 14: 97–105.

Shugart, H. H. l984. *A Theory of Forest Dynamics: The Ecological Implications of Forest Succession Models.* Springer-Verlag, New York.

Shugart, H. H. 1987. Dynamic ecosystem consequences of tree birth and death patterns. *BioScience* 37: 596–602.

Shugart, H. H. 1997. Plant and ecosystem functional types. In: T. M. Smith, H. H. Shugart and F. I. Woodward (eds.). *Plant Functional Types: Their Relevance to Ecosystem Properties and Global Change.* Cambridge University Press, Cambridge.

Shugart, H. H. and W. R. Emanuel. 1985. Carbon dioxide increase: the implications at the ecosystem level. *Plant, Cell and Environment* 8: 381–386.

Shugart, H. H. and S. B. McLaughlin. 1986. Modelling SO_2 effects on forest growth and community dynamics (pp. 478–491). In: W. E. Winner, H. A. Mooney and R. A. Goldstein (eds.). *Sulfur Dioxide and Vegetation: Physiology, Ecology, and Policy Issues.* Stanford Press, Stanford, CA.

Shugart, H. H. and I. R. Noble. 1981. A computer model of succession and fire response to the high altitude eucalyptus forest of the Brindabella Range, Australian Capital Territory. *Australian Journal of Ecology* 6: 149–164.

Shugart, H. H. and R. V. O'Neill (eds.). 1979. *Systems Ecology.* Dowden, Hutchinson and Ross, Stroudsburg, PA.

Shugart, H. H. and B. C. Patten. 1972. Niche quantification and the concept of niche pattern (pp. 284–326). In: B. C. Patten (ed.). *Systems Analysis and Simulation in Ecology,* Vol. II. Academic Press, New York.

Shugart, H. H. and D. L. Urban. 1988. Scale, synthesis and ecosystem dynamics (pp. 279–290). In: L. R. Pomeroy and J. J. Alberts (eds.). *Essays in Ecosystems Research: A Comparative Review.* Springer-Verlag, New York.

Shugart, H. H. and D. L. Urban. 1989. Factors affecting the relative abundance of forest tree species (pp. 249–274). In: P. J. Grubb and J. B. Whittaker (eds.). *Toward a More Exact Ecology.* Jubilee Symposium of the British Ecological Society, Blackwell, Oxford.

Shugart, H. H. and D. C. West. 1977. Development of an Appalachian deciduous forest succession model and its application to assessment of the impact of the chestnut blight. *Journal of Environmental Management* 5: 161–179.

Shugart, H. H. and D. C. West. 1979. Size and pattern in simulated forest stands. *Forest Science* 25: 120–122.

Shugart, H. H. and D. C. West. 1980. Forest succession models. *BioScience* 30: 308–313.

Shugart, H. H. and D. C. West. 1981. Long-term dynamics of forest ecosystems. *American Scientist* 69: 647–652.

Shugart, H. H., T. R. Crow and J. M. Hett. 1973. Forest succession models: a rationale and methodology for modeling forest succession over large regions. *Forest Science* 19: 203–212.

Shugart, H. H., R. A. Goldstein, R. V. O'Neill and J. B. Mankin. 1974. TEEM: a terrestrial ecosystem energy model for forest. *Oecologica Plantarium* 9: 251–284.

Shugart, H. H., M. S. Hopkins, I. P. Burgess and A. T. Mortlock. 1980a. The development of a succession model for subtropical rain forest and its application to assess the effects of timber harvest at Wiangaree State Forest, New South Wales. *Journal of Environmental Management* 11: 243–265.

Shugart, H. H., W. R. Emanuel, D. C. West, and D. L. DeAngelis. 1980b. Environmental gradients in a beech–yellow poplar stand simulation model. *Mathematical Biosciences* 50: 163–170.

Shugart, H. H., D. C. West and W. R. Emanuel. 1981. Patterns in the long-term dynamics of forests: an application of simulation models (pp. 74–94). In: D. C. West, H. H. Shugart and D. B. Botkin (eds.). *Forest Succession: Concepts and Application*. Springer-Verlag, New York.

Shugart, H. H., M. Ja. Antonovsky, P. G. Jarvis and A. P. Sandford. 1986. CO_2, climatic change and forest ecosystems: assessing the response of global forests to the direct effects of increasing CO_2 and climatic change (pp. 475–521). In: B. Bolin, B. R. Döös, J. Jager and R. A. Warrick (eds.). *The Greenhouse Effect, Climatic Change and Ecosystems, SCOPE 29*. Wiley, New York.

Shugart, H. H., G. B. Bonan and E. B. Rastetter. 1988. Niche theory and community organization. *Canadian Journal of Botany* 66: 2634–2639.

Shugart, H. H., R. Leemans and G. B. Bonan (eds.). 1992a. *A Systems Analysis of the Global Boreal Forest*. Cambridge University Press, Cambridge.

Shugart, H. H., T. M. Smith and W. M. Post. 1992b. The application of individual-based simulation models for assessing the effects of global change. *Annual Reviews in Ecology and Systematics* 23: 15–38.

Shukla, J. and Y. Mintz. 1982. Influence of land-surface evapotranspiration on the earth's climate. *Science* 215: 1498–1501.

Shukla, J., C. Nobre and P. J. Sellers. 1990. Amazon deforestation and climate change. *Science* 247: 1498–1501.

Siegenthaler, U., H. Friedli, H. Loetscher, E. Moor, A Neftel, H. Oeschger and B. Stauffer. 1988. Stable isotope ratios and concentration of CO_2 in air from polar ice cores. *Annals in Glaciology* 10: 1–6.

Silvertown, J. and B. Smith. 1988. Gaps in the canopy: the missing dimension in vegetation dynamics, *Vegetatio* 77: 57–60.

Simberloff, D. and T. Dayan. 1991. The guild concept and the structure of ecological communities. *Annual Reviews in Ecology and Systematics* 22: 115–143.

Sinclair, A. R. E. 1989. Population regulation in animals (pp. 197–241). In: J. M. Cherrett (ed.). *Ecological Concepts*. Blackwell Scientific, Oxford.

Singh, G., A. P. Kershaw and R. Clark. 1981. Quaternary vegetation and fire history in Australia (pp. 23–54). In: A. M. Gill, R. H. Groves and I. R. Noble (eds.). *Fire and the Australian Biota*. Australian Academy of Sciences, Canberra.

Sinko, J. W. and W. Streifer. 1967. A new model for age-size structure of a population. *Ecology* 48: 910–918.

Sirois, L., G. B. Bonan and H. H. Shugart. 1994. Development of a simulation model of the forest–tundra transition zone of northeastern Canada. *Canadian Journal of Forest Research* 24: 697–706.

Skellam, J. G. 1973. The formulation and interpretation of mathematical models of diffusionary processes in population biology (pp. 63–85). In: M. S. Bartlett and R. W. Hiorns (eds.). *The Mathematical Theory of the Dynamics of Biological Populations*. Academic Press, London.

Slatyer, R. O. 1961. Methodology of a water balance study conducted on a desert woodland community in central Australia (pp. 15–26). In: *UNESCO Arid Zone Research 16: Plant-water relationships in arid and semi-arid conditions*. UNESCO, Paris.

Slatyer, R. O. 1990. Alpine and valley bottom treelines (pp. 169–184). In: R. Good (ed.). *The Scientific Significance of the Australian Alps*. Australian Alps National Parks Liaison Committee and Australian Academy of Science, Canberra, Australia.

Slingo, A. 1989. A CGM parametrization for the shortwave radiative properties of water clouds. *Journal of Atmospheric Science* 46: 1419–1427.

Smalley, G. W. and R. L. Bailey. 1974. *Yield Tables and Stand Structure for Loblolly Pine Plantations in Tennessee, Alabama, and Georgia Highlands. Research Paper 50–96.* US Department of Agriculture-Forest Service, Washington, DC.

Smith, J. B. and D. A. Tirpak (eds.). 1989. *The potential effects of global climate change on the U.S. : Appendix D – Forests.* Office of Policy, Planning and Evaluation, USEPA, Washington, DC.

Smith, J. M. 1974. *Models in Ecology.* Cambridge University Press, Cambridge.

Smith, O. L. 1980. The influence of environmental gradients on ecosystem stability. *American Naturalist* 116: 1–24.

Smith, R. L. 1980. *Ecology and Field Biology*, 3rd edn. Harper and Row, New York.

Smith, T. M. 1986. Habitat simulation models: integrating habitat-classification into forest-simulation models (pp. 389–394). In: J. Verner, M. L. Morrison and C. J. Ralph (eds.). *Wildlife 2000: Modeling Habitat Relationships of Terrestrial Vertebrates.* University of Wisconsin Press, Madison, WI.

Smith, T. M. and M. Huston. 1989. A theory of the spatial and temporal dynamics of plant communities. *Vegetatio* 83: 49–69.

Smith, T. M. and H. H. Shugart. 1987. Territory size variation in the ovenbird: the role of habitat structure. *Ecology* 68: 695–704.

Smith, T. M. and H. H. Shugart. 1993. The transient response of terrestrial carbon storage to a perturbed climate. *Nature* 361: 523–526.

Smith, T. M. and D. L. Urban. 1988. Scale and the resolution of forest structural pattern. *Vegetatio* 74: 143–150.

Smith, T. M., H. H. Shugart and D. C. West. 1981a. FORHAB. A forest simulation model to predict habitat structure for nongame bird species (pp. 114–121). In: D. E. Capen (ed.). *The Use of Multivariate Statistics in Studies of Wildlife Habitat. General Technical Report RM–87.* Rocky Mountain Forest and Range Experimental Station, US Department of Agriculture-Forest Service, Ft. Collins, CO.

Smith, T. M., H. H. Shugart and D. C. West. 1981b. The use of forest simulation models to integrate timber harvest and nongame bird habitat management (pp. 501–510). *46th North American Wildlife and Natural Resources Conference 1981.* Wildlife Management Institute, Washington, DC.

Smith, T. M., H. H. Shugart, G. B. Bonan and J. B. Smith. 1992a. Modeling the potential response of vegetation to global climate change. *Advances in Ecological Research* 22: 93–116.

Smith, T. M., R. Leemans and H. H. Shugart. 1992b. Sensitivity of terrestrial carbon storage to CO_2 induced climate change: comparison of five scenarios based on general circulation models. *Climatic Change* 21: 367–384.

Smith, T. M., H. H. Shugart, F. I. Woodward, P. J. Burton. 1993. Plant functional types (pp. 272–292). In: A. M. Solomon and H. H. Shugart (eds.). *Vegetation Dynamics and Global* Change. Chapman & Hall, New York.

Smith, T. M., P. N. Halpin, H. H. Shugart and C. M. Secrett. 1995. Global forests (pp. 146–179). In: K. M. Strzepek and J. B. Smith (eds.). *If Climate Changes: International Impacts of Climate Change.* Cambridge University Press, Cambridge.

Smith, T. M., H. H. Shugart and F. I. Woodward (eds.). 1997. *Plant Functional Types: Their Relevance to Ecosystem Properties and Global Change.* Cambridge University Press, Cambridge.

Smith, W. H. 1981. *Air Pollution and Forests.* Springer Verlag, New York.

Solbrig, O. T. 1994. Biodiversity: an introduction (pp. 13–20). In: O. T. Solbrig, H. M. van Emden and P. G. W. J. van Oordt (eds.). *Biodiversity and Global Change.* CAB International, Wallingford, UK.

Solomon, A. K. 1949. Equations for tracer experiments. *Journal of Clinical Investigation* 28: 1297–1307.

Solomon, A. M. 1986a. Comparison of taxon calibrations, modern analog techniques, and forest-stand simulation models for the quantitative reconstruction of past vegetation: a critique. *Earth Surface Processes and Landforms* 11: 681–685.

Solomon, A. M. 1986b. Transient response of forests to CO_2-induced climate change: simulation experiments in eastern North America. *Oecologia* 68: 567–79.

Solomon, A. M. and Shugart, H. H. 1984. Integrating forest-stand simulations with paleoecological records to examine long-term forest dynamics (pp. 333–357). In: G. I. Ågren (ed.). *State and Change of Forest Ecosystems: Indicators in Current Research.* Report Number 13, Swedish University of Agricultural Science, Uppsala, Sweden.

Solomon, A. M. and T. Webb III. 1985. Computer-aided reconstruction of late Quaternary landscape dynamics. *Annual Reviews of Ecology and Systematics* 16: 63–84.

Solomon, A. M., H. R. Delcourt, D. C. West and T. J. Blasing. 1980. Testing a simulation model for reconstruction of prehistoric forest-stand dynamics. *Quaternary Research* 14: 275–293.

Solomon, A. M., D. C. West and J. A. Solomon. 1981. Simulating the role of climate change and species immigration in forest succession (pp. 154–177). In: D. C. West, H. H. Shugart and D. B. Botkin (eds.). *Forest Succession: Concepts and Application.* Springer-Verlag, New York.

Solomon, A. M., M. L. Tharp, D. C. West, G. E. Taylor, J. M. Webb and J. L. Trimble. 1984. *Response of Unmanaged Forests to CO_2-induced Climate Change: Available Information, Initial Tests and Data Requirements. Tech.* Report *TR009.* US DOE Carbon Dioxide Research Division, Washington, DC.

Solomon, D. S. 1974. Simulation of the development of natural and silviculturally treated stands of even-aged northern hardwoods (pp. 327–352). In: J. Fries (ed.). *Growth Models for Tree and Stand Simulation*. Res. Notes 30. Department of Forest Yield Research, Royal College of Forestry, Stockholm, Sweden.

Song, J. 1988. *Population System Control*. China Academic Press, Beijing.

Song, J. and J. Yu. 1981. On stability theory of population systems and critical fertility rates. *Mathematical Modelling* 2: 109–121.

Song, J., C. Taun and J. Yu. 1985. *Population Control in China – Theory and Application*. Preager, New York.

Sorenson, L. H. 1981. Carbon–nitrogen relationships during the humification of cellulose in soil containing different amounts of clay. *Soil Biology and Biochemistry* 13: 313–321.

Sousa, W. P. 1979. Disturbance in marine intertidal boulder fields: the nonequilibrium maintenance of species diversity. *Ecology* 60: 1225–1239.

Sousa, W. P. 1984. The role of disturbance in natural communities. *Annual Reviews in Ecology and Systematics* 15: 353–91.

Southwood, T. R. E. 1977. Habitat, the template for ecological strategies. *Journal of Animal Ecology* 46: 337–365.

Spies, T. A. and J. F. Franklin. 1989. Gap characteristics and vegetation response in coniferous forests of the Pacific Northwest. *Ecology* 70: 543–545.

Spies, T. A., J. F. Franklin and M. Klopsch. 1990. Gap characteristics and vegetation response in coniferous forests of the Pacific Northwest. *Ecology* 70: 543–545.

Sprugel, D. G. 1976. Dynamic structure of wave-generated *Abies balsamea* forests in northeastern United States. *Journal of Ecology* 64: 889–911.

Sprugel, D. G., and F. H. Bormann. 1981. Natural disturbance and the steady state in high-altitude balsam fir forests. *Science* 211: 390–393.

Spurr, S. H. 1952. *Forest Inventory*. Ronald Press, New York.

Staebler, G. R. 1951. Growth and spacing in an even-aged stand of Douglas-fir. MF Thesis, University of Michigan, Ann Arbor MI.

Stafford-Smith, M. and G. Pickup. 1993. Out of Africa, looking in. Understanding vegetation change and its implications for management in Australian rangelands (pp. 196–226). In: R. H. Benke, I. Scoones and C. Kerven (eds.). *Range Ecology at Disequilibrium*. Overseas Development Institute, Commonwealth Secretariat, London.

Stenger, J. 1958. Food habits and available food of ovenbirds in relation to territory size. *Auk* 75: 125–140.

Stenger, J. and J. B. Falls. 1959. The utilized territory of the ovenbird. *Wilson Bulletin* 71: 125–140.

Stephenson, N. T. 1990. Climatic control of vegetation distribution: the role of the water balance. *American Naturalist* 135: 649–670.

Stimson, J. 1973. The role of territory in the ecology of the intertidal limpet *Lottia gigantea* (Gray). *Ecology* 54: 1020–1030.

Strain, B. R. 1992. Atmospheric carbon dioxide: A plant fertilizer? *New Biologist* 4: 87–89.

Sugihara, G. 1980. Minimal community structure: an explanation of species abundance patterns. *American Naturalist* 116: 770–787.

Sukachev, V. 1968a. Basic concepts in forest biogeocoenology (pp. 4–59). In: V.

Sukachev and N. Dylis (eds.). *Fundamentals of Forest Biogeocoenology.* Oliver and Boyd, Edinburgh. (A translation of: *Osnovy lesnoi biogeotsenologii.* 1964. Nauka, Moscow.)

Sukachev, V. 1968b. Dynamics of forest biogeocoenoses (pp. 538–571). In: V. Sukachev and N. Dylis (eds.). *Fundamentals of Forest Biogeocoenology.* Oliver and Boyd, Edinburgh. (A translation of: *Osnovy lesnoi biogeotsenologii.* 1964. Nauka, Moscow.)

Sullivan, A. D. and J. L. Clutter. 1972. A simultaneous growth and yield model for loblolly pine. *Forestry Science* 18: 76–86.

Sutherland, W. J. and P. M. Dolman. 1994. Combining behaviour and population dynamics with applications for predicting consequences of habitat loss. *Proceedings of the Royal Society of London, Series B* 255: 133–138.

Sutherland, W. J. and P. Koene. 1992. Field estimates of the strength of interference between Oystercatchers *Haematopus ostraegus. Oecologia* 55: 108–109.

Suzuki, T. and T. Umemura. 1967a. Forest transition as a stochastic process. III. *Journal of the Japanese Forestry Society* 49: 208–210.

Suzuki, T. and T. Umemura. 1967b. Forest transition as a stochastic process. IV. *Journal of the Japanese Forestry Society* 49: 402–404.

Suzuki, T. and T. Umemura. 1974. Forest transition as a stochastic process. V. *Journal of the Japanese Forestry Society* 56: 195–204.

Swain M. D. and J. B. Hall. 1988. The mosaic theory of forest regeneration and the determination of forest composition in Ghana. *Journal of Tropical Ecology* 4: 253–269.

Swift, L. W., W. T. Swank, J. B. Mankin, R. J. Luxmore and R. A. Goldstein. 1979. Simulation of evapotranspiration and drainage from mature and clearcut deciduous forests and young pine plantation. *Water Resources Research* 11: 667–673.

Tanemura, M. and M. Hasegawa. 1980. Geometric models of territory. I. Models for synchronous and asynchronous settlement of territories. *Journal of Theoretical Biology* 82: 477–496.

Tans, P. P., I. Y. Fung and T. Takahashi. 1990. Observational constraints on the global CO_2 budget. *Science* 247: 1431–1438.

Tans, P. P., J. A. Berry and R. F. Keeling. 1993. Oceanic $^{13}C/^{12}C$ observations: a new window on ocean CO_2 uptake. *Global Biogeochemical Cycles* 7: 353–368.

Tansley, A. G. 1935. The use and abuse of vegetational concepts and terms. *Ecology* 16: 284–307.

Taylor, P. R. and M. M. Littler. 1982. The roles of compensatory mortality, physical disturbance, and substrate retention in the development and organization of a sand-influenced, rocky-intertidal community. *Ecology* 63: 135–146.

Teorell, T. 1937. Kinetics of distribution of substances administered to the body. *Archives in International Pharmacodynamics* 57: 205–240.

Terborgh, J. 1989. *Where Have All the Birds Gone?* Princeton University Press, Princeton, NJ.

Tharp, M. L. 1978. *Modeling major perturbations on a forest ecosystem.* MS Thesis. University of Tennessee, Knoxville.

Thienemann, A. 1918. *Lebengemeinschaft und Lebensraum. Naturwissenshaft Wocheschrift,* N. F. 17: 282–290; 297–303.

Thiéry, J. M., J.-M. D'Herbès and C. Valentin. 1995. A model simulating the genesis of banded vegetation patterns in Niger. *Journal of Ecology* 83: 497–507.

Thornthwaite, C. W. 1931. The climates of North America according to a new classification. *Geographical Review* 21: 633–655.

Tiessen, H. and J. W. B. Stewart. 1983. Particle-size fractions and use in studies of soil organic matter: II. Cultivation effects on organic matter composition in size fractions. *Soil Society of America Journal* 47: 509–514.

Tiessen, H., J. W. B. Stewart and J. R. Bettany. 1982. Cultivation effects on the amounts and concentration of carbon, nitrogen and phosphorus in grassland soils. *Agronomical Journal* 74: 831–835.

Tilman, D. 1982. *Resource Competition and Community Structure*. Princeton University Press, Princeton, NJ.

Tilman, D. 1985. The resource-ratio hypothesis of plant succession. *American Naturalist* 125: 827–852.

Tilman, D. 1988. *Plant Strategies and the Dynamics and Structure of Plant Communities*. Princeton University Press, Princeton, NJ.

Timothy, L. K. and E. Bona. 1968. *State Space Analysis: An Introduction*. McGraw-Hill, New York.

Tongway, D. J. and J. A Ludwig. 1990. Vegetation and soil patterning in semi-arid mulga lands of Eastern Australia. *Australian Journal of Ecology* 15: 23–34.

Tucker, C. J. 1979. Red and photographic infrared linear combinations for monitoring vegetation. *Remote Sensing of the Environment* 8: 127–150.

Tucker, C. J. and P. J. Sellers. 1986. Satellite remote sensing of primary production. *International Journal of Remote Sensing* 7: 1395–1416.

Tucker, C. J., C. L. Vanpraet, M. J. Sharman and G. Van Ittersum. 1985a. Satellite remote sensing of total herbaceous biomass production in the Senegalese Sahel: 1980–1984. *Remote Sensing of the Environment* 17: 233–249.

Tucker, C. J., J. U. Hielkema and J. Roffey. 1985b. The potential of satellite remote sensing of ecological conditions for survey and forecasting desert-locust activity. *International Journal of Remote Sensing* 7: 1395–1416.

Tucker, C. J., J. R. G. Townshend and T. E. Goff. 1985c. African land-cover classification using satellite data. *Science* 227: 369–375.

Tucker, C. J., I. Y. Fung, C. D. Keeling and R. H. Gammon. 1986. Relationship between atmospheric CO_2 variations and a satellite derived vegetation index. *Nature* 319: 195–199.

Turner, M. G., V. H. Dale and R. H. Gardner. 1989a. Predicting across scales: theory development and testing. *Landscape Ecology* 3: 245–252.

Turner, M. G., R. V. O'Neill, R. H. Gardner and B. T. Milne. 1989b. Effects of changing spatial scale on the analysis of landscape pattern. *Landscape Ecology* 2: 153–162.

Urban, D. L. 1981. Habitat relationships of birds and small mammals in second-growth forests. Master's thesis. Southern Illinois University, Carbondale, IL.

Urban, D. L. and H. H. Shugart. 1989. Forest response to climate change: a simulation study for southeastern forests. (pp. 3–1 to 3–45). In: J. Smith and D. Tirpak (eds.). *The Potential Effects of Global Climate Change on the United States*. EPA–230–05–89–054, US Environmental Protection Agency, Washington, DC.

Urban, D. L. and H. H. Shugart. 1992. Individual-based models of forest succession (pp. 249–293). In: D. C. Glenn-Lewin, R. K. Peet and T. T. Veblen (eds.). *Plant Succession: Theory and Prediction.* Chapman & Hall, London.

Urban, D. L. and T. M. Smith. 1989. Microhabitat pattern and the structure of forest bird communities. *American Naturalist* 133: 811–829.

Urban, D. L., R. V. O'Neill and H. H. Shugart. 1987. Landscape ecology. *BioScience* 37: 119–27.

Urban, D. L., G. B. Bonan, T. M. Smith and H. H. Shugart. 1991. Spatial applications of gap models. *Forest Ecology and Management* 42: 95–110.

Usher, M. B. 1973. *Biological Management and Conservation: Ecological Theory, Application and Planning.* Chapman & Hall, London.

Usher, M. B. 1981. Modelling ecological succession, with particular reference to Markovian models. *Vegetatio* 46: 11–18.

Usher, M. B. 1992. Statistical models of succession (pp. 215–248). In: D. C. Glenn-Lewin, R. K. Peet and T. T. Veblen (eds.). *Plant Succession: Theory and Prediction.* Chapman & Hall, London.

Van Cleve, K. and L. A. Vierick. 1981. Forest succession in relation to nutrient cycling in the boreal forest of Alaska (pp. 185–221). In: D. C. West, H. H. Shugart, and D. B. Botkin (eds.). *Forest Succession: Concepts and Application.* Springer-Verlag, New York.

van Daalen, J. C. and H. H. Shugart. 1989. OUTENIQUA – A computer model to simulate succession in the mixed evergreen forests of the southern Cape, South Africa. *Landscape Ecology* 24: 255–267.

van der Pijl, L. 1972. *Principles of Dispersal in Higher Plants.* Springer-Verlag, Berlin.

van Horne, B. 1983. Density as a misleading indicator of habitat quality. *Journal of Wildlife Management* 47: 893–901.

van Hulst, R. 1979. On the dynamics of vegetation: succession in model communities. *Vegetatio* 39: 85–96.

van Tongeren, O. and I. C. Prentice. 1986. A spatial simulation model for vegetation dynamics. *Vegetatio* 65: 163–173.

Van Veen, J. A., J. H. Ladd and M. J. Frissel. 1984. Modeling C and N turnover through the microbial biomass in soil. *Plant Soil* 76: 257–274.

Van Voris, P., R. V. O'Neill, W. R. Emanuel and H. H. Shugart. 1980. Functional complexity and ecosystem stability. *Ecology* 61: 1352–1360.

Van Wagner, C. E. 1973. Height of crown scorch in forest fires. *Canadian Journal of Forestry Research* 3: 373–378.

Varley, G. C. 1957. Ecology as an experimental science. *Journal of Animal Ecology* 26: 251–261.

Veblin, T. T. 1979. Structure and dynamics of *Nothofagus* forests near timberline in south-central Chile. *Ecology* 60: 937–945.

Veblin, T. T., D. H. Ashton and F. M. Schlegel 1979. Tree regeneration strategies in a lowland *Nothofagus* forest in south-central Chile. *Journal of Biogeography* 6: 329–340.

Veblin, T. T., Z. C. Donoso, F. M. Schlegel, and R. B. Escobar. 1981. Forest dynamics in south-central Chile. *Journal of Biogeography* 8: 211–247.

VEMAP [J. Borchers, J. Chaney, S. Fox, A Haxeltine, A. Janetos, D. W. Kicklighter, T. G. F. Kitel, A. D. McGuire, R. McKeown, J. M. Melillo, R. Neilson, R. Nemani, D. S. Ojima, T. Painter, Y. Pan, W. J. Parton, L. Pierce, L. Pitelka, C. Prentice, B. Rizzo,

N. A. Rosenboom, S. Running, D. S. Schimel, S. Sitch, T. M. Smith, F. I. Woodward]. 1995. Vegetation/Ecosystems Modeling and Analysis Project (VEMAP): Comparing biogeography and biogeochemistry models in a continental-scale study of terrestrial ecosystems responses to climate change and CO_2 doubling. *Global Biogeochemical Cycles* 9: 407–437.

Verner, J., M. L. Morrison and C. J. Ralph (eds.). 1986. *Wildlife 2000: Modeling Habitat Relationships of Terrestrial Vertebrates.* University of Wisconsin Press, Madison, WI.

Volterra, V. 1926. Fluctuations in the abundance of a species considered mathematically. *Nature* 118: 588–56.

Volz, A. 1983. Studie über die Auswirkungen von Kohlendioxidemissionen auf das Klima. Kernforschungsanlage Jülicich, F. R. Germany.

Von Foerster, G. 1959. Some remarks on changing populations (pp. 382–407). In: *Kinetics of Cellular Proliferation.* Grune and Stratton, New York.

Waggoner, P. E. and G. R. Stephens. 1971. Transition probabilities for a forest. *Nature* 225: 93–114.

Waldrop, T. A., E. R. Buckner, H. H. Shugart and C. E. McGee. 1986. FORCAT: a single tree model of stand development on the Cumberland Plateau. *Forest Science* 32: 297–317.

Walker, B. H., D. Ludwig, C. S. Holling and R. M. Peterman. 1981. Stability of semi-arid savanna grazing systems. *Journal of Ecology* 69: 473–498.

Walker, D. and Y. Chen. 1987. Palynological light on rainforest dynamics. *Quarterly Science Review* 6: 77–92.

Walker, J., P. J. Sharpe, L. K. Pendridge and H. Wu. 1989. Ecological field theory: the concept and field tests. *Vegetatio* 83: 81–95.

Wallin, D. O. 1990. Habitat dynamics of an African weaver-bird: the red-billed quelea (*Quelea quelea*). PhD Dissertation, Department of Environmental Sciences, University of Virginia, Charlottesville, VA.

Wallin, D. O., C. C. H. Elliott, H. H. Shugart, C. J. Tucker and F. Wilhelmi. 1992. Satellite remote sensing of breeding habitat for an African weaver-bird. *Landscape Ecology* 2: 87–99.

Walsh, S. J., L. Bian, D. G. Brown, D. R. Butler and G. P. Malanson. 1989. Image enhancement of Landsat Thematic Mapper digital data for terrain evaluation, Glacier National Park, Montana. *Geocarta International* 4: 55–58.

Walter, H. and S. W. Breckle. 1985. *Ecological Systems of the Geobiosphere.* Springer-Verlag, Berlin.

Wang, B., 1986. A study of spatial pattern in mixed broadleaved-*Pinus koraiensis* (Korean pine) forest on Changbaishan mountain (pp. 164–170). In: H. Yang, Z. Wang, J. N. R. Jeffers and P. A. Ward (eds.). *The Temperate Forest Ecosystem. Proceedings of the ITE symposium No. 20*, Antu, China. 5–11 July 1986. Institute of Terrestrial Ecology, Grange-over- Sands, UK.

Ward, P. 1965a. Feeding ecology of the black-faced dioch, *Quelea quelea*, in Nigeria. *Ibis* 107: 173–214.

Ward, P. 1965b. The breeding biology of the black-faced dioch, *Quelea quelea*, in Nigeria. *Ibis* 107: 326–349.

Ward, P. 1971. The migration patterns of *Quelea quelea* in Africa. *Ibis* 113: 275–297.

Ward, P. 1973. *Manual of Techniques Used in Research on Quelea Birds.* APG: RAF/67/087, UNDP/FAO, Rome.

Ward, R. T. 1956. The beech forests of Wisconsin – changes in forest composition and the nature of the beech border. *Ecology* 37: 407–409.

Waring, R. M. and J. Major. 1964. Some vegetation of the California coastal redwood region in relation to gradients of moisture, light and temperature. *Ecological Monographs* 34: 167–215.

Warming, E. 1909. *Oecology of Plants: An Introduction to the Study of Plant Communities.* Humphrey Milford and Oxford University Press, Oxford.

Warrick, R. A., H. H. Shugart, M. Ja. Antonovsky, J. R. Tarrant and C. J. Tucker. 1986. The effects of increased CO_2 and climatic change on terrestrial ecosystems (pp. 363–392). In: B. Bolin, B. R. Döös, J. Jager and R. A. Warrick (eds.). *The Greenhouse Effect, Climatic Change and Ecosystems (SCOPE 29).* John Wiley, New York.

Watson, R. T., H. Rodhe, H. Oeschger and U. Siegenthaler. 1990. (pp. 1–40). In: J. T. Houghton, G. J. Jenkins and J. J. Ephraums (eds.). *Climate Change, The IPCC Scientific Assessment.* Cambridge University Press, Cambridge.

Watt, A. S. 1925. On the ecology of British beech woods with special reference to their regeneration. II. The development and structure of beech communities on the Sussex Downs. *Journal of Ecology* 13: 27–73.

Watt, A. S. 1947. Pattern and process in the plant community. *Journal of Ecology* 35: 1–22.

Watts, W. A. 1988. Europe (pp. 155–192). In: B. Huntley and T. Webb, III (eds.). *Vegetation History.* Kluwer, Dordrecht.

Webb, D. S. 1984. Ten million years of mammal extinctions in North America (pp. 189–210). In: P. S. Martin and R. G. Klein (eds.). *Quaternary Extinctions: A Prehistoric Revolution.* University of Arizona Press, Tucson, AZ.

Webb, L. J., J. G. Tracey, W. T. Williams and G. N. Lance. 1970. Studies in the numerical analysis of complex rain-forest communities. V. A comparison of the properties of floristic and physiognomic-structural data. *Journal of Ecology* 58: 203–232.

Webb, T., III. 1987. The appearance and disappearance of major vegetational assemblages: long-term vegetation dynamics in eastern North America. *Vegetatio* 69: 177–187.

Webb, T., III. 1988. Glacial and Holocene vegetation history: eastern North America (pp. 385–414). In: B. Huntley and T. Webb, III (eds.). *Vegetation History.* Kluwer Academic, Dordrecht.

Webb, T., III, P. J. Bartlein and J. E. Kutzbach. 1987. Climatic change in eastern North America during the past 18,000 years; comparisons of pollen data with model results (pp. 441–462). In: W. E. Ruddiman and H. E. Wright, Jr (eds.). *North America and Adjacent Oceans during the Last Deglaciation. The Geology of North America.* Vol. K-3. US Geological Society, Boulder, CO.

Weiner, J. 1986. How competition for light and nutrients affects size variability in *Ipooea tricolor* populations. *Ecology* 67: 1425–1427.

Weinstein, D. A. and H. H. Shugart. 1983. Ecological modelling of landscape dynamics (pp. 29–45). In: H. A. Mooney and M. Godron (eds.). *Disturbance and Ecosystems.* Springer-Verlag, New York.

Weinstein, D. A., H. H. Shugart and D. C. West. 1982. *The Long-term Nutrient Retention Properties of Forest Ecosystems: A Simulation Investigation.* ORNL/TM–8472. Oak Ridge National Laboratory, Oak Ridge, TN.

Weinstein, D. A., H. H. Shugart and C. C. Brandt. 1983. Energy flow and the persistence of a human population: a simulation analysis. *Human Ecology* 11: 201–225.

Weishampel, J. F., D. L. Urban, H. H. Shugart and J. B. Smith. 1992. A comparison of semivariograms from a forest transect gap model and remotely sensed data. *Journal of Vegetation Science* 3: 521–526.

West, D. C., S. B. McLaughlin and H. H. Shugart. 1980. Simulated forest response to chronic air pollution stress. *Journal of Environmental Quality* 9: 43–49.

West, R. G. 1961. Interglacial and interstadial vegetation in England. *Proceedings of the Linnean Society of London* 172: 81–89.

West, R. G. 1977. *Pleistocene Geology and Biology.* Longman, London.

Wetherald, R. T. and S. Manabe. 1990. Section in chapter, Processes and modeling (by U. Cubasch and R. D. Cess, pp. 69–91): In: J. T. Houghton, G. I. Jenkins and J. J. Ephraums (eds.). *Climate Change: The IPCC Scientific Assessment.* Cambridge University Press, Cambridge.

White, L. P. 1970. 'Brousse tigrée' patterns in southern Niger. *Journal of Ecology* 58: 549–553.

White, L. P. 1971. Vegetation stripes on sheet wash surfaces. *Journal of Ecology* 59: 615–622.

White, P. S. 1979. Pattern, process and natural disturbance in vegetation. *Botanical Review* 45: 229–299.

Whitehead, F. H. 1954. A study of the relation between growth form and exposure on Monte Maiella, Italy. *Journal of Ecology* 42: 180–186.

Whitehead, F. H. 1959. Vegetational change in response to alterations of surface roughness on Monte Maiella, Italy. *Journal of Ecology* 47: 603–606.

Whitford, P. B. and P. J. Salmun. 1954. An upland forest survey of the Milwaukee area. *Ecology* 35: 533–540.

Whitmore, T. C. 1974. Change in time and the role of cyclones in the tropical rain forest on Kolombangara, Solomon Islands. *Commonwealth Forestry Institute Paper 46.*

Whitmore, T. C. 1975. *Tropical Rain Forests of the Far East.* Clarendon Press, Oxford.

Whitmore, T. C. 1982. On pattern and process in forests (pp. 45–59). In: E. I. Newman (ed.). *The Plant Community as a Working Mechanism.* Special Publ. No. 1, British Ecological Society. Blackwell Scientific, Oxford.

Whitmore, T. C. 1989. Canopy gaps and the two major groups of tree species. *Ecology* 70: 536–538.

Whitmore, T. C. and G. T. Prance (eds.). 1987. *Biogeography and Quaternary History in Tropical America. Oxford Monographs on Biogeography No. 3.* Oxford Science Publications, Oxford.

Whittaker, R. H. 1953. A consideration of climax theory. The climax as a population and a pattern. *Ecological Monographs* 23: 41–78.

Whittaker, R. H. 1960. *Ecological Monographs* 30: 279–338.

Whittaker, R. H. 1967. Gradient analysis of vegetation. *Biological Review* 42: 207–264.

Whittaker, R. H. 1975. *Communities and Ecosystems*, 2nd edn. MacMillan, New York.

Whittaker, R. H. and D. Goodman. 1979. Classifying species according to their demographic strategy. 1. Population fluctuations and environmental heterogeneity. *American Naturalist* 113: 185–200.

Whittaker, R. H. and S. A. Levin. 1977. The role of mosaic phenomena in natural communities. *Theoretical Population Biology* 12: 117–139.

Whittaker, R. H. and G. E. Likens. 1973. The primary production of the biosphere. *Human Ecology* 1: 299–369.

Whittaker, R. H. and G. E. Likens. 1975. The biosphere and man (pp. 305–328). In: H. Leith and R. H. Whittaker (eds.). *Primary Productivity of the Biosphere*. Springer Verlag, New York.

Whittaker, R. H. and P. L. Marks. 1975. Methods of assessing terrestrial productivity (pp. 55–118). In: H. Leith and R. H. Whittaker (eds.). *Primary Productivity of the Biosphere*. Springer Verlag, Berlin.

Whittaker, R. H. and W. A. Niering. 1965. Vegetation of the Santa Catalina Mountains, Arizona. II. A gradient analysis of the south slope. *Ecology* 46: 429–452.

Whittaker, R. H., S. A. Levin and R. B. Root. 1973. Niche, habitat and ecotope. *American Naturalist* 109: 479–482.

Wiegert, R. G. 1988. The past, present and future of ecological energetics (pp. 29–56). In: L. R. Pomeroy and J. J. Alberts (eds.). *Concepts of Ecosystem Ecology*. Springer-Verlag, New York.

Wiens, J. A. and M. I. Dyer. 1977. Assessing the potential impact of granivorous birds in ecosystems (pp. 205–266). In: J. Pinowski and S. C. Kendeigh (eds.). *Granivorous Birds in Ecosystems*. Cambridge University Press, Cambridge.

Wiens, J. A. and R. F. Johnston. 1977. Adaptive correlates of granivory in birds (pp. 301–340). In: J. Pinowski and S. C. Kendeigh (eds.). *Granivorous Birds in Ecosystems*. Cambridge University Press, Cambridge.

Wilcox, B. A. and D. D. Murphy. 1985. Conservation strategy: the effects of fragmentation on extinction. *American Naturalist* 125: 879–887.

Williams, B. K. 1983. Some observations on the used of discriminant function analysis. *Ecology* 64: 1283–1291

Williamson, D. L., J. T. Kiehl, V. Ramanathan, R. E. Dickinson and J. J. Hack. 1987. *Description of the NCAR community climate model (CCM1). NCAR Technical Note TN–285+ STR*. National Center for Atmospheric Research, Boulder, CO.

Williamson, M. 1972. *The Analysis of Biological Populations*. Edward Arnold, London.

Wilson, C. A. and J. F. B. Mitchell. 1987. A doubled CO_2 climate sensitivity experiment with a global climate model including a simple ocean. *Journal of Geophysical Research* 92: 13 315–13 343.

Wilson, E. O. 1961. The nature of the taxon cycle in the Melanesian ant Fauna. *American Naturalist* 95: 169–193.

Wilson, E. O. 1975. *Sociobiology*. Belknap, Cambridge, MA.

Wilson, J. B. 1990. Mechanisms of species coexistence: twelve explanations for Hutchinson's Paradox of the Plankton: evidence from New Zealand plant communities. *New Zealand Journal of Ecology* 13: 17–42.

Wilson, J. B., J. C. E. Hubbard and G. L. Rapson. 1988. A comparison of the realised niche relations of species in New Zealand and Britain. *Journal of Biogeography* 19: 183–193.

Wistar, C. 1799. An account of the bones deposited, by the President, in the museum of the Society, and represented in the annexed plates. *Transactions of the American Philosophical Society* 4: 525–531.

Wong. S. C. and F. X. Dunin. 1987. Photosynthesis and transpiration of trees in a eucalypt forest stand: CO_2, light and humidity responses. *Australian Journal of Plant Physiology* 14: 619–632.

Wood, D. L. 1982. The role of pheromones, kaironmones, and allomones in the host selection and colonization behavior of bark beetles. *Annual Reviews in Entomology* 27: 411–446.

Woods, K. D. and R. H. Whittaker. 1981. Canopy-understory interaction and the internal dynamics of mature hardwood and hemlock-hardwood forests (pp. 305–323). In: D. C. West, H. H. Shugart and D. B. Botkin (eds.). *Forest Succession: Concepts and Application.* Springer-Verlag, New York.

Woodward, F. I. 1987a. Stomatal numbers are sensitive to increases in CO_2 from pre-industrial levels. *Nature* 327: 617–618.

Woodward, F. I. 1987b. *Climate and Plant Distribution.* Cambridge University Press, Cambridge.

Woodward. F. I. 1988. Temperature and the distribution of plant species (pp. 59–75). In: S. P. Long and F. I. Woodward (eds.). *Plants and Temperature.* Company of Biologists, Cambridge.

Woodward, F. I. 1993. Plant responses to past concentrations of CO_2. *Vegetatio* 104/105: 145–155.

Woodward, F. I. and B. G. Williams. 1987. Climate and plant distributions at global and local scales. *Vegetatio* 69: 189–197.

Woodward, F. I., T. M. Smith and W. R. Emanuel. 1995. A global land primary productivity and phytogeography model. *Global Biogeochemical Cycles* 9: 471–490.

World Climate Programme. 1981. *On the Assessment of the Role of CO_2 on Climate Variations and their Impact.* Report of a WMO/UNEP/ICSU meeting of experts in Villach, Austria, November 1980. World Meteorological Organization, Geneva.

Worrall, G. A. 1959. The Butana grass patterns. *Journal of Soil Science* 10: 34–53.

Wright, H. E., Jr. 1977. Quaternary vegetation history – some comparisons between Europe and America. *Annual Reviews Earth Plant Sciences* 5: 123–158.

Wu, H.-I, P. J. H. Sharpe, J. Walker and L. K. Penridge. 1985. Ecological field theory: a spatial analysis of resource interference among plants. *Ecological Modelling* 29: 215–243.

Wu, H.-I., E. J. Rykiel, Jr, T. Hatton and J. Walker. 1993. *Multi-Factor Growth Rate Modelling Using an Intergrated Rate Methodology (IRM).* Technical Memorandum 93/4 Division of Water Resources, CSIRO, Canberra, Australia.

Wu, J. and S. I. Levin. 1994. A spatial patch dynamic modeling approach to pattern and process in an annual grassland. *Ecological Monographs* 64: 447–464.

Wullschleger, S. D. 1993. Biochemical limitations to carbon assimilation in C_3 plants – a retrospective analysis of the A/C_i curves from 109 species. *Journal of Experimental Botany* 44: 907–920.

Xue, Y. and J. Shukla. 1993. The influence of land surface properties on Sahel climate: Part I. Desertification. *Journal of Climate* 6: 2232–2245.

Yamamura, N. 1976. A mathematical approach to spatial distribution and temporal succession in plant communities. *Bulletin of Mathematical Biology* 38: 517–526,

Yan, X. and S. Zhao. 1995. Simulating the carbon storage dynamics of temperate broadleaved coniferous mixed forest ecosystems: Dynamics of the tree layer of broadleaved Korean Pine forests in the Changbai Mountains. *Chinese Journal of Ecology* 14: 6–12.

Yang, H. and Y. Wu. 1986. Tree composition, age structure and regeneration strategy of the mixed broadleaved-*Pinus koraiensis* (Korean pine) forest in Changbaishan Mountain Reserve (pp. 12–20). In: H. Yang, Z. Wang, J. N. R. Jeffers and P. A. Ward (eds.). *The Temperate Forest Ecosystem.* Proceedings of the ITE symposium No. 20, Antu, China. 5–11 July 1986. Institute of Terrestrial Ecology, Grange-over-Sands, UK.

Yodzis, P. 1989. *Introduction to Theoretical Ecology*. Harper and Row, New York.

Zabinski, C. and M. B. Davis. 1989. Hard times ahead for Great Lakes forests; a climate threshold model predicts responses to CO_2-induced climate change (pp. 5–1 to 5–19). In: J. B. Smith and D. A. Tirpak (eds.). *The Potential Effects of Global Climate Change on the United States* (EPA–230–05–89–054). US Environmental Protection Agency, Washington, DC.

Zach, R. and J. B. Falls. 1976a. Ovenbird (Aves: Parulidae) hunting behavior in a patchy environment: an experimental study. *Canadian Journal of Zoology* 54: 1863–1879.

Zach, R. and J. B. Falls. 1976b. Foraging behavior, learning and exploration by captive ovenbirds (Aves: Parulidae). *Canadian Journal of Zoology* 54: 1880–1893.

Zach, R. and J. B. Falls. 1976c. Do ovenbirds (Aves: Parulidae) hunt by expectation? *Canadian Journal of Zoology* 54: 1894–1903.

Zach, R. and J. B. Falls. 1977. Influence of capturing a prey on subsequent search in ovenbirds (Aves: Parulidae). *Canadian Journal of Zoology* 55: 1958–1969.

Zach, R. and J. B. Falls. 1979. Foraging and territoriality of male ovenbirds (Aves: Parulidae) in a heterogeneous habitat. *Journal of Animal Ecology* 48: 33–52.

Ziegler, P. and O. Kandler. 1980. Tonoplast stability as a critical factor in frost injury and hardening of spruce (*Picea abies* L. Karst.) needles. *Zeitschrift für Pflanzenphysiologie* 105: 229–239.

Zilversmit, D. B., C. Entenmann and M. C. Fishler. 1943. On the calculation of turnover time and turnover rate from experiments involving the use of labeling agents. *Journal of General Physiology* 26: 325–331.

Zinke, P. J. 1962. The pattern of influence of individual trees on soil properties. *Ecology* 43: 130–133.

Zobel, M. 1992. Plant species coexistence – the role of historical, evolutionary and ecological factors. *Oikos* 65: 314–320.

Zobler, L. 1986. *A World Soil File for Global Climate Modeling. NASA Technical Memorandum 87802.* National Aeronautics and Space Administration, Washington, DC.

Zyabchenko, S. S. 1982. Age dynamics of scotch pine forests in the European North. *Lesovedenie* 2: 3–10.

Index

536 · Index

FRANKLIN PIERCE COLLEGE LIBRARY

00110581